Gloria Straub

GEOMETRY   A GUIDED INQUIRY

The Stein Series

Elementary Algebra:
A Guided Inquiry

Geometry: A Guided Inquiry

Intermediate Algebra with
Trigonometry: A Guided Inquiry

HOUGHTON MIFFLIN COMPANY BOSTON

NEW YORK   ATLANTA   GENEVA, ILLINOIS   DALLAS   PALO ALTO

Instructor's Edition

GEOMETRY A GUIDED INQUIRY

The Stein Series

G. D. Chakerian

Professor of Mathematics

University of California, Davis

Calvin D. Crabill

Department of Mathematics

Davis High School

Sherman K. Stein

Professor of Mathematics

University of California, Davis

Copyright © 1972 by Houghton Mifflin Company.

Copyright © 1970 by G. D. Chakerian, Calvin D. Crabill, and Sherman K. Stein.

All rights reserved. No part of this work may be reproduced or transmitted in any form or by any means, electronic or mechanical, including photocopying and recording, or by any information storage or retrieval system, without permission in writing from the publisher.

Printed in the United States of America

Library of Congress Catalog Card Number: 71-179132

ISBN: 0-395-13148-0

ISBN: 0-395-13149-9 (Instructor's Edition)

This book is dedicated

to the students and teachers

who experimented with

the preliminary editions

and

to our wives and children,

who lived through

the numerous revisions.

A Note to the Student

In using the small-group method, it is especially important that the students understand what is expected of them. Thus a teacher-directed discussion of "A Note to the Student" is essential before the groups are formed for the first time. See also "Suggestions for Using the Small-Group Learning Method" in the Preface for the Instructor.

Have you ever been confused by mathematical symbols and words? Have you ever been afraid to ask questions in class? This text is designed to prevent some of that confusion and fear as you learn geometry through small-group discussions. Technical language and symbols are kept to a minimum; everyday language is used wherever possible. The exercises encourage you to ask questions and to talk about mathematics. We believe that the questions *you* ask are most important in helping you to learn.

In this kind of geometry course, your teacher will be able to spend more time with individual students and groups who need help. You will have the chance to assume more responsibility for learning than you probably are used to, and you will do more of your own thinking. Everyone is encouraged to voice his ideas, but you should not accept an idea offered by someone in your group without trying to understand the reasons behind it.

Each chapter is made up of three major sections. The Central section presents basic concepts. The Review section will help you remember what you have learned in previous chapters, and it will reinforce your understanding of the current material. There are self tests, with answers, at the end of the Central and Review sections. Chapters 1–7 also have separate Algebra Review sections. The Projects will enable you to go more deeply into some of the interesting byways and applications of geometry. Some of these are easier than the exercises in the Central section; others are more challenging. The Projects should deepen your understanding and enjoyment of geometry.

The composition of the groups is important. The most successful grouping consists of four students arranged with their desks close together. This grouping permits six different dialogue patterns; each student can easily converse with any other student in his group. Having an odd num-

Four is a stable number in group dynamics; no one gets left out. Each person has a partner for dialogue. If a four group is not possible, a five group is generally preferable to a three group.

ber of students in a small group often does not work well, since students tend to pair off, and one student may be left out of discussions.

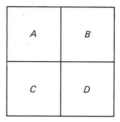

Just as the roles of individuals constantly change in activities outside the classroom, so do the roles of individuals constantly change in the small-group method of learning. However, at any moment the ideal group has individuals playing the following roles (any one person may play one or more roles during a class period):

The teacher is a catalyst, making sure these roles are informally assumed in each group. If roles are assigned, a great deal of spontaneity is lost.

the *assured leader* (he is not afraid to speak up)
the *idea person* (he offers a mathematical idea)
the *challenger* (he questions ideas)
the *synthesizer* (he reconciles opposing ideas)
the *ego-builder* (he praises others in the group)

The groups work best when the individuals in the classroom alternately assume these roles as the situation requires. While the groups are working, your teacher will be moving about the room, encouraging discussion and answering questions that a group cannot resolve on its own—being like a roving coach. At times, he may call the class together and lecture in the traditional style, such as when introducing new material, summarizing, or discussing a common difficulty.

In order that the teacher or a student may have an opportunity to present a mathematical idea to the entire class, the following rule should be observed:

Follow this crucial rule from the very first day!

If anyone wishes to speak to the entire class, he should be given courteous attention.

When your questions have been answered in class, when you have completed the exercises in the Central and Review sections and you have done the self tests, you should then feel confident, for what you have found by your own effort and spoken in your own words will remain a part of your permanent knowledge.

Preface for the Instructor

The student of today, as part of the heritage of his generation, tends to challenge existing ideas more than ever before. Students themselves and also social pressures demand that the teaching-learning process be more flexible. The small-group learning approach, which we advocate, permits students to take an active role in the learning process. Through playing the different roles in a functioning group, every student can learn mathematics in a context of discovery and mutual reinforcement. This text is part of a new series that is designed to provide students with materials geared to small-group study and to provide the teacher with all the encouragement and guidance we can give in using the small-group method to best advantage.

Our Philosophy

Who usually gets the most practice speaking in the classroom? Who needs it most? *The student* needs dialogue, because often he cannot separate in his mind what he does understand from what he does not understand. If he can state his confusions clearly, he will probably be able to help himself. A student is handicapped if he cannot express his difficulties, yet little is done in the typical classroom to help him become more articulate and more confident.

Because the teacher has long been considered the central figure and generating force in the classroom, he may easily miss the fact that students learn most easily from their own efforts and from each other. Formal classroom teaching with the instructor lecturing to the entire class should not be the only method of instruction, for it imposes a single pattern of learning; it ignores the fact that students can teach each other. Generally speaking, students of any age have a higher natural affinity for each other than for the teacher, and this attraction can be channeled constructively.

Students' interest in each other is the prime source of inattention in the conventional classroom; students want to talk and in other ways relate to each other. For this reason, communication between students during class traditionally has been discouraged. But a classroom so tightly controlled that dialogue between students is prohibited tends to invite an atmosphere of apathy which can paralyze curiosity and stifle creativity. On the other hand, when students learn in small groups, guided by a text that encourages discussion, they are less apt to become bored. They find that learning and human relationships are compatible, and they retain more of the information presented.

As classes get larger and larger, the problems of maintaining discipline while encouraging communication increase. In a small class, it is possible to maintain at least the illusion that the teacher has complete control and is teaching all of the students; this is probably impossible in a large class. Large classes challenge us to break from the academic lockstep and somehow get more students involved in the learning process.

Ideally, every student should have an opportunity to express and develop his own ideas. Generally the student will speak up freely in a small group, whereas he may be reluctant to talk before the whole class. He should be encouraged to ask questions so that erroneous notions can be exposed and corrected. Because he is in a vulnerable position, great sensitivity is required by everyone in the classroom, including the teacher. In a conventional classroom it is a rare student who willingly and openly shares his ignorance and confusion; yet, this is the very essence of learning in the small-group approach. The teacher who uses this method finds his most effective role as a director and motivator of the learning activities and a supplier of information not readily available to the student. He should be something like a roving coach—a stimulator, a questioner, an encourager, a morale-builder, and quite often a listener. The most important thing the teacher can do is to create an atmosphere in which the sharing of ideas not only takes place but is the predominant pattern.

This text is based on the conviction that learning almost comes naturally when the situation is right for it, when proper materials are available and the teacher knows how to guide the students in helping each other to learn from them. A conventional text generally does not work well in small-group learning, for the spirit of discovery is often smothered by excessive explanations and by exercises that are not designed to invite dialogue between students. This text was written specifically for use by students in small groups, with the help of a roving teacher. It has an abundance of exercises that provide both drill in mathematical concepts and practice in oral expression.

We have tried to show that mathematics can be learned without the usual heavy dose of technical jargon. The use of symbols and jargon in mathematics has greatly increased in recent years, and often confuses the student. We have found that when symbols and technical words are kept to a minimum, the student understands the mathematics better because the ideas are not hidden by esoteric language. By having students work with mathematical ideas expressed in everyday language before they learn the names of those concepts, we hope to avoid intimidating students who lack confidence in their mathematical ability. Through student dialogue, misconceptions can be cleared up, new concepts tried, and confidence strengthened. Although our approach is informal, the student is encouraged to develop habits of precise thought. Occasionally a definition is made more precise after its introduction. The point is this: Total precision too early in the course is usually futile and often reduces the students' enthusiasm for learning.

For example, the student is encouraged to discover geometric properties experimentally. As the course progresses, he is asked to justify his steps and to give reasons. Gradually he learns how to prove a theorem, using previously established theorems and given axioms. However, the idea of an axiomatic system is not emphasized, and the axiomatic method is not discussed until Chapter 4. It is hoped that the student will acquire some ability to do proofs through example and practice early in the course, and that he will simultaneously acquire a knowledge of geometry sufficient for use in everyday life. After he has this background, he will be ready to understand the axiomatic basis of geometry, and, indeed, of algebra or of mathematics generally. He will have a reservoir of experience to draw from when the logical basis of geometry is discussed. In the meantime, he will have acquired a knowledge of the fundamental facts of geometry and the knack of problem solving.

The basic philosophy of this series can be summarized as follows:

1. Present movements for reform in mathematics should focus not on changing the mathematical content but on changing the mood of the classroom. This can be accomplished only by presenting the material in a fresh and provocative manner, specifically, in a way that actively involves the student in the learning process.

2. Learning based on experiment and discovery is more exciting and probably motivates most students better than the axiomatic approach.

3. Excessive use of technical vocabulary and symbols often hinders learning. Thus mathematical concepts should be presented in a meaningful context and in everyday language whenever possible.

4. Students need to talk mathematics in order to clarify their thoughts. Therefore, we have tried to use exercises that are discussion-provoking. Mathematics is not only a skill but also a language.

5. The classroom activities should be structured to permit students to learn essential ideas, yet give them the freedom to experiment.

6. The self-image of almost any student can be improved; when a student understands what he is doing and experiences success, he becomes more self-confident and self-respecting.

7. Students of all abilities have a part to play in the teaching-learning process, and that part is never static; sometimes the "slow" student can help the more able one.

8. The teacher can be more helpful to the students by departing from complete dependence on the lecture style.

9. The teaching materials should offer the teacher the greatest latitude in choice of methods to suit the needs of the students and his own personality.

Suggestions for Using the Small-Group Learning Method

This section contains some suggestions for using the small-group learning method, which has been employed successfully by many who have taught with preliminary editions of the books in this series. You may find it helpful to return to this section from time to time.

The composition of the groups is important. The most successful grouping consists of four students arranged with their desks close together. This grouping permits six different dialogue patterns; each student

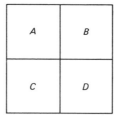

can easily converse with any other student in his group. Having an odd number of students in a small group often does not work well, since students tend to pair off, and one student may be left out of discussions. Admittedly, we are speaking now of ideal arrangements. The teacher will have to adapt these suggestions to the actual classroom situation on any given day. Generally a five group is preferable to a three group. In a six group some students become rather passive while others become "permanent teachers," thus stifling the creativity of the less aggressive students.

There are certain key roles to be played in each group:

the *assured leader* (he is not afraid to speak up)
the *idea person* (he offers a mathematical idea)
the *challenger* (he questions ideas)
the *synthesizer* (he reconciles opposing ideas)
the *ego-builder* (he praises others in the group)

Each may be played by any or all members of the group from time to time during a class period. You may become aware of other important roles, but the five cited are the essential ones needed for a group to function smoothly.

It is an exciting experience to observe a group of students debating mathematics for an entire class period! When the group approach is working, a class of 32 students working in groups of four can be monitored by a teacher almost as easily as a class of eight; in some cases more easily, since the teacher has 32 assistants if everyone is involved in the teaching-learning process. The so-called slower student may originate ideas, and all students quickly realize that playing teacher is fun and an efficient way of learning. The teacher can spend time with an individual student during the regular class period, at no expense to the rest of the class.

But NOTE: *Although the teacher is seemingly on the sidelines, he cannot afford to retire to his desk or withdraw from the students.* He should stay away from his desk as much as possible. Almost instant feedback results when he moves about the room and is alert to the needs and progress of each group. He can be a catalyst for any group that lacks leadership. He can sense when one or more of the groups are stymied and take the steps listed below to identify and remove obstacles. The formal lecture is still necessary when many students will benefit, such as when the teacher is introducing new material or summarizing, or when the students are meeting a common difficulty.

Classroom control by the teacher is imperative. However, it should be made clear that students as well as the teacher may initiate a "lecture" or general discussion. We suggest the rule:

If anyone wishes to speak to the entire class, he should be given courteous attention.

The amount of noise in the classroom may be irritating at first. However, if the students are talking about mathematics, what more can a teacher ask? What the student discovers by his own efforts and says in his own words he will not soon forget.

The extent of student discussion is a good indicator of the level of learning with this method. If almost constant dialogue is not present in some groups, then the following possibilities should be considered:

1. *Students in a group may become bored with each other.* Changing the membership of the groups with every chapter is a successful approach, although some students may not like it at first. Consider changing the manner in which groups are formed, perhaps using a completely random selection. Or, in a class of 32 students, you might pick eight students who have proven abilities to make a group work and let the rest of the class form groups of four around them as they wish.

2. *Each group should have adequate student leadership.* All the leaders may have formed a clique in the same group.

3. *The students may feel abandoned.* Move constantly among the groups so you can quickly detect any common difficulty. You may need to give short lectures from time to time.

4. *Students may be defeated by a difficult exercise.* In this case, return to a more traditional, teacher-directed approach for a time, or let a student explain the problem to the entire class.

5. *The students may need a change of pace.* Allow a day or two for Projects at the end of the Central section or at the end of the Review section, or both. Encourage students to bring in their own projects, if they wish.

6. *Too much accuracy may be required at a given stage.* Students should be allowed to stumble and make mistakes, yet recover without stigma. This helps build self-esteem.

7. *Everyone should constantly strive to build the students' confidence.* Several recent studies show that self-image is the most important single factor in learning. The "ego-builder" may not be present in every small group. Or, students may be used to a society where it is acceptable to "put the other person down." Examine yourself as a teacher to see if you are doing this to individual students. Some are so fearful of failure that they scarcely try at all.

8. *The teacher may feel uneasy because he is not the central figure in the classroom.* This feeling may affect the students. Are you lecturing to a class unnecessarily? If you are not sure, ask the students—they will tell you!

9. *The teacher may be getting mixed feedback.* A good method is to let the students struggle with a new idea briefly in the groups, then lecture briefly, then let them discuss again in groups. That way immediate feedback tells you where the problems lie. You might lecture again briefly if necessary; some classes require very little lecturing, others need more.

Features of This Textbook

The basic features of the text, developed after several years of experience with preliminary editions, are as follows:

1. The mathematical content has been changed very little; traditional material is still needed by students who will be going on to study more mathematics.

2. Symbols and technical vocabulary have been kept to a minimum. The student is encouraged to talk about mathematics in everyday language.

3. The text is deliberately constructed to get students more involved in the teaching-learning process. Students with varying abilities can learn together in the same classroom by performing activities that enhance their self-images. In so doing, they teach each other.

4. An Algebra Review section is included in each of the first seven chapters.

5. Each chapter is divided into the following major sections, to provide the teacher several options in teaching:

Central (presents essential ideas)
Review (reviews prerequisite material and essential ideas)
Projects (optional, for enrichment)
Self Tests and Self Quizzes (for student self-evaluation)
Algebra Review (in Chapters 1–7 only)

6. The Central Self Quiz, Review Self Test, Review, and Cumulative Review sections often have all the answers given so that the student may check his work immediately.

7. Throughout the book, there are boxes, colored arrows, and boldface type to focus the student's attention on the important ideas.

8. Definitions are preceded by exercises that provide the student with the necessary background.

9. Examples are often used to help the student avoid common errors.

10. This annotated Instructor's Edition contains solutions and teaching suggestions. (These always appear in black.)

11. There is also a *Solutions Key* which provides additional solutions, often more detailed, that could not be included in the text or Instructor's Edition.

Organization of This Text

Chapters 1–10 may be considered the basic text, with the chapters taken in numerical order.

Chapters 11 and 12 may be considered optional.

Many chapters are rich in projects, which are optional. If time permits, you may wish to return to earlier Projects sections and do some of the exercises that the students have not yet done.

Acknowledgments

The authors wish to acknowledge the following people for their help and encouragement:

Dr. Donald McKinley, Auburn, Calif.; Mrs. Sharon Frost, Whittier, Calif.; Mr. Harold Oxsen, Mr. Arthur Dull, Diablo Valley College; Mr. Edward Harper, Mr. Richard Detar, American River College; Mrs. Mary Gorse, Miss Elaine Kasimatis, Mr. William Raski, Mrs. Idella Shideler, Mrs. Maria Wong, Mrs. Storm Gelinske, Mr. E. A. White, Mr. Arnold Manke, Mr. William Watkins, Mr. Joshua Stein, Davis, Calif.; Mr. Edward Perry, Orinda, Calif.; Mr. James Nudelman, Mr. Amar Chadda, Adm. William O'Regan, Mr. Conrad Schwarze, Mountain View, Calif.; Mr. Lyle Fisher, Mr. Richard Newman, Mr. Hart Smith, Mill Valley, Calif.; Mr. Milton Dennison, Dixon, Calif.; Miss Nancy Morlock, Hugson, Calif.; Mrs. Barbara Bohannan, Miss Marion Phillips, Sacramento, Calif.; Mrs. Marjorie Hauge, Los Angeles, Calif.; Mr. Robert Greisman, Walnut Creek, Calif.; Mr. Joseph Rizzo, Mr. Norman Siever, Los Angeles Valley College; Mr. Larry Luck, Minneapolis, Minn.; Mr. Stanley Paumer, Mr. Wayne Boyer, Thousand Oaks, Calif.

Contents

Chapter 1   **The Shortest Path**

　　　　　　Central / 1
　　　　　　Central Self Quiz / 16
　　　　　　Review / 18
　　　　　　Review Self Test / 22
　　　　　　Projects / 23
　　　　　　Algebra Review / 26

Chapter 2   **Tiling the Plane**

　　　　　　Central / 33
　　　　　　Central Self Quiz / 55
　　　　　　Review / 56
　　　　　　Review Self Test / 64
　　　　　　Projects / 65
　　　　　　Algebra Review / 71

Chapter 3   **Triangles**

　　　　　　Central / 81
　　　　　　Central Self Quiz / 111
　　　　　　Review / 112
　　　　　　Review Self Test / 121
　　　　　　Cumulative Review / 123
　　　　　　Projects / 125
　　　　　　Algebra Review / 129

Chapter **4** **What is a Proof?**

> **Central** / 137
> Central Self Quiz / 159
> **Review** / 160
> Review Self Test / 166
> **Projects** / 167
> **Algebra Review** / 171

Chapter **5** **Constructions with Straightedge and Compass**

> **Central** / 179
> Central Self Quiz / 208
> **Review** / 209
> Review Self Test / 216
> **Projects** / 217
> **Algebra Review** / 224

Chapter **6** **Area and Volume**

> **Central** / 231
> Central Self Quiz / 249
> **Review** / 250
> Review Self Test / 257
> **Projects** / 258
> **Algebra Review** / 266

Chapter **7** **The Pythagorean Theorem**

> **Central** / 273
> Central Self Quiz / 285
> **Review** / 287
> Review Self Test / 297
> **Cumulative Review** / 300
> **Projects** / 302
> **Algebra Review** / 311

Chapter **8** **Similar Figures**

Central / 317
Central Self Quiz / 341
Review / 342
Review Self Test / 349
Projects / 351

Chapter **9** **Perimeter, Area, and Volume of Similar Figures**

Central / 365
Central Self Quiz / 378
Review / 379
Review Self Test / 382
Cumulative Review / 383
Projects / 389

Chapter **10** **Circles**

Central / 399
Central Self Quiz / 431
Review / 432
Review Self Test / 438
Projects / 440

Chapter **11** **Coordinates**

Central / 463
Central Self Quiz / 492
Review / 493
Review Self Test / 497
Projects / 498

Chapter **12** **The Conic Sections**

Central / 507
Central Self Quiz / 523
Review / 524

Review Self Test / 530
Projects / 531
Cumulative Review / 538

Table of Squares and Approximate Square Roots / 549
Photo Credits / 551
Index / 553

GEOMETRY   A GUIDED INQUIRY

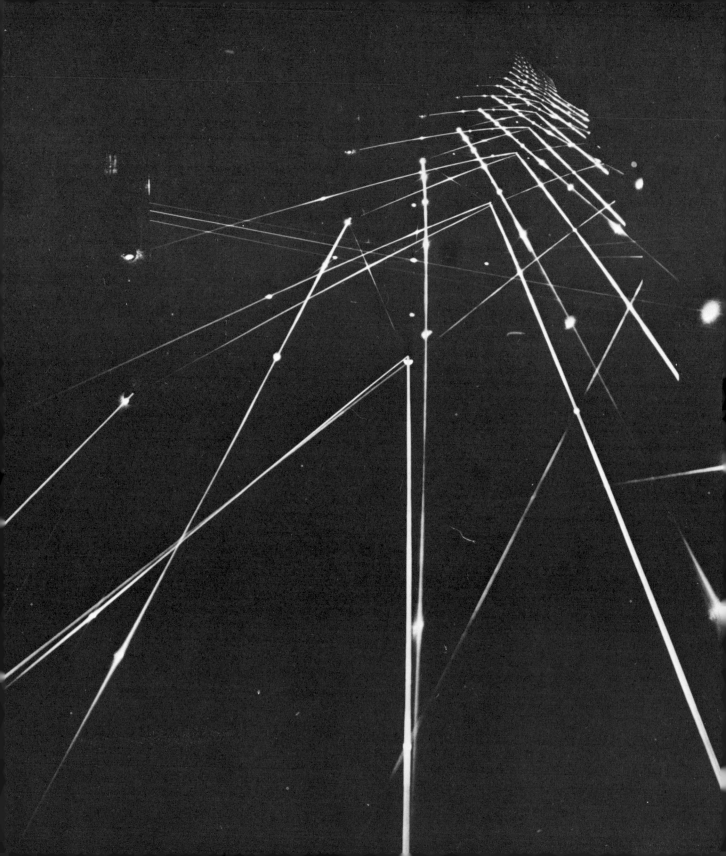

Chapter 1 The Shortest Path

Chapter 1 should provide the student with an enjoyable and relaxed introduction to the geometry course. The problem of the shortest path, while interesting, is not the main point of the chapter from a pedagogical standpoint. The problem only serves as a framework for introducing in a natural way the basic concepts used throughout the course. The student is also being quietly prepared for the idea of proof (see Central Exercises 13, 18, and 20).

Read through the entire chapter before the first class meeting, and then resist the temptation to tell the class "how to solve it by reflection." Only when the Central section is finished, should you summarize the basic ideas. Maintain a pace that is conducive to discussion, for communication between students is a very efficient and satisfying learning method. Review the suggestions in the Preface concerning the small-group method.

This chapter is important for several reasons:

(a) *The student meets such basic concepts as* line, angle, perpendicular, *and* triangle inequality.
(b) *The student develops skill in using ruler and protractor.*
(c) *The student has a chance to be self-reliant.*
(d) *The student sees that geometry is useful.*
(e) *The large problem approach makes the learning of geometric ideas more meaningful.*
(f) *The classes who are using the small-group learning method are solving a significant problem that encourages a great deal of dialogue (along with teacher attitude, a very important prerequisite for success in this method).*

Central

The shortest path problem is a gold mine for geometric concepts, as this chapter will show.

This chapter is based upon a famous problem, about 2000 years old, whose solution involves several important geometric ideas about points, lines, distances, and angles. The idea behind this problem has many applications in our physical world, but we state it simply in terms of a camper, his tent, and a nearby river.

A man has a campsite in a large flat clearing next to a straight river. He is at point *A* (600 feet from the river), and his tent is at point *B* (300 feet from the river). He sees a large spark leap from his campfire and set his tent aflame. The man has an empty bucket already in his hand. *At what point P on the river should he fill his bucket in order to make the shortest possible path to put out the fire?*

Have the proper tools ready for this first problem: compasses, centimeter rulers, and protractors. Graph paper may be helpful.

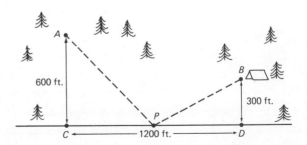

1

The scale drawing should be made carefully with a sharp pencil.

1. Using a protractor and a centimeter ruler, make a scale drawing of the situation in the problem above (scale: 1 centimeter = 100 feet).

Distance Between Two Points

Make sure they understand the notation for distance between two points.

Corresponding to any two points, there is a number called the "distance" between the two points. Since this idea occurs so often in geometry, it will be convenient to introduce the following notation for the distance between points.

If P and Q are two points, we represent the *distance* from P to Q by the symbol PQ.

For example, in the problem we are studying, the distance from the point A to the point C is 600 feet. We write

$$AC = 600 \text{ feet.}$$

Similarly, $BD = 300$ feet and $CD = 1200$ feet.

Note that the order of the points does not matter in the notation for distance. In other words, for any points P and Q,

$$PQ = QP.$$

Again, careful drawing and measurements are called for when using both ruler and protractor. Encourage students to discuss and to compare answers as they work Exercise 2.

2. (a) Use your scale drawing in Exercise 1 and a centimeter ruler to estimate the total distance the man must go, $AP + PB$, when P is chosen to be exactly at C.   *about 1837 feet*

(b) Record your estimate from part (a) in the table where $CP = 0$.

(c) Do the same with point P chosen successively between C and D so that $CP = 100$ ft., 200 ft., ..., 1200 ft.

(d) In each case, also use your protractor to measure angle 1 (the angle at which he approaches the river) and angle 2 (the angle at which he leaves the river).

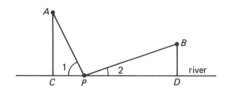

1: The Shortest Path

Copy and fill in the table showing the results of your measurements:

CP	Total distance AP + PB	Angle of approach angle 1	Angle of leaving angle 2
0	1837	90°	14°
100	1748	80°	15°
200	1676	71°	17°
300	1620	64°	18°
400	1576	57°	20°
500	1543	50°	23°
600	1519	45°	26°
700	1505	40°	31°
800	1500	37°	37°
900	1506	33°	44°
1000	1527	31°	55°
1100	1569	28°	70°
1200	1642	26°	90°

Do not tell the solution of the problem at this point, even if a student has a fine conjecture.

3. (a) From your table in Exercise 2, guess how P should be chosen in order to make $AP + PB$ as small as possible.   *800 ft. from point C*

(b) How do angle 1 and angle 2 compare when $AP + PB$ is as small as possible?   *Most students will say ∡1 ≐ ∡2.*

(c) Compare your conclusions in (a) and (b) with those of your classmates.   *Encourage student dialogue.*

In the previous exercise we estimated, experimentally, how to choose P in order to give the shortest path to the burning tent by way of the river. We will next develop a method for finding P exactly, without any measuring and experimenting. Before doing this, we need to introduce some new ideas. For example, we must agree on what we mean by "the distance from a point to a line" and "perpendicular lines."

Distance From a Point to a Line

When we speak of "the distance from a point to a line," we mean the distance from that point to the *nearest* point on the line. Thus, if ℓ is a

The distance from P to ℓ is the "shortest distance." The experiment in Exercise 4 suggests that the distance from P to ℓ is also the "perpendicular distance." See Review Exercise 1 concerning this and the original camper problem.

straight line and if A is any point not on ℓ, then *the distance from the point A to the line ℓ is AQ, where Q is the point on ℓ nearest to point A*.

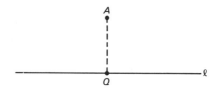

Perpendicular Lines

Two straight lines that intersect to form a 90° angle we call *perpendicular lines*.

Perpendicular lines are often labeled with a small square, as shown in the second figure below.

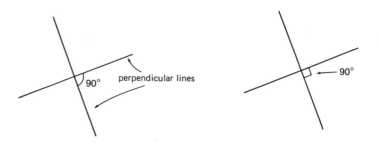

Exercise 4 is very important. Make sure they understand (c).

4. Imagine a point Q moving from left to right on the straight line ℓ in the figure below. The point A is fixed.

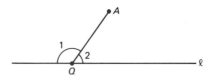

(a) Which angle do you think will be larger (angle 1 or angle 2) when Q is nearest A? ∡1 = ∡2

(b) If Q is the point on ℓ nearest A, then what is the size of angle 1? of angle 2? ∡1 = ∡2 = 90°

(c) Complete the following statement, and compare your answer with those of your classmates.

> If **A** is any point not on the line ℓ, and if **Q** is the point on ℓ nearest **A**, then the line joining **A** to **Q** is _____ to the line ℓ.
> *perpendicular*

Reflection of a Point Across a Line

Concerning the definition of reflection of a point across a line: in a later chapter it will be pointed out that ℓ is the perpendicular bisector of the line segment joining B to B*. At this point, you may want to lecture and stress that there are two conditions for the reflection of a point across a line.

The idea of "reflection" will later help us find the point P that gives the shortest path to the burning tent in a simple and exact way.

When you look into an ordinary mirror, you see your reflection at the same distance from the mirror, but on the opposite side:

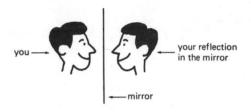

Note that we have the student "doing things" immediately, showing him it is a learning workshop in which he is an active participant, not just a lecture course where he can be passive. The teacher should lecture — but only when it is called for.

Now suppose that ℓ is a straight line and that B is a point not on ℓ:

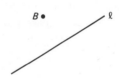

Then, with the idea of "mirror reflection" in mind, we associate with B another point B^* ("B-star") called *the reflection of B across ℓ*:

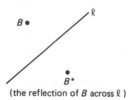

(the reflection of B across ℓ)

We require B^* to satisfy the following two conditions.

DEFINITION OF REFLECTION OF A POINT ACROSS A LINE	If ℓ is a line and if B is a point not on ℓ, then *the reflection of B across ℓ* is that point B^* such that (*i*) the distance from B^* to ℓ is equal to the distance from B to ℓ, and (*ii*) the line joining B to B^* is perpendicular to ℓ.
Make sure that both conditions of reflection are understood.	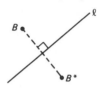

Students should not be allowed to skip Ex. 5 (b).

Make sure N* satisfies conditions (i) and (ii) for reflection. Do not let them guess at the right angle.

5. (*a*) Draw a line ℓ on a sheet of paper, and mark a point B not on ℓ. Now find B^*, the reflection of B across ℓ, *by folding the paper* (this will be easier if you use very thin paper, such as tracing paper). Compare your method with those of your classmates. B^* coincides with B.

(*b*) Use your ruler and protractor to check that the point B^* you found in part (*a*) satisfies conditions (*i*) and (*ii*) in the definition of reflection across a line.

6. (*a*) Draw a line ℓ on a sheet of paper, and mark a point N not on ℓ. Now find N^*, the reflection of N across ℓ, *using your ruler and protractor*. Compare your method with those of your classmates.

(*b*) Did you use both conditions (*i*) and (*ii*) in the definition of reflection across a line to find N^*?

7. If B^* is the reflection of B across ℓ, then what is the reflection of B^* across ℓ? B

8. Suppose B^* is the reflection of B across ℓ. Discuss with your classmates how you could use B and B^* to find the point on ℓ nearest B.
Draw the line connecting B and B^*.

9. Draw a line ℓ. Mark a point B not on ℓ and its reflection B^* across ℓ.

(*a*) Choose a point P on ℓ and measure PB and PB^*. Do this for at least three different choices of P on ℓ. What do you observe?

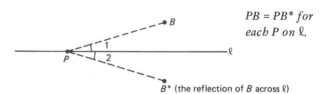

$PB = PB^*$ for each P on ℓ.

The observations in Exercise 9 will be proved in a later chapter, using congruent triangles. For the moment, students should accept these properties on faith.

(b) If B^* is the reflection of B across ℓ, then for any point P on ℓ, PB _____ PB^*. (Compare your answer with those of your classmates.) *equals*

(c) Choose a point P on ℓ and measure angle 1 and angle 2 (picture in part (a)). Do this for at least three different choices of P on ℓ. What do you observe? *∡1 = ∡2 for each P on ℓ.*

(d) If B^* is the reflection of B across ℓ, then for any point P on ℓ angle 1 _____ angle 2. (Compare your answer with those of your classmates.) *equals*

In the preceding exercise we discovered experimentally certain properties of reflection of a point across a line. Of course experiments only suggest that these properties hold. It will not be until a later chapter that we *prove* that these properties are valid. However, right now a simple diagram suggests the reason. Connect B to B^* with a line, as in the picture below.

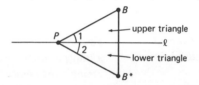

Do not teach congruent triangles at this stage. Allow the student to accept the concept intuitively.

Now observe that if the paper is folded on line ℓ, then the upper triangle folds over onto the lower triangle. Discuss with your classmates why this suggests that $PB = PB^*$ and angle 1 = angle 2.

10. In the picture below, B^* is the reflection of B across line ℓ. Why is $AP + PB = AP + PB^*$? [*Hint:* Exercise 9(b)] $PB = PB^*$

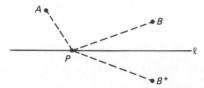

The Triangle Inequality

The next exercise deals with a fundamental property of distance. It is the basis for the intuitive idea that "the shortest path between two points is the straight line joining the points."

7 Central

Encourage students to experiment in Exercise 11.

11. (a) Draw three triangles of different shapes. (Since you will need to measure their sides, do not draw them too small. Make their sides between 3 and 10 centimeters long.)

(b) In each case in part (a), compare the sum of the lengths of two sides with the length of the third side. What do you observe?
side 1 + side 2 > side 3

Your observation in the preceding exercise will be crucial in solving the problem of the camper and the burning tent.

THE TRIANGLE INEQUALITY

Point out that there are three inequalities satisfied for each triangle.

The length of any side of a triangle is less than the lengths of the other two sides.

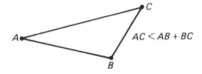

Do not rush Exercises 12 and 13. Allow as much student discussion as is needed.

12. (a) If P is any point on ℓ and if Q is the point where the straight line joining A to B^* intersects ℓ, why is $AQ + QB^* < AP + PB^*$?
Triangle Inequality

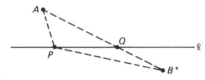

(b) Draw a line ℓ and mark points A and B^* on opposite sides of ℓ as in the picture below.

Point P is where the line joining A to B* crosses ℓ.

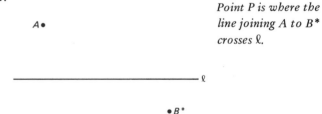

Find that point P on ℓ which makes $AP + PB^*$ as small as possible.

(c) Explain, using part (a), why your choice of P in part (b) is correct.
Triangle Inequality

8   1: The Shortest Path

Using Reflection to Find the Shortest Path

We will now use the reflection of a point across a line to find the point on the river which gives the shortest path to the burning tent.

Exercise 13 should be summarized in a total class discussion, as it is the key exercise of the chapter. The exercise may be difficult for some students. Be sure to allow discussion time when it is completed.

13. In the picture below, B^* is the reflection of B across ℓ and Q is the point where the line joining A to B^* intersects ℓ. P is *any* point on ℓ other than Q.

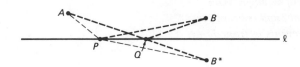

$AP + PB = AP + PB^*$

$AQ + QB = AQ + QB^*$

$AP + PB^* > AQ + QB^*$

$AP + PB > AQ + QB$

(a) Compare $AP + PB$ and $AP + PB^*$ (recall Exercise 10).

(b) Compare $AQ + QB$ and $AQ + QB^*$.

(c) Compare $AP + PB^*$ and $AQ + QB^*$ (recall Exercise 12).

(d) Compare $AP + PB$ and $AQ + QB$. [*Hint:* Remember what you just did in parts (a), (b), and (c)!]

(e) Let ℓ be a line and A and B points on one side of ℓ. Let B^* be the reflection of B across ℓ and Q the point where the line joining A to B^* intersects ℓ. If P is any point on ℓ other than Q, then

$$AP + PB \underline{\qquad} AQ + QB. \quad \textit{greater than}$$

Fill in the blank with $>$ (greater than), $=$ (equal to), or $<$ (less than).

Reflect B across the river to obtain B^*, draw the line joining A to B^*.

14. (a) Use what you learned in Exercise 13 to find the point on the river in the opening problem which gives the shortest path to the burning tent. Compare your method with those of your classmates.

(b) How does your solution compare with the result you obtained in Exercise 2? *They should be approximately equal.*

(c) Which method of solving the problem do you feel is more satisfactory —the method you just used or the experimental method used in Exercise 2? Why? *In Exercise 13 we have a precise method for finding the point we are looking for.*

Some Properties of Angles

Concerning our notation and terminology for angles: A mathematical conscience might beg us to define "angle" in full precision, and carefully distinguish between "angle" and "measure of an angle." Our moral conscience and concern for the student prevents us from doing this at this stage.

You may remember our comparison of angle 1 and angle 2 in Exercise 3(*b*), when *P* is chosen to give the shortest path to the burning tent. In order to explain the result of that comparison, we need to consider some fundamental properties of angles.

The following notation for angles will save us much writing in the future. We will use the symbol "∡" in place of the word "angle." For example, we will refer to the angle indicated in the figure below as ∡1 ("angle one").

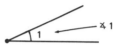

We will use this symbol in more than one way. When it is clear what we mean, we will also denote the *size* of angle 1 by the symbol ∡1. For example, if the two lines in this figure are perpendicular, then we will write ∡1 = 90°.

The concept of straight angle will not be new to many students.

When the two sides of an angle form a straight line, we have a *straight angle*:

![straight angle]

15. (*a*) Using your protractor, measure a straight angle.

(*b*) Compare your result in part (*a*) with those of your classmates, and complete the following.

The size, in degrees, of any straight angle is ___180°___.

---

16. Use what you learned in the preceding exercise to find *x*, in degrees, in each of the following pictures, without measuring.

(a) 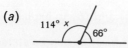 (b)

(c) 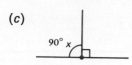 (d)

Vertical Angles

A short presentation on vertical angles may be useful to the class.

Consider the four angles formed by two intersecting straight lines:

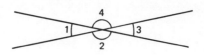

 The angles opposite each other are called *vertical angles*. In other words,

∡1 and ∡3 are vertical angles, and
∡2 and ∡4 are vertical angles.

17. (*a*) In the preceding picture, compare the *sizes* of ∡1 and ∡3; of ∡2 and ∡4. Draw some other pairs of intersecting lines and check your observation by measuring other vertical angles with a protractor.

(*b*) What do you think is true about the size of vertical angles?   *equal*

The next exercise gives some practice in using the notation for angles, and also shows *why* vertical angles are equal.

18. Consider the two intersecting lines in this figure.

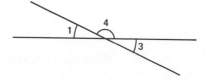

Do not emphasize "proof" at this stage. We present the concept of proof over several chapters. Do not expect too much of the student too soon.

(*a*) ∡1 + ∡4 = _____ (in degrees).   180°

(*b*) ∡3 + ∡4 = _____ (in degrees).   180°

11 Central

(c) Use parts (a) and (b) to explain why ∡1 = ∡3.   ∡1 + ∡4 = ∡3 + ∡4
Subtract ∡4
(d) The result of part (c) can be stated as follows: from both sides.

**Vertical angles are **___*equal*___.

---

19. In the following picture we have two intersecting straight lines:

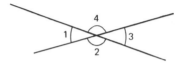

Find the other three angles, without measuring, if

(a) ∡1 = 30°   (b) ∡2 = 135°   (c) ∡3 = 10°
                         ∡4 = 135°   ∡1 = 10°
(d) ∡4 = 110°.   ∡1 = ∡3 = 45°   ∡2 = ∡4 = 170°
     ∡2 = 110°
     ∡1 = ∡3 = 70°

The Angle of Approach and the Angle of Leaving

We are now in a position to explain a result we discovered earlier with experiments (Exercise 3(b)) concerning the angle of approach (∡1) and the angle of leaving (∡2).

20. In the picture below, B^* is the reflection of B across the straight line ℓ, and Q is the point where the line joining A to B^* intersects ℓ.

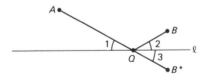

What must be true about

(a) ∡2 and ∡3? [*Hint:* Exercise 9(d)]   ∡2 = ∡3
(b) ∡1 and ∡3? Why?   ∡1 = ∡3, vertical angles
(c) ∡1 and ∡2? Discuss with your classmates how you obtain your answer, using parts (a) and (b).   ∡1 = ∡2

21. (*a*) Draw a line ℓ and mark two points A and B on one side of ℓ. Using the preceding exercise, find that point P on ℓ which makes $\angle 1 = \angle 2$.

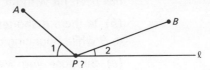

(*b*) Using Exercise 13, find the point P on ℓ which makes $AP + PB$ as small as possible.   *same point as in (a)*

(*c*) What is the reason behind the result you observed in Exercise 3(*b*)? Discuss with your classmates.   *The point P which minimizes AP + PB is the same point which makes $\angle 1 = \angle 2$.*

The preceding exercises prove what is often called "the reflection principle." Namely, the shortest path from A to B via ℓ is given by a path which bounces off ℓ, as a light ray is reflected by a mirror. Exercise 22 should bring this physical analogy to the attention of the students.

22. John and Bill are arguing about a light ray, reflected by a mirror, traveling from A to B:

JOHN: The light ray travels along a path such that the angle of approach equals the angle of leaving.

BILL: The light ray travels along the path which takes the least time.

Who is right? Discuss with your classmates.   *Both are right; the paths are the same.*

23. John and Bill stand before a mirror. Where does John look in the mirror in order to see Bill's image? Discuss with your classmates.
at the point where the angle of approach equals the angle of leaving

24. A billiards player makes a shot from point A bouncing off a cushion and hitting a ball at B.

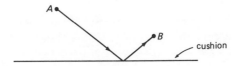

(a) What is true about the "angle of approach" versus the "angle of leaving" when a billiard ball bounces off a cushion (assuming the ball has been hit without "English," or spin)?   *equal*

(b) Is there a shorter path which the ball could follow in going from A to B while bouncing once off the cushion?   *no*

(c) Suppose you want to shoot the ball from A, bounce once off the cushion, and hit the ball at B. Describe where you would aim.
the mirror image of B, i.e., B*

Important Ideas in This Chapter

A lecture summarizing the Central section is valuable for most students at this point, using perhaps Important Ideas in This Chapter as an outline, as well as the Summary.

Perpendicular Lines Two straight lines which intersect to form a 90° angle are called perpendicular lines.

The Distance From a Point to a Line If ℓ is a straight line and A a point not on ℓ, then by "the distance from A to ℓ," we mean the distance from A to the nearest point on ℓ.

A Property of Nearest Point on a Line Let ℓ be a straight line and A a point not on ℓ. If Q is the nearest point to A on ℓ, then the line joining A to Q is perpendicular to the line ℓ.

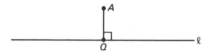

The Reflection of a Point Across a Line If ℓ is a line and B a point not on ℓ, then the reflection of B across ℓ is that point B^* such that
(i) the distance from B^* to ℓ is equal to the distance from B to ℓ, and
(ii) the line joining B to B^* is perpendicular to ℓ.

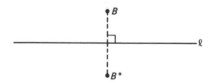

The Triangle Inequality The length of any side of a triangle is less than the sum of the lengths of the other two sides.

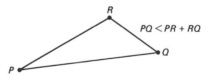

Straight Angle The size of a straight angle is 180°.

Vertical Angles are equal.

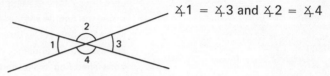

∡1 = ∡3 and ∡2 = ∡4

Symbols   ∡ (angle),   PQ (the distance from P to Q)

Summary

In this chapter we met some fundamental ideas involving points, lines, distances, and angles. These ideas were used in solving the shortest path problem: that is, in finding that point P on the line ℓ in the diagram below so that $AP + BP$ is as small as possible.

If B* is the reflection of B across ℓ, and if P is the point where ℓ intersects the line joining A to B*, then we saw that P is the point giving the shortest path. Furthermore, we showed that this same choice of P makes ∡1 = ∡2.

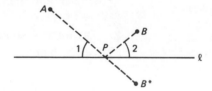

We will return to the shortest path problem in later chapters.

Historical Note

The solution of the problem of the shortest path originated with Heron of Alexandria (also known as Hero), a Greek geometer who lived around A.D. 100. However, Heron is probably better known for the ingenious mechanical inventions described in his books, among them a primitive steam-powered engine. The following passage from Morris Kline's *Mathematics, A Cultural Approach* (Reading, Mass.: Addison-Wesley,

1962), gives a colorful description of Heron's machines:

> Another science whose beginning may be found in Alexandria is the study of gases. The Alexandrians, notably Heron (about first century A.D.), a famous mathematician and engineer, learned that the steam created by heating water seeks to expand and that compressed air can also exert force. Heron is responsible for many inventions which used these forces. Temple doors opened automatically when a coin was deposited. Inside the temple another coin inserted in a machine blessed the donor by automatically sprinkling holy water upon him. Fires lit under the altar created steam, and the mystified and awe-struck audience observed gods who raised their hands to bless the worshippers, gods shedding tears, and statues pouring out libations. Doves rose and descended under the unobservable action of steam. Guns similar to the toy bee-bee gun were operated by compressed air. Steam power was used to drive automobiles in the annual religious parade along the streets of Alexandria.

"The Tunnel of Eupalinus," by June Goodfield, Scientific American, June 1964, pp. 104-112, is an interesting article.

In his book *Dioptra,* Heron describes how the famous tunnel of Eupalinus on the island of Samos may have been built. The tunnel was a striking engineering achievement of ancient times. It was 3400 feet long and constructed by digging in from the opposite sides of a mountain. The workers met in the middle with an error of only 30 feet horizontally and 10 feet vertically.

Central Self Quiz

This self quiz enables the student to examine himself. He has a second chance to remedy his weaknesses in the Review section. Answers are given at the end of the quiz.

1. Suppose in the opening problem about the camper and his burning tent that $AC = 250$ feet, $BD = 500$ feet, and $CD = 1000$ feet. Using an accurate scale drawing (scale: 1 centimeter = 100 feet), apply what you learned in Exercise 13 to estimate the length of the shortest path from A to the river, and then to the burning tent at B.

2. Do the preceding exercise for $AC = 500$ feet, $BD = 700$ feet, and $CD = 500$ feet.

3. (a) Use what you learned in Exercise 20 to quickly locate that point P on line ℓ in the following diagram which makes $\angle 1 = \angle 2$.

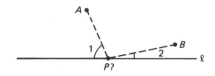

(b) Using a protractor, check that your choice of P in part (a) works.

After the Central Self Quiz, you may wish to spend a day or two on the Projects section to give a change of pace.

4. In each case, find x without measuring. You should be able to explain how you arrived at your answer.

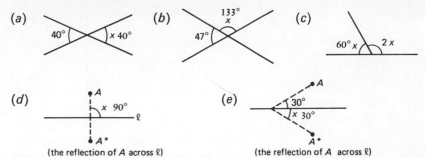

(the reflection of A across ℓ)   (the reflection of A across ℓ)

5. A map of Loco County shows that the distance from Arlington to Bonzo is 17 miles, from Bonzo to Coaltar is 23 miles, and from Coaltar to Arlington is 42 miles. Explain why this is strange.

6. Suppose a billiards player must hit ball A to make it hit ball B, first bouncing ball A off a side. Can he do this by using *any* side he chooses? Explain, and show with a picture, how he should aim in each case.

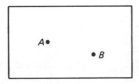

Answers

1. about 1250 ft.

2. 1300 ft.

3. (a) Join A to the reflection of B across ℓ. The point P where this line intersects ℓ is the point we are looking for.

4. (a) x = 40° (vertical angles are equal)   (b) x + 47° = 180°, therefore x = 133°   (c) 3x = x + 2x = 180°, therefore x = 60°.   (d) x = 90° (see definition of reflection of a point across a line)   (e) x = 30° (see Exercise 9 (d))

5. The Triangle Inequality doesn't hold.

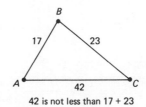

42 is not less than 17 + 23

6. To bounce off a side and hit *B*, aim at *B**, the reflection of *B* across that side. The figure illustrates this for the bottom side.

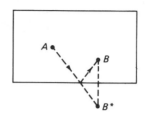

Review

Selected answers are given at the end of the Review section.

His shortest path will be from B to the nearest point Q on the river and back to B. But for the shortest path we have seen that ∡1 = ∡2. Thus if Q is the nearest point to B on ℓ, then ∡1 = ∡2 = 90°. This link between the main problem of the chapter and the concept of "distance from a point to a line" is very illuminating (as a bit of reflection will show).

Note. Do not look at the answer until you have made a good attempt to work the exercise.

1. Suppose our camper is in the tent when it catches fire. He runs to the nearest point *Q* on the river, fills his bucket, and runs back to the tent.

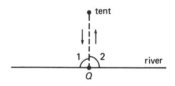

(a) Discuss with your classmates how this problem is a special case of the shortest path problem.

(b) What is the size of ∡1? of ∡2?

(c) How is your answer to part (b) related to our study of "angle of approach" and "angle of leaving" in the shortest path problem?

2. In the figure below, use your protractor and ruler

(a) to find the point on ℓ nearest to *P*

(b) to find *P**, the reflection of *P* across ℓ.

3. If B^* is the reflection of B across line ℓ, then the line joining B to B^* and the line ℓ are _____ lines.

4. In the figure below, B^* is the reflection of B across ℓ.

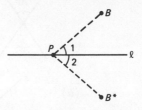

(a) If $PB^* = 5$, what is PB?   (b) If $\angle 2 = 40°$, how large is $\angle 1$?

5. State three different inequalities involving XY, YZ, and XZ.

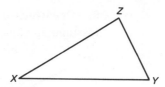

6. Q is the point where the line joining A to B^* intersects the line ℓ. Give a complete explanation of why $AQ + QB < AP + PB$, where P is any point on ℓ other than Q.

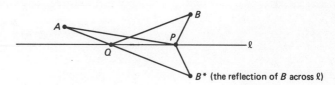

7. Use an accurate scale drawing (1 cm. = 10 ft.) to measure the length of the shortest path starting at A, touching ℓ, and ending at B.

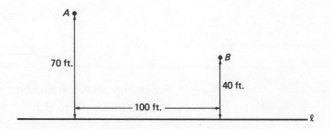

8. In each case, find x without measuring. You should be able to explain your answers (assume that lines that appear to be straight lines actually *are* straight lines).

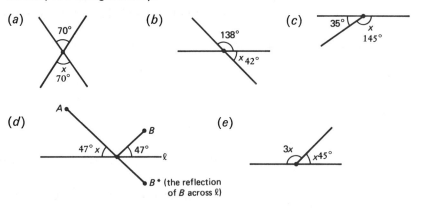

9. In each case, explain what is wrong in the picture without measuring (assume that lines which look like straight lines actually *are* straight lines).

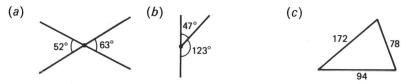

10. Describe a method for finding the point P on ℓ which makes $\angle 1 = \angle 2$.

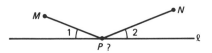

After the Review section, you may wish to spend a day or two on the Projects section before giving an exam on the chapter.

11. If one were to go from A to B to C to D, it would be a greater distance than going directly from A to D.

Prove the above statement, *using the Triangle Inequality.* [*Hint:* Draw the line joining A to C.]

Review Answers

1. (a) This is the shortest path problem with $A = B$.

(b) $\angle 1 = 90°$ and $\angle 2 = 90°$, since Q is the nearest point.

(c) The shortest path, in this case, is from the tent to Q and back to the tent. We can think of $\angle 1$ as the "angle of approach" and $\angle 2$ as the "angle of leaving." For the shortest path, $\angle 1 = \angle 2$. But notice that $\angle 1 + \angle 2 = 180°$. Thus both $\angle 1$ and $\angle 2$ equal 90°, as we have already seen in part (b).

2. (a) You simply need to find the point Q on ℓ such that the line joining P to Q is perpendicular to ℓ.

(b) You need to locate P^*, on the opposite side of ℓ, such that the line joining P to P^* is perpendicular to ℓ and the distance from P^* to ℓ is the same as the distance from P to ℓ. Notice that P^* lies on the line passing through P and Q, where Q is the nearest point to P on ℓ.

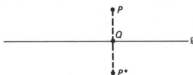

3. perpendicular

4. (a) $PB = 5$ (b) $\angle 1 = 40°$, see Central Exercise 9.

5. $XY < XZ + YZ$, $XZ < XY + YZ$, $YZ < XY + XZ$

6. $AQ + QB = AQ + QB^* < AP + PB^* = AP + PB$. Each step is explained as follows: $QB = QB^*$, since B^* is the reflection of B across ℓ; hence $AQ + QB = AQ + QB^*$ (we have simply added the same number to QB and QB^*). Now note that $AQ + QB^* = AB^*$; hence $AQ + QB^* < AP + PB^*$ by the Triangle Inequality. Finally, $AP + PB^* = AP + PB$, since $PB^* = PB$.

7. Find the reflection B^* of B across ℓ, and let Q be the point where the line joining A to B^* intersects the line ℓ. Then $AQ + QB$ is the length of the shortest path. Note that since $AQ + QB = AB^*$ (why?), the length of the shortest path is AB^*, which can be obtained with a single measurement. The result is about 148.7 feet (it is exactly $10\sqrt{221}$ feet).

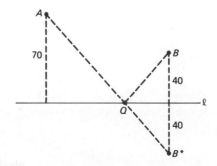

8. (a) $x = 70°$ (vertical angles)   (b) $x = 42°$ $(x + 138° = 180°)$
   (c) $x = 145°$ (d) $x = 47°$ (e) $x = 45°$ $(4x = x + 3x = 180°)$

9. (a) vertical angles not equal   (b) $123° + 47° \neq 180°$
   (c) The Triangle Inequality does not hold.

10. Let N^* be the reflection of N across ℓ. Then the desired point P is the point where the line joining M to N^* intersects the line. The reason this works was given in Central Exercise 20. Notice that we could also obtain the desired point P by joining N to M^*, the reflection of M across ℓ (why?).

11.

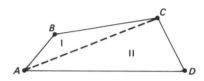

The Triangle Inequality, using triangle I, shows that

$$AC < AB + BC.$$

From triangle II,

$$AD < AC + CD.$$

Combining these, we have

$$AD < AC + CD < (AB + BC) + CD = AB + BC + CD.$$

[*Suggestion:* Always try to explain things using ideas which you have previously encountered, and facts which have previously been established, without introducing ideas or facts which have not been discussed.]

Review Self Test

1. Is there any triangle whose sides have lengths 13, 51, and 37 respectively? Explain.

2. What is wrong with the picture below?

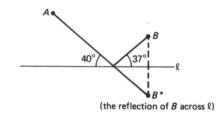

(the reflection of B across ℓ)

3. If the dotted path is the shortest path from A to B touching line ℓ, then what is x?

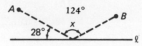

Answers

1. Since $13 + 37$ is less than 51, such a triangle, if it existed, would violate the Triangle Inequality. Hence there is no such triangle.

2. $40° \neq 37°$ (Why should these angles be equal in the picture?)

3. Since there are $180°$ in a straight angle, $28° + x + 28° = 180°$, hence $x = 124°$.

Projects

Students should attempt selected problems from the Projects section. Even the very best students will find interesting and challenging problems here. Moreover, the Projects are usually not so difficult that average students cannot enjoy doing some of them.

Projects provide a break in the routine after the Central section, Review section, or both.

Some students may wish to devise projects of their own.

Here are three references for further reading about the shortest path problem, or Heron's Problem (see the Historical Note).

R. Courant and H. Robbins, *What is Mathematics?* (New York: Oxford University Press, 1941), chap. 7.

G. Polya, *Induction and Analogy in Mathematics* (Princeton: Princeton University Press, 1954), chap. 9.

N. D. Kazarinoff, *Geometric Inequalities* (New York: Random House, 1961), chap. 3.

1. The following inspiration might lead us to a solution of Heron's Problem. We think of ℓ as the edge of a piece of paper folded in half with A on the front fold and B on the back fold:

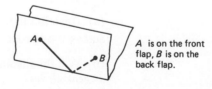

A is on the front flap, B is on the back flap.

Heron's Problem is then equivalent to finding the shortest path from A to B drawn on the surface of the paper.

(a) What is the easiest way to draw that path? *Unfold the paper and draw the line segment from A to B.*
(b) Discuss how this is related to the solution of Heron's Problem using "reflection." *Unfolding the paper corresponds to reflecting B across ℓ.*

2. Find the shortest path connecting A to B and touching first line ℓ and then line m. *Reflect A across ℓ and B across m, then draw the line joining A* to B*.*

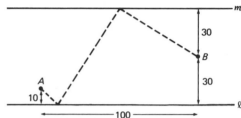

[*Hint:* Make an accurate scale drawing and use *two* reflections to find the shortest path. Determine the length of the path by measurement.]

Students may be surprised to learn that (a) and (b) are not necessarily the same distance.

3. A man is going from A to B on a horse, but first he wants to stop at the river to let the horse drink and then at the pasture to let the horse graze.

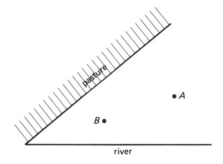

(a) Describe the shortest path he can follow to do this. [*Hint:* Reflect B across the border of the pasture and reflect A across the river.]

(b) Will the distance be different if the horse is to graze before he drinks? *in general, yes*

4. Here is a shortest path problem of Henry Dudeney, an Englishman famous for his construction of ingenious mathematical puzzles: A water

The books of Martin Gardner are an excellent source for material of a recreational, but very instructive, nature. Chapter 3 of The 2nd Scientific American Book of Mathematical Puzzles and Diversions (New York: Simon and Schuster, 1961), discusses some of Dudeney's best puzzles, and gives another interesting shortest path problem.

glass is 4 inches high and has a 6-inch circumference. On the *outside* of the glass, 1 inch from the bottom, sits a spider. On the *inside* of the glass and "diametrically opposite" the spider, 1 inch from the top, sits a fly.

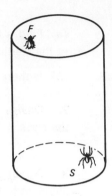

The spider sees the fly and wants to follow the shortest path on the glass to catch the fly. Find that path. [*Hint:* Replace the glass by a paper cylinder. Imagine cutting along a vertical line and flattening out the cylinder.]

Some Billiards Problems

5. Two balls, *A* and *B*, rest on a billiards table.

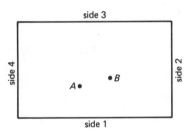

Find the path ball *A* must follow to hit ball *B* if it *first* must touch

(a) side 1   (b) side 1, then side 2   (c) side 1, then side 3

(d) side 1, then side 2, then side 3.

(Assume *A* gets no "English," or spin, when it is hit or bounces off a side.)

6. A ball rests on a square billiards table as in the following figure.

(a) Find a path the ball can follow to touch each side once and return to the starting point.

(b) Is there more than one such path?

A creative student will learn a great deal from a problem he devises.

7. Devise and solve a billiards problem of your own, then relate it to the class.

Some Mirror Problems

8. A man is in a rectangular room walled with mirrors. Where can he stand and see the back of his head?   *anywhere but dead center*

Exercise 9 should be intriguing to many students, although some of them may need a hint.

9. Can A see B's image after it has been reflected in all three mirrors?

Reflect B across #3 and A across #1. Find the shortest path from B* to A* by way of #2.

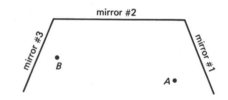

If not, why not? If so, where does A look in the mirror nearest him?

Algebra Review (Optional)

There is an Algebra Review at the end of each of the first seven chapters. These sections are optional, but most students in geometry seem to need a review of elementary algebra.

1. Add.

(a) 5
    7

(b) 5
    −7

(c) −5
     7

(d) −5
    −7

2. Subtract the lower number from the upper number in each part of Exercise 1. (What must be added to the lower number to get the upper number?)

Exercises 1-4 show the student's knowledge of the fundamental operations in using negative numbers.

3. Write as a single integer.

(a) $9 - 4 + 3 - 6 - 4$ (b) $17 - (-11)$ (c) $(-8)(-4)$

(d) $(-6)(5)$ (e) $\dfrac{18}{-6}$ (f) $\dfrac{-48}{-16}$

4. Compute.

(a) $(-1)(-1)(-1)(-1)$ (b) $(-2)(-2)(-2)(-2)(-2)$

(c) $\dfrac{(-45)(3)(-4)(0)}{(-10)(6)}$

Exponents

If x is any number and n is any positive integer, then x^n means x multiplied by itself n times:

$$x^n = \underbrace{x \cdot x \cdots x}_{n \text{ times}}$$

We call n an *exponent.* For example,

$$2^5 = 2 \cdot 2 \cdot 2 \cdot 2 \cdot 2 = 32.$$

Recall the following rules (where x and y are any numbers different from 0, and m and n are integers).

Rules for Exponents

If the student understands the following examples and Exercise 5, don't force him to memorize the general rules for exponents. He obviously understands.

I. $x^m \cdot x^n = x^{m+n}$ (the "Basic Law" for exponents)

II. $x^0 = 1$ (where $x \neq 0$)

III. $x^{-n} = \dfrac{1}{x^n}$

IV. $\dfrac{x^m}{x^n} = x^{m-n}$

V. $(x^m)^n = x^{mn}$

VI. $\left(\dfrac{x}{y}\right)^n = \dfrac{x^n}{y^n}$

VII. $(x \cdot y)^n = x^n \cdot y^n$

Algebra Review

For example,

$$2^2 \cdot 2^3 = 2^5 \quad \text{the "Basic Law"}$$

$$3^0 = 1$$

$$5^{-2} = \frac{1}{5^2} = \frac{1}{25}$$

$$\frac{4^3}{4^2} = 4^{3-2} = 4^1 = 4$$

$$(7^2)^3 = 7^6$$

$$\left(\frac{2}{3}\right)^4 = \frac{2^4}{3^4}$$

$$(2 \cdot 3)^5 = 2^5 \cdot 3^5.$$

This exercise may call for some additional examples, depending on class ability.

5. Find x.

(a) $(3^2)(3^4) = 3^x$ (b) $(2^5)^3 = 2^x$ (c) $\dfrac{5^7}{5^3} = 5^x$

(d) $\left(\dfrac{1}{2}\right)^0 = x$ (e) $\dfrac{4^3}{7^3} = \left(\dfrac{4}{7}\right)^x$ (f) $(2 \cdot 5)^4 = 2^x \cdot 5^x$

(g) $\dfrac{1}{81} = 3^x$ (h) $(2^3)(2^{-4}) = 2^x$ (i) $(3^{-2})^2 = 3^x$

6. Multiply.

(a) $(3x)(-2x^2)$ (b) $(-5ax)(2a^2)$ (c) $(-4x^2y)(-5xy^4)$

(d) $(-x)(x^2)$

7. Divide.

(a) $\dfrac{16x^{10}}{8x^2}$ (b) $\dfrac{-10a^4b^2}{-2a^3b}$ (c) $\dfrac{-5x^3}{-5x^3}$ (d) $\dfrac{x^2}{-x}$

Order of Operations

Some classes may need additional exercises on order of operations.

To simplify an expression such as

$$2(3)^3 - 4\left(\frac{1}{5}\right)^2 + 7,$$

follow these steps:

(1) First, consider the exponents and multiply as they indicate,

$$2(27) - 4\left(\frac{1}{25}\right) + 7.$$

(2) Then multiply (or divide) within each term left to right,

$$54 - \frac{4}{25} + 7.$$

(3) Finally, add (or subtract) the remaining terms,

$$61 - \frac{4}{25} = 60\frac{21}{25}.$$

8. Write as a single number.

(a) $16\left(\frac{1}{2}\right)^3 - 6\left(\frac{1}{2}\right)^2 + 9$ (b) $3(-4)^2 - 5(-2)^4 + 6(3)^3$

(c) $(2)^3(3) - (5)^2(6) + 7$

9. Fill in the table for $y = 2x^2 + 3x - 1$. [*Suggestion:* Use parentheses when you substitute for x as you see below when $x = 2$.]

x	$2x^2 + 3x - 1$	y
2	$2(2)^2 + 3(2) - 1$ $2(4) + 3(2) - 1$ $8 + 6 - 1$ 13	13
1		
0		
$-\frac{3}{4}$		
-1		
-2		
-3		

The remaining exercises provide an opportunity to use order of operations in a more meaningful context.

Encourage the use of ( ) when substituting for x.

The data of Exercise 9 will be used again in Exercise 11.

Plotting Points in the (x, y)-Plane

Recall how we plot points in the (x, y)-coordinate plane. For example, on the next page we plot the points $A = (2, 3)$, $B = (-3, 5)$, $C = (-2, -4)$, $D = (5, -3)$, and the point $(0, 0)$, which we call the origin.

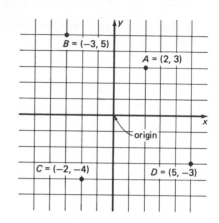

10. Plot the following points in the (x, y)-plane.

(a) (0, 2) (b) (−1, −7) (c) (3, −5) (d) (−2, 6)

11. (a) Plot the points (x, y) from the table in Exercise 9 in the (x, y)-coordinate plane.

(b) Draw a smooth curve through the points in (a).

Students often confuse $-x^2$ with $(-x)^2$.

In a later Algebra Review we state the general form of the equation of a line and a parabola. Do not stress them at this stage.

12. (a) Fill in the table below for $y = -x^2 - 3x + 7$.
[Careful! Observe that $-x^2 = -(x)^2$, not $(-x)^2$.]

x	$-x^2 - 3x + 7$	y
1	$-(1)^2 - 3(1) + 7$	
0		
−1		
$-\frac{3}{2}$		
−2		
−3		
−4		

(b) Plot the points (x, y) from your table in (a).

(c) Draw a curve through the points in (b).

13. (a) Fill in the table for $y = 2x - 3$.

Make sure they use ( ) in substituting for x in order to minimize errors.

x	2x − 3	y
3		
2		
1		
0		
−1		
−2		
−3		

(b) Plot the points (x, y) from your table and draw a line through them.

14. Do Exercise 13 for $y = -x + 4$.

Answers

Answers are given to encourage students to check their work.

1. (a) 12   (b) −2   (c) 2   (d) −12

2. (a) −2   (b) 12   (c) −12   (d) 2

3. (a) −2   (b) 28   (c) 32   (d) −30   (e) −3   (f) 3

4. (a) 1   (b) −32   (c) 0

5. (a) 6   (b) 15   (c) 4   (d) 1   (e) 3   (f) 4
   (g) −4   (h) −1   (i) −4

6. (a) $-6x^3$ (b) $-10a^3x$ (c) $20x^3y^5$ (d) $-x^3$

7. (a) $2x^8$ (b) $5ab$ (c) 1   (d) $-x$

8. (a) $9\frac{1}{2}$ (b) 130   (c) −119

9.

x	2	1	0	$-\frac{3}{4}$	−1	−2	−3
y	13	4	−1	$-2\frac{1}{8}$	−2	1	8

12. (a)

x	1	0	−1	$-\frac{3}{2}$	−2	−3	−4
y	3	7	9	$9\frac{1}{4}$	9	7	3

13. (a)

x	3	2	1	0	−1	−2	−3
y	3	1	−1	−3	−5	−7	−9

14. (a)

x	3	2	1	0	−1	−2	−3
y	1	2	3	4	5	6	7

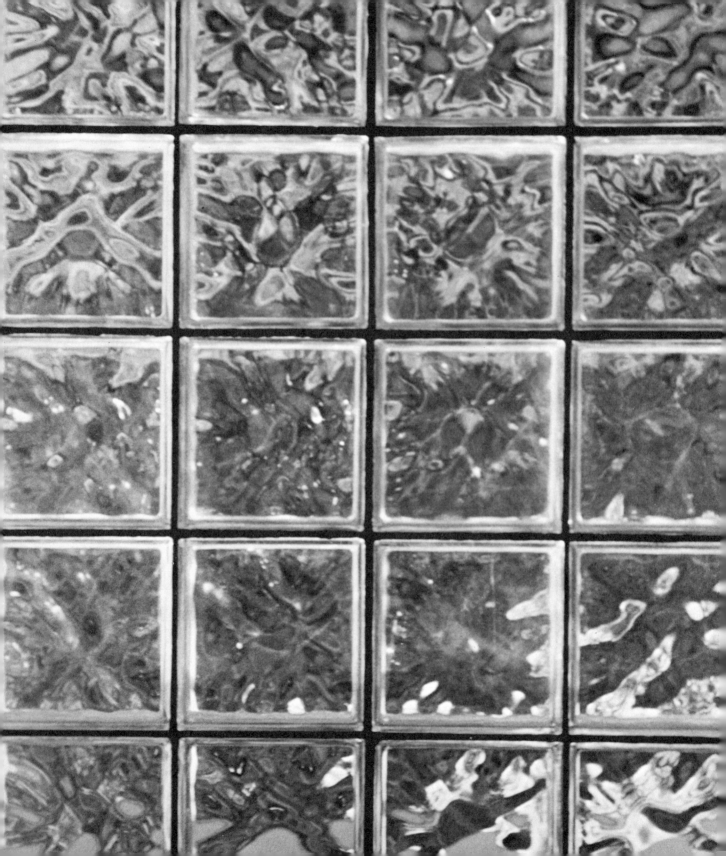

Chapter 2 Tiling the Plane

You may be surprised at our avoidance of the axiomatic technique. Our reasons for the more informal approach are many:

(a) Axiomatic Euclidean geometry requires so many axioms as to be inappropriate at this level. If one wishes to teach the "axiomatic method," the simple eleven axioms of a field are more effective, as there are more models (rational field, real number field, finite fields).

(b) The concept of proof can be learned without ever knowing about axiomatic systems, seldom does a working mathematician list "axioms." This chapter has proofs (e.g., Central Exercises 14, 17, and 28); these are "local proofs," the type students and professors meet most, not proofs set so much in a "global" axiomatic setting.

(c) Students who study mathematics beyond geometry need the concepts of geometry far more than knowledge of logical structure.

(d) Students who do not go on with mathematics have every right to enjoy the learning of geometry. We have tried to use interesting "grabber" questions (such as tiling the plane) to interest them. For most of these students, the axiomatic method is not enthralling.

Work with scissors and paper helps students learn much better than emphasizing axiomatic systems and is more effective than listening to lectures. Some students may believe that working with their hands is undignified, but they should enjoy it when they experience the mathematics first hand by actually creating their own model. Scissors and heavy paper are needed in this chapter.

Central

In tiling the plane, we do not mean we only go out so far. Stress that if a figure tiles the plane we could just keep going out farther and farther.

A figure "tiles the plane" if we can fit copies of that figure together to fill up the plane, without overlapping or leaving gaps. As the picture below shows, a square tiles the plane.

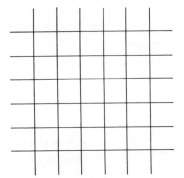

You can imagine a floor that goes on and on in all directions, covered by squares. In this chapter we will find other figures which tile the plane, but our main interest is in the basic properties of angles and polygons.

We will do some experimental tiling with various shapes. It will be convenient to have scissors and heavy paper available.

It is generally a surprise that any triangle will tile the plane. They should experiment with odd-shaped ones as well as the special triangles. Let them learn for themselves; do not tell them.

1. (*a*) Find a triangle which tiles the plane. (To test if a figure tiles the plane, cut it from heavy paper and trace its outline in different positions. Be careful not to "flip over" the figure as you move it in different positions.)

(*b*) Can you find a triangle which does *not* tile the plane? Discuss with your classmates.   *No, any triangle will tile the plane.*

Do not rush the experimental exercises. Much time will be saved later by allowing dialogue and a spirit of discovery now.

2. Which of the figures below tile the plane? Make paper models and experiment by tracing them on a sheet of paper.

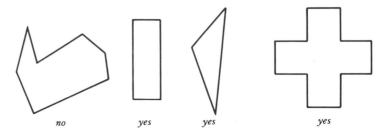

We pause to make some useful definitions.

DEFINITION   Consider the straight line passing through two points A and B. The part of that line between A and B (including both A and B) is called the *line segment* joining A to B. The points A and B are called the *endpoints* of the line segment. We denote this line segment by $\overline{AB}$.

Note that AB is a number and $\overline{AB}$ is a line segment. However, it is perhaps wise to not *make a fetish of these two notations.*

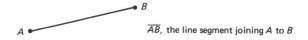

Note.   Recall from Chapter 1 that the distance from A to B, or the length of $\overline{AB}$, is denoted by the symbol AB. Keep in mind that AB is a *number*, the length of the line segment $\overline{AB}$.

An important type of figure in geometry is a polygon, defined as follows:

DEFINITION   A *polygon* is a figure formed by fitting line segments together end to end in such a way that each segment meets two other segments, one at each

You may wish to give a short presentation on polygon *and* n-gon. *We are implicitly assuming that the polygon lies in a plane; however, this is a valid definition in space also. Throughout this chapter it is understood that polygons lie in a plane.*

endpoint. The line segments are called the *sides* of the polygon. A point where two sides meet is called a *vertex* (pl. *vertices*).

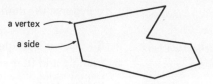

DEFINITION

An *n-gon* is a polygon with *n* sides. For example,

a 3-gon is usually called a *triangle*

a 4-gon is usually called a *quadrilateral*

a 5-gon is usually called a *pentagon*

a 6-gon is usually called a *hexagon*

a 7-gon is usually called a *heptagon*

an 8-gon is usually called an *octagon.*

Few students will guess that any *quadrilateral tiles the plane. Again, do not tell them; let them discover this. (See Ex. 45.)*

3. (a) Which of the quadrilaterals below will tile the plane? [*Hint:* Label the sides *a, b, c,* and *d,* and place copies of the figure so that sides with the same label fit together. Do not "flip over" the figure as you move it in different positions.]

(b) Can you find a quadrilateral which does *not* tile the plane? Discuss with your classmates.   *no*

4. (a) Find a 5-gon which tiles the plane.   *a particular non-regular pentagon*
(b) Find a 5-gon which does *not* tile the plane. Compare your examples with those of your classmates.   *regular pentagon*

Creative students should devise some interesting examples in Exercise 5.

5. Find a figure which tiles the plane, but is *not* a polygon. Compare your example with those of your classmates.

35 Central

The Sum of the Angles of a Triangle

We can learn something about tiling the plane by studying the angles of polygons. We start with an important property of the angles of any triangle.

The need for experimentation here varies from class to class.

6. Cut a triangle out of a sheet of paper. Tear the corners from your triangle and fit them together so that the vertices meet at a point, as shown below. What do you think is true about the sum of the angles in your triangle? *They sum to 180°.*

 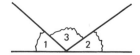

7. Use your protractor to measure the angles in three different triangles, and check the statement below for your triangles.

The sum of the angles of any triangle is 180°.

Your experiments in the preceding exercises suggested that the sum of the angles of any triangle is 180°. But is this really true for all triangles? Perhaps there is a triangle which you have not measured whose angles add up to only 179°; or worse, perhaps your measurements were not accurate! How can you be sure that what you have discovered by measuring a few triangles is true for *all* triangles? The next few exercises deal with this question. First, we need some facts about parallel lines which will help us prove this statement about the sum of the angles of a triangle.

Parallel Lines

DEFINITION

Students may not realize the importance of "in the same plane." Skew lines in space do not meet, yet they are not parallel.

We say that two straight lines are *parallel* if they lie in the same plane and do not intersect, no matter how far they are extended.

The next exercise examines the angles formed by a line crossing two parallel lines.

8. Draw two parallel lines and label them ℓ and m (ordinary ruled paper

In Exercise 8 the student should intuitively recognize which angles are equal when parallel lines ℓ and m are crossed by line t.

provides parallel lines to work with). Draw a line *t* crossing ℓ and *m*, as in the following picture:

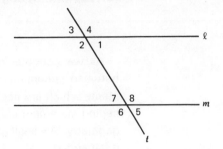

Students often have difficulty because of excess jargon. For this reason we have delayed using the names "corresponding angles," etc.

All should do part (d).

(a) Measure ∡3 and ∡7 in your picture. What do you observe?   equal

(b) Measure ∡2 and ∡8 in your picture. What do you observe?   equal

(c) Find as many pairs of equal angles as you can in your picture.
∡3 = ∡7 = ∡1 = ∡5 and ∡4 = ∡2 = ∡8 = ∡6

(d) Do the above experiment with another choice for the line *t*.

(e) Without measuring or looking at your work in parts (a)–(d), tell a classmate which angles you think are equal in the above picture.

It is convenient to have names for various pairs of angles formed by a line crossing two other lines. Hence we make the following definition.

DEFINITION Suppose a line *t* crosses two lines ℓ and *m* as in the picture below:

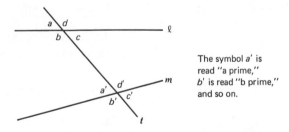

The symbol *a'* is read "a prime," *b'* is read "b prime," and so on.

Then we say the pair of angles ∡*a* and ∡*a'* are *corresponding angles*. Similarly, ∡*b* and ∡*b'* are corresponding angles, ∡*c* and ∡*c'* are corresponding angles, and ∡*d* and ∡*d'* are corresponding angles.

9. Complete the following, suggested by the experiment in Exercise 8:

> Suppose line t crosses lines ℓ and m. If ℓ is parallel to m, then corresponding angles are ___equal___.

As we continue our study of geometry, we will learn to distinguish between statements which can be proved, called *theorems*, and statements which are accepted as true without proof, called *axioms*. The preceding statement about corresponding angles is an axiom in our study of geometry. We shall assume it is true, and use it in proving theorems when it is needed.

You may wish to teach the notation "∥" for parallel. Thus, $\ell \parallel m$.

Thus, if t is a line crossing parallel lines ℓ and m:

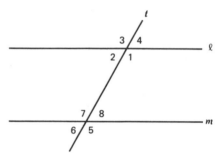

then we will assume that each pair of corresponding angles are equal; that is,

$$\angle 1 = \angle 5$$
$$\angle 2 = \angle 6$$
$$\angle 3 = \angle 7$$
$$\angle 4 = \angle 8.$$

Besides corresponding angles, certain other pairs of angles formed by a line crossing two other lines are of interest.

DEFINITION Suppose a line t crosses two other lines ℓ and m as in the picture below:

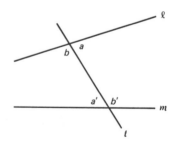

38   2: Tiling the Plane

Then the pair of angles ∡a and ∡a' are *alternate interior angles*. Similarly, ∡b and ∡b' are alternate interior angles.

Your experiment in Exercise 8 may have suggested that something special is true about alternate interior angles formed by a line *t* crossing two *parallel* lines ℓ and m. The next exercise is concerned with this.

10. Let ℓ and m be parallel lines, and let *t* be a line crossing ℓ and m:

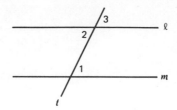

(a) Why is ∡1 = ∡3?   *corresponding angles*

(b) Why is ∡2 = ∡3?   *vertical angles*

(c) Discuss with your classmates why this proves ∡1 = ∡2.

(d) Complete the statement:

Suppose line *t* crosses lines ℓ and m. If ℓ is parallel to m, then alternate interior angles are ___equal___ .

It is the rare student who cannot tell you which angles are equal in Exercise 11. However, if too many technical words are used, many students become confused.

11. Assume that ℓ is parallel to m in the picture below:

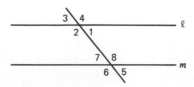

Use vertical, corresponding, and alternate interior angles to show ∡3 = ∡1 = ∡7 = ∡5 and ∡4 = ∡2 = ∡8 = ∡6.

List all those angles which are equal to each other, and give reasons why the angles are equal. (For example, ∡1 = ∡3, since they are vertical angles, and ∡3 = ∡7, since they are corresponding angles formed by a line crossing two parallel lines. Why is ∡1 = ∡7? Why is ∡4 = ∡6?) Discuss your answers and reasons with your classmates.

Encourage student dialogue as they work.

Difficulties often arise in the small-group approach if the teacher remains at his desk for long periods of time, as many students hesitate to ask questions.

12. Assume that ℓ is parallel to m in each picture, and find angles r, s, u, v, w, x, and y in each case without measuring.

(a)   (b)

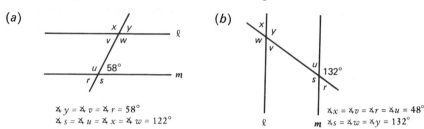

∡y = ∡v = ∡r = 58°
∡s = ∡u = ∡x = ∡w = 122°

∡x = ∡v = ∡r = ∡u = 48°
∡s = ∡w = ∡y = 132°

Exercise 13 sets up the proof in Exercise 14.

13. Line ℓ is parallel to line m in the figure. Which angles are equal?

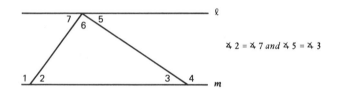

∡2 = ∡7 and ∡5 = ∡3

A Proof that the Sum of the Angles of a Triangle is 180°

We now use what we know about angles and parallel lines to *prove* that the sum of the angles of every triangle must be 180°.

14. Let ABC be any triangle. Let ℓ be a line through C parallel to the opposite side $\overline{AB}$.

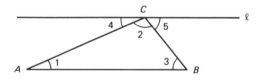

(a) What is true about ∡1 and ∡4? about ∡3 and ∡5? Why?
Alternate interior angles are equal.
(b) Using part (a), show that ∡1 + ∡2 + ∡3 = ∡4 + ∡2 + ∡5.

(c) ∡4 + ∡2 + ∡5 = _____ . 180°

Students should be aware we have proved the general case: The sum of the angles of any triangle is 180°.

(d) Using parts (b) and (c), show that the sum of the angles of the triangle ABC is 180°.   ∡1 + ∡2 + ∡3 = ∡4 + ∡2 + ∡5 = 180°

The Parallel Postulate

Too much rigor regarding the Parallel Postulate is both dull and ineffective for most students at this stage. We mention it here merely so that they know it exists.

In the preceding exercise we proved that the sum of the angles of any triangle is 180°. In a proof, we should use only those things which have been previously proved or accepted as true. You may or may not have noticed that our proof about the sum of the angles of a triangle was not quite complete, because we used a certain statement which we have not proved or previously accepted as true. If you look back at the proof, you will see that we *assumed* there actually is a line ℓ through C parallel to side $\overline{AB}$ of triangle ABC. The assumption we used is known as the *parallel postulate*. (The word *postulate* means the same thing as the word *axiom*. We use the new word in this case because the axiom we are about to state has come to be known by this name.)

THE PARALLEL POSTULATE **Given any line m and any point P not on m, there is one and only one line ℓ passing through P and parallel to m.**

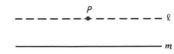

In a later chapter we will discuss the Parallel Postulate further, as it is a crucial assumption in our geometry, with a very interesting history.

Exercises Using the Sum of the Angles of a Triangle

You will often meet exercises in which you are asked to find some quantity shown in a figure. Unless otherwise instructed, you should not try to find the answer by measurement, because the drawings are often intentionally inaccurate. You should find the answer by using concepts you have learned earlier.

Throughout the course exercises are presented similar to the types in Exercises 15 and 16.
Students are encouraged not merely to get an answer, but to discuss with each other how they get that answer.

15. Find *x*. Explain to a classmate your reasoning in each case.

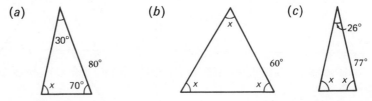

41   Central

You may wish to stress labeling all angles in a figure.

16. In each case, find x, and explain your reasoning to a classmate.

(a)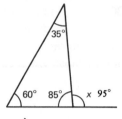

(b)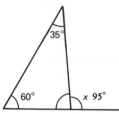

(c)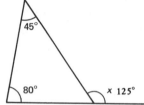

(d)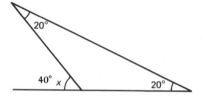

17. Consider this figure:

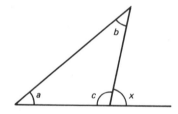

Copy and complete each of the following steps:

(a) $a + b + c =$ _____ $180°$

(b) $x + c =$ _____ $180°$

(c) Explain why $a + b + c = x + c$.

(d) $x =$ _____ (in terms of a and b) $a + b$

DEFINITION In the figure below, ∡x is called an *exterior angle* of the triangle, and ∡a and ∡b are the *opposite interior angles*.

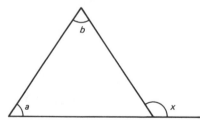

42   2: Tiling the Plane

18. Exercise 17 proves:

| THE EXTERIOR ANGLE THEOREM | Any exterior angle of a triangle is equal to the ___sum___ of the opposite interior angles. |

Our approach to proofs is a gentle and gradual process continuing throughout the course. The concept of proof is spread over several chapters.

Recall that a theorem is a statement which can be proved using previously established or accepted facts. In this course we will develop skill in proving theorems, in addition to building up a knowledge of geometry. Learning how to write a correct proof is an important skill that requires much practice and experience. We will learn more about proofs in Chapters 3 and 4.

19. What were the "previously established or accepted facts" used in proving the Exterior Angle Theorem? Discuss with your classmates.

The sum of the angles of a triangle equals 180°; a straight angle equals 180°.

20. Find x.

(a)

(b)

(c)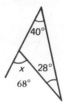

(d)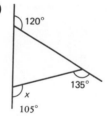

21. Draw a line and label it t. Next draw lines ℓ and m crossing t in such a way that a pair of corresponding angles are equal (use your protractor).

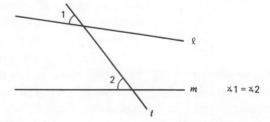

(a) What appears to be true about ℓ and m in your picture? *parallel*

(b) Do the experiment using a different line t and a different pair of corresponding angles. Complete the statement:

> **Suppose line t crosses lines ℓ and m, lying in the same plane. If corresponding angles are equal, then ℓ is _parallel_ to m.**

More will be said about converses in Chapter 4.

If you compare the above statement with the statement in Exercise 9, you will notice they are very similar. Each statement can be obtained from the other by reversing the "if" and "then" parts of the statements. Such statements are called *converses* of each other and will be studied further in Chapter 4.

Students should not panic by fearing that this is a model for proofs. It simply is an argument to justify the statement in Exercise 22 (a).

You should lead a discussion of the argument presented.

We can prove the statement in Exercise 21(b) using what we have learned so far. Here is a proof of the statement:

Suppose line t crosses lines ℓ and m in such a way that a pair of corresponding angles are equal. We want to prove that ℓ must be parallel to m. If ℓ were not parallel to m, then ℓ would intersect m and we would have a triangle formed, as in the picture below,

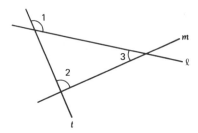

where $\angle 1$ and $\angle 2$ are equal corresponding angles. But then $\angle 1 = \angle 2 + \angle 3$ (by the Exterior Angle Theorem), and $\angle 1 = \angle 2$ (by assumption), so $\angle 3 = 0°$. This is impossible, since then there would be no triangle; hence ℓ cannot intersect m. That is, ℓ is parallel to m, as we wanted to prove.

Label the vertical angle opposite $\angle 1$. Then use the same type of argument as above.

Actually, our proof is not quite complete, since we did not consider the possibility that ℓ and m intersect "on the other side" of t with $\angle 1 = \angle 2$, as in the following picture. Discuss with your classmates why this is also impossible.

44 2: Tiling the Plane

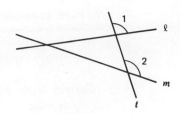

22. (*a*) Complete the statement:

> **Suppose line *t* crosses lines *ℓ* and *m*, lying in the same plane. If alternate interior angles are equal, then *ℓ* is ___parallel___ to *m*.**

(*b*) What is the relationship between the preceding statement and the statement in Exercise 10(*d*)?   *converse*

Use vertical angles and the statement in Exercise 21 (b).

(*c*) Give an argument which shows why the statement in part (*a*) is true. Compare your argument with those of your classmates. (You may use the theorem in Exercise 21 (*b*) which we have just proved in your argument.)

Types of Angles

The concept of a *ray* often occurs in discussing angles:

You may wish to give a short presentation of the material between Exercises 22 and 23.

The arrow indicates that the ray does not end. The point *P* is called the *endpoint* of the ray. A ray is often called a *half-line,* for obvious reasons. Note that an angle is formed by two rays going out from a common point. This common point is called the *vertex* of the angle. The rays are often referred to as the *sides* of the angle.

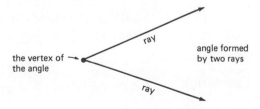

45 Central

We have a straight angle when the two rays form a straight line:

Certain types of angles are encountered often enough to give them special names:

DEFINITION   An *obtuse angle* is an angle larger than 90°.

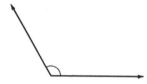

An *acute angle* is an angle less than 90°.

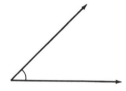

A *right angle* is an angle equal to 90°.

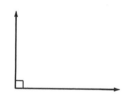

23. Can a triangle have more than one

(a) right angle?   *no*   (b) obtuse angle?   *no*

Explain your answers, using the fact that the sum of the angles of any triangle is 180°.

DEFINITION   A *right triangle* is a triangle having a 90° angle.

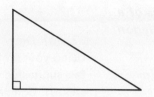

24. (*a*) Explain why in any right triangle the angles different from the 90° angle must both be acute.   *Their sum is 90°.*

(*b*) Explain why $a + b = 90°$: $90° + a + b = 180°$

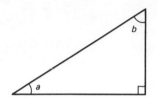

(*c*) Complete:

The sum of the two acute angles of any right triangle is   90° .

---

25. Find *x*.

(*a*)   (*b*)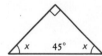

(*c*)   (*d*)

DEFINITION   If the sum of two angles is 90°, then the angles are called *complementary angles*.

26. The two acute angles of a right triangle are _____ angles.
complementary

The Sum of the Angles of a Polygon

Exercise 27 begins a series of important exercises. Be sure to encourage experimentation and discussion.

We exclude quadrilaterals such as this:

27. (a) Draw three arbitrary quadrilaterals, such as the samples below. Using your protractor, measure the angles of a quadrilateral and add them (as suggested in the picture, you should measure the *interior* (inside) angles). Compare your result with the experiments of your classmates. 360°

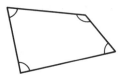 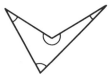

(b) What do you think is true about the sum of the interior angles of any quadrilateral?   *sum = 360°*

In the preceding exercise you reached a conclusion based on experiments. In the next exercise we prove that this conclusion is correct.

28. Using the fact that the sum of the angles of any triangle is 180°, prove that the sum of the angles of an arbitrary quadrilateral is (2)(180°) = 360°. [*Hint:* In the figure below, the dotted line is a *diagonal* of the quadrilateral. Label the angles of the two triangles which are thus formed. Keep in mind what you want to prove!]

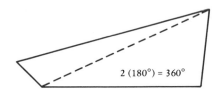

29. (a) Notice how the 5-gon in the figure below is cut up into 3 triangles by two diagonals.

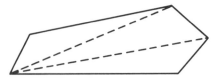

Use this to prove that the sum of the interior angles of the 5-gon must be (3)(180°) = 540°.

48   2: Tiling the Plane

Lead a discussion summarizing Exercises 28 and 29. Some students may have difficulty in accepting the generalization in Exercise 29(e).

(b) The sum of the interior angles of any 6-gon is _____. 720°

(c) The sum of the interior angles of any 7-gon is _____. 900°

(d) The sum of the interior angles of any 1000-gon is _____. 998 (180°)

(e) Explain why the following statement is true:

> **The sum of the angles of any n-gon is $(n-2)(180°)$.**

30. Find x, using what you learned in the preceding exercises.

(a)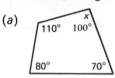

(b)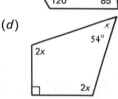

(c)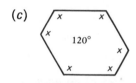

(d)

Regular Polygons

DEFINITION If a polygon has all angles equal and all sides of equal length, then it is called a *regular polygon*. Examples:

regular 3-gon   regular 4-gon   regular 6-gon

"equilateral triangle"   "square"   "regular hexagon"

Students sometimes try to define regular polygons by omitting the angle condition. Sticks nailed together (or straws with pins) will dispel this idea.

Exercise 31 builds on Exercise 29. Again, part (e) may cause difficulty.

31. Using what you have learned so far, find the size of each interior angle of a *regular*

(a) 3-gon 60°   (b) 4-gon 90°   (c) 5-gon 108°   (d) 6-gon. 120°

[*Hint:* You know the sum of the angles, and all the angles are equal.]

(e) Explain why the size of each interior angle of a regular *n*-gon is

$$\frac{(n-2)(180°)}{n}.$$

Students who have trouble with algebra may need help in Exercise 32.

32. Is there a regular polygon, each of whose angles is

(a) 120°?   (b) 108°?   (c) 72°?   (d) 10°?   (e) 176°?

    6-gon   5-gon   no   no   90-gon

33. (a) The picture below shows five copies of a triangle which fit "around a point" in a plane.

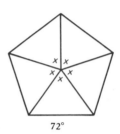

What is the size of ∡*x*? [*Hint:* There are 360° "around a point," since the total angle around a point is the sum of two straight angles: 180° + 180°.]

(b) How many copies of an *equilateral* triangle will fit "around a point" in a plane?   6

34. Three copies of a certain regular polygon fit "around a point" in a plane.

(a) What is the size of each angle of the polygon?   120°

(b) What kind of polygon is it?   regular 6-gon

Student discussion is important in Exercise 35. Encourage it.

35. If a regular polygon tiles the plane, then the size of each angle must be a number which divides the number _____ . [*Hint:* Consider copies of the polygon "around a point."]   360

Students should look at Exercises 31 and 35 again if they have trouble here.

A regular 5-gon does not tile the plane because 108 does not divide 360.

36. (a) Bill tells the class: "Any regular *n*-gon, with $3 \leq n \leq 6$, tiles the plane." Is he correct? Discuss with your classmates.   no

(b) Which *regular n*-gons will tile the plane? Discuss with your classmates.   3-gon, 4-gon, 6-gon

Parallelograms

The parallelogram and the rectangle, familiar figures which occur frequently in geometry, are defined as follows:

DEFINITION   A *parallelogram* is a quadrilateral whose opposite sides are parallel.

Students tend to use other properties of a parallelogram to define it at this stage.

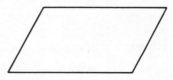

DEFINITION   A *rectangle* is a parallelogram whose angles are all right angles.

Every rectangle is a parallelogram, not vice versa.

37. Suppose ℓ and m are parallel lines, and t crosses ℓ and m, as in the figure below:

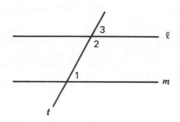

(a) Why is $\angle 3 = \angle 1$?   *corresponding angles*

(b) $\angle 3 + \angle 2 =$ _____.   180°

(c) Explain why $\angle 1 + \angle 2 = 180°$. Compare your explanation with those of your classmates.   *Substitute the equality in (a) into the equation in (b).*

38. If s is parallel to t and ℓ is parallel to m, find x, y, and z on p. 52.

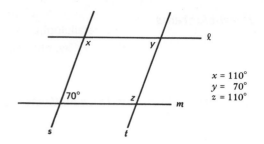

$x = 110°$
$y = 70°$
$z = 110°$

DEFINITION

If the sum of two angles is 180°, then the angles are called *supplementary angles*.

Students sometimes confuse the names complementary and supplementary. An easy mnemonic: c comes before s in the alphabet and 90 comes before 180 on the number line.

39. Suppose ℓ is parallel to m, and s is parallel to t, in the picture below.

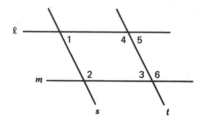

(a) ∡1 and ∡2 are _____ angles. (Recall Exercise 37 and the definition of supplementary angles.) *supplementary*

(b) ∡2 and ∡5 are _____ angles. *supplementary*

(c) ∡1 and ∡3 are _____ angles. *equal*

(d) ∡2 and ∡6 are _____ angles. *equal*

(e) ∡1 and ∡6 are _____ angles. *supplementary*

40. Consider the parallelogram in the picture below.

(a) Find all pairs of equal angles. ∡1 = ∡3, ∡2 = ∡4

(b) Find all pairs of supplementary angles. ∡1 and ∡2, ∡1 and ∡4, ∡4 and ∡3, ∡2 and ∡3

52   2: Tiling the Plane

41. If one angle of a parallelogram is 30°, find the other three angles.
150°, 30°, 150°

42. (*a*) If one angle of a parallelogram is 90°, find the other three angles.
90°, 90°, 90°
(*b*) If a parallelogram has a right angle, then it is a _____.
rectangle

43. With a careful drawing, show how any parallelogram tiles the plane.

44. (*a*) Show how two copies of any triangle can be put together to form a parallelogram.

(*b*) Use part (*a*) to show how any triangle can be used to tile the plane.

Most students are surprised that every quadrilateral will tile the plane.

45. (*a*) Verify that a quadrilateral such as the one below tiles the plane.

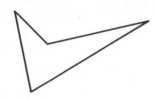

(*b*) Check that in your tiling pattern there are four copies of the quadrilateral "around each vertex," and each angle is represented once.

(*c*) Discuss with your classmates the relationship between part (*b*) and the fact that the sum of the angles of any quadrilateral is 360°.

46. Report to the class on some interesting tiling patterns you have observed in everyday life.

Important Ideas in This Chapter

Definitions

Give a presentation summarizing the Central section. Important Ideas in This Chapter *and* Summary *provide a useful outline for this.*

Line segment, endpoints of a line segment

Polygon, sides of a polygon, *vertices* of a polygon

An *n-gon* is a polygon with *n* sides.

Triangle, quadrilateral, pentagon, hexagon, heptagon, octagon

A *regular polygon* is a polygon all of whose angles are equal and all of whose sides have equal length.

An *equilateral triangle* is a regular 3-gon; a *square* is a regular 4-gon.

Two straight lines are *parallel* if they lie in the same plane and do not intersect (no matter how far they are extended).

A *parallelogram* is a quadrilateral whose opposite sides are parallel.

A *rectangle* is a parallelogram whose angles are all 90°.

Ray, or *half-line*

Vertex of an angle, *sides* of an angle

An *obtuse angle* is an angle larger than 90°.

An *acute angle* is an angle less than 90°.

A *right angle* is an angle equal to 90°.

A *right triangle* is a triangle having a right angle.

Corresponding angles (formed by a line crossing two other lines)

Alternate interior angles (formed by a line crossing two other lines)

Two angles are called *complementary angles* if their sum is 90°.

Two angles are called *supplementary angles* if their sum is 180°.

A *theorem* is a statement that can be proved using previously established or accepted facts.

An *axiom* is a statement which is accepted as true, without proof.

Axioms   If a line t crosses two parallel lines ℓ and m, then each pair of corresponding angles are equal.

The Parallel Postulate   Given any line m and a point P not on m, there is one and only one line through P parallel to m.

Theorems   The sum of the angles of any triangle is 180°.

The sum of the acute angles of a right triangle is 90°.

The Exterior Angle Theorem   Any exterior angle of a triangle is equal to the sum of the opposite interior angles.

If a line t crosses two parallel lines ℓ and m, then each pair of alternate interior angles are equal.

If a line t crosses two lines ℓ and m and a pair of corresponding angles are equal, then ℓ is parallel to m.

If a line t crosses two lines ℓ and m and a pair of alternate interior angles are equal, then ℓ is parallel to m.

The sum of the angles of any n-gon is $(n - 2)(180°)$.

Each interior angle of a regular n-gon is $\dfrac{(n - 2)(180°)}{n}$.

Symbols

$\overline{PQ}$ (the line segment with endpoints P and Q)

Summary

In this chapter tiling problems introduced the study of angles and polygons. We also considered a number of properties of parallel lines, and used some of these properties to prove that the sum of the angles of any triangle is 180°. From this we deduced several consequences; for example, we found the sum of the angles of any n-gon and used it to find the size of each angle of a regular n-gon, which in turn helped us decide which regular polygons tile the plane.

Historical Note

Many of the rudimentary facts of geometry, especially those involving measurement of length and area, were known to the ancient Egyptians and Babylonians as early as 2000 B.C. The brilliant achievement of Greek mathematicians was to develop these facts into an organized mathematical structure, giving logical arguments to prove that the results were correct. The man credited with taking the first steps in this direction is Thales (thā′lēz) of Miletus, who lived about 600 B.C. Thales was one of the "seven sages" of antiquity, and there are many stories about his wisdom and the breadth of his achievements. It is said that he spent his early years becoming wealthy as a merchant and devoted the latter part of his life to the pursuit of knowledge. When asked to describe the most remarkable thing he had ever seen, Thales replied, "an aged tyrant."

We will meet a useful theorem named after Thales in Chapter 10.

Central Self Quiz

1. The sum of the angles of a triangle is _____.

2. Find x.

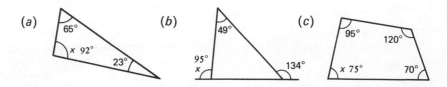

(a) 65°, x, 92°, 23°

(b) 49°, 95°, x, 134°

(c) 95°, 120°, x, 75°, 70°

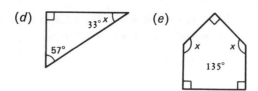

3. Can we tile the plane with any

 (a) triangle? (b) quadrilateral? (c) hexagon?

4. If line m is parallel to line ℓ, solve for the other letters.

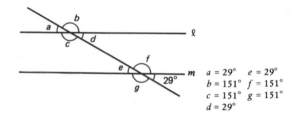

$a = 29°$ $e = 29°$
$b = 151°$ $f = 151°$
$c = 151°$ $g = 151°$
$d = 29°$

5. One of the angles of a parallelogram is 125°. Find the other angles.

Answers

1. 180°
2. (a) 92° (b) 95° (c) 75° (d) 33° (e) 135°
3. (a) yes (b) yes (c) no
4. $a = 29°, b = 151°, c = 151°, d = 29°, e = 29°, f = 151°, g = 151°$
5. 55°, 125°, 55°

Review

1. In Loco County there is a lake (known as Polygon Lake) in the shape of a *regular* polygon. Surveyors measured one of the "exterior" angles of the lake and found it to be 20°, as indicated in the picture below.

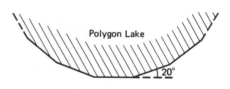

The student may recall that each angle has size $\frac{(n-2)(180°)}{n}$ and equate this with $160°$. Alternatively, he may follow the hint and write

$$(n-2)(180°) = n(160°)$$

and solve for n.

How many sides does Polygon Lake have? [*Hint:* If the polygon has n sides, then the sum of all its angles is $(n-2)(180°)$. Can you deduce from the picture above how large each angle of the polygon is? Use this to write the sum of the angles in another way.]

2. In the following exercises, find x in degrees (without measuring). Explain your answer in each case. (Beware: the figures may not be accurately drawn.)

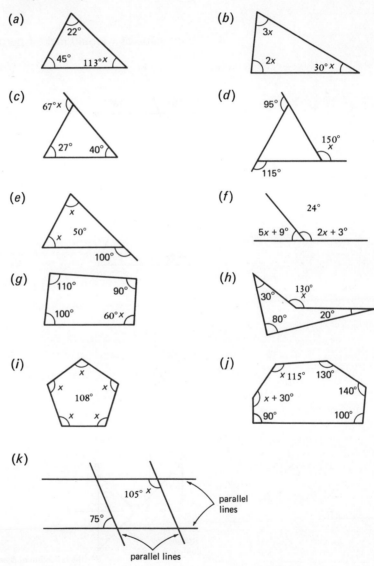

3. Find x and y (without measuring, of course).

(a)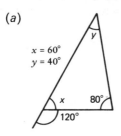
x = 60°
y = 40°

(b)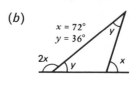
x = 72°
y = 36°

4. What value of x would make ℓ perpendicular to m?

(a)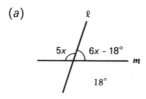

(b)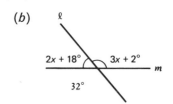

5. In each case, explain what is wrong in the figure.

(a)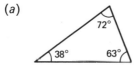

(b)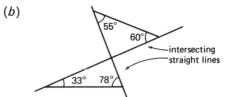

(c)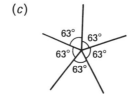

(d)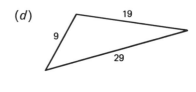

(e)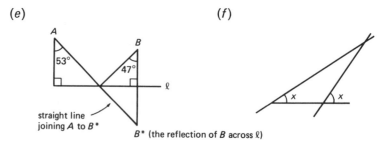

(f)

58   2: Tiling the Plane

(g)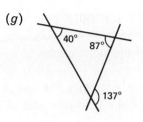

(h)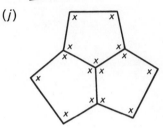

Exercise 5 (j) may suggest reopening the question: "Does a regular pentagon tile the plane?"

(i)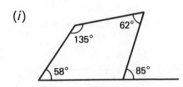

(j)

6. (a) Using the fact that the sum of the angles of a triangle is 180°, explain why the sum of the angles of any 7-gon is 900°.

(b) What is the size of each angle of a *regular* 7-gon?

7. The sum of the interior angles of any n-gon is _____.

8. Find the size of each interior angle of a regular

(a) 3-gon (b) 4-gon (c) 5-gon (d) 6-gon

(e) 100-gon (f) n-gon.

9. Can we tile the plane with a regular

(a) 3-gon? (b) 4-gon? (c) 5-gon? (d) 6-gon?

(e) 7-gon? (f) 8-gon?

(If so, show how. If not, explain why not.)

10. (a) For which n will any n-gon tile the plane?

(b) For which n will any regular n-gon tile the plane?

11. How many sides does a regular polygon have if each of its interior angles is

(a) 120°? (b) 135°? (c) 90°? (d) $128\frac{4}{7}$°?

(e) 108°?   (f) 60°?

12. The sum of the acute angles of a right triangle is ____. Explain.

13. Suppose ℓ is a straight line and A is a point not on ℓ. Can there be more than one line passing through A perpendicular to ℓ? Explain, using a picture.

14. Lines ℓ and m are parallel, and line t intersects both of them:

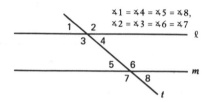

(a) Name the angles that are the same size.

(b) Name all pairs of supplementary angles.

15. Find the other angles of a parallelogram

(a) if one angle is 4°   (b) if one angle is 57°.

16. If ℓ is parallel to m, what is x? [Hint: Angle x is an exterior angle of the triangle formed when $\overline{AB}$ is extended until it meets line m.]

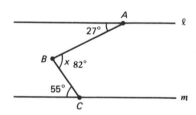

17. Use a protractor and ruler to help you accurately draw a regular

(a) 3-gon (equilateral triangle)   (b) 4-gon (square)

(c) 5-gon (pentagon)   (d) 6-gon (hexagon)   (e) 8-gon (octagon).

18. The following figures are distorted. Make an accurate drawing of each figure, where each segment is drawn with length 5.

(a) (b) (c)

19. Which of the figures in the preceding exercise tile the plane (when correctly drawn !) ?

You may wish to summarize some of the more interesting Review exercises when the students have completed them.

20. What kind of triangle must the large triangle be ? Explain. [*Hint:* $2x + 2y = $ ____]

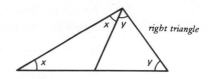

Review Answers

1. Each interior angle of the polygon is 160°. Thus if the polygon has n sides, then the sum of its angles is $(n)(160°)$. But the sum of the angles is also $(n-2)(180°)$. Hence
$$(n)(160°) = (n-2)(180°).$$
It follows that $n = 18$.

2. (a) $x = 113°$, since the sum of the angles in a triangle is 180°.

(b) $x + 2x + 3x = 180°$; that is, $6x = 180°$; hence $x = 30°$.

(c) $x = 27° + 40° = 67°$

(d) The angles opposite the exterior angle x have sizes 85° and 65°; hence $x = 85° + 65° = 150°$.

(e) $x + x = 100°$; hence $x = 50°$.

(f) $(5x + 9°) + (2x + 3°) = 180°$, so $x = 24°$.

(g) $x = 60°$, since the sum of the angles of any quadrilateral is 360°.

(h) The sum of the interior angles of the quadrilateral is 360°. Hence the interior angle (other side of x) is 230°. Then $x + 230° = 360°$, so $x = 130°$.

(i) The sum of the angles of any pentagon is 540°; hence $5x = 540°$, so $x = 108°$.

(j) The sum of the angles of the hexagon is 720°. It follows that $x = 115°$.

(k) $x + 75° = 180°$; hence $x = 105°$.

3. (a) $x + 120° = 180°$, so $x = 60°$. By the Exterior Angle Theorem, $y + 80° = 120°$, so $y = 40°$. (Once we have $x = 60°$, we could also find y from $y + 80° + 60° = 180°$.)

(b) By the Exterior Angle Theorem, $x = 2y$. Also, $2x + y = 180°$. Solving the two equations, we obtain $x = 72°$, $y = 36°$.

4. (a) With $x = 18°$, we get $5x = 90°$ and $6x - 18° = 90°$; hence ℓ perpendicular to m.

(b) No value of x will make both $2x + 18°$ and $3x + 2°$ equal to 90° (why?). We must have $(2x + 18°) + (3x + 2°) = 180°$. Solving this for x gives $x = 32°$.

5. (a) The sum of the angles is not 180°.

(b) Consider the vertical angles formed by the two intersecting straight lines. Since the sum of the angles of a triangle is 180°, one of these vertical angles is 65° and the other is 69°. But this contradicts the fact that vertical angles are equal!

(c) The five angles should add up to 360°, but they do not.

(d) The Triangle Inequality does not hold!

(e) Our work in Chapter 1 shows that certain angles should be equal, but they are not (why?).

(f) If a line t crosses two other lines ℓ and m in such a way that corresponding angles are equal, then ℓ must be parallel to m. Also, notice that the Exterior Angle Theorem cannot hold in the picture.

(g) The Exterior Angle Theorem does not hold.

(h) From the bottom right triangle we obtain $a + b = 90°$; hence $2a + 2b = 180°$. But then the sum of the angles of the big triangle must be greater than 180°.

(i) The sum of the angles of a quadrilateral is 360°; hence the unmarked angle is 105°. But $105° + 85° \neq 180°$.

(j) The sum of the angles in a pentagon is 540°. It follows that $x = 108°$. But looking at the three angles around the point in the center, we obtain $3x = 360°$, or $x = 120°$.

6. (a) Any 7-gon can be cut into 5 triangles. The sum of the angles of the 7-gon is equal to the sum of all the angles of the 5 triangles (draw a picture!). This sum is $(5)(180°) = 900°$.

(b) There are 7 equal angles with sum 900°. Hence each angle is $\dfrac{900°}{7} = 128\tfrac{4}{7}°$.

7. $(n - 2)(180°)$

8. (a) 60°   (b) 90°   (c) 108°   (d) 120°   (e) 176.4°

    (f) $\dfrac{(n - 2)(180°)}{n}$

9. (a) yes   (b) yes   (c) no   (d) yes   (e) no   (f) no

10. (a) $n = 3, 4$ (any triangle or any quadrilateral will tile the plane)

(b) $n = 3, 4, 6$

11. (a) 6   (b) 8   (c) 4   (d) 7   (e) 5   (f) 3

12. 90°. If a and b are the acute angles, then $a + b + 90° = 180°$ (since the sum of the angle of any triangle is 180°). Hence $a + b = 90°$.

13. If the lines s and t in the figure below were both perpendicular to ℓ, then the triangle formed would have two right angles, which we have already seen is impossible (why?).

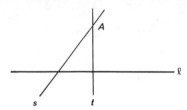

14. (a) $\angle 1 = \angle 4 = \angle 5 = \angle 8$, $\angle 2 = \angle 3 = \angle 6 = \angle 7$

(b) For example, $\angle 1$ and $\angle 2$, $\angle 5$ and $\angle 7$, $\angle 4$ and $\angle 6$, $\angle 3$ and $\angle 5$, $\angle 3$ and $\angle 8$. There are many more pairs.

15. (a) 176°, 4°, 176°   (b) 123°, 57°, 123°

16. $x = 82°$. One way to show this is to extend $\overline{AB}$ until it intersects line m at point D:

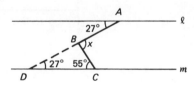

The acute angle at D is then 27° (alternate interior angles). Then the Exterior Angle Theorem, applied to triangle BDC, gives $x = 27° + 55° = 82°$. Another way is to draw a line through B parallel to ℓ and m and add the alternate interior angles.

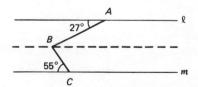

63 Review

19. All will tile the plane.

20. A right triangle! Since $2x + 2y = 180°$, $x + y = 90°$, that is, the top angle of the large triangle is a right angle.

Review Self Test

1. What is the sum of the angles in any

(a) triangle? (b) quadrilateral? (c) pentagon?

(d) hexagon?

2. What is the size of each angle in a regular n-gon, if

(a) $n = 3$? (b) $n = 4$? (c) $n = 5$? (d) $n = 6$?

(e) $n = 7$?

3. Find x.

(a)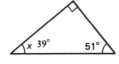

(b)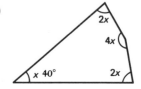

(c)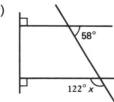

(d)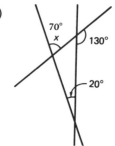

4. One of the angles of a certain parallelogram is twice the size of another angle of the parallelogram. Find the angles of this parallelogram.

Answers

1. (a) 180° (b) 360° (c) 540° (d) 720°

2. (a) 60° (b) 90° (c) 108° (d) 120° (e) $128\frac{4}{7}°$

3. (a) 39° (b) 40° (c) 122° (d) 70°

4. 60°, 120°, 60°, 120°

Projects

Some Tiling Problems

Exercises 1 and 2 are easy and interesting for most students.

1. Find a non-regular hexagon which will tile the plane.
   Lengthen two parallel sides of a regular hexagon.

2. A "Greek Cross" is obtained by fitting together five squares:

   Any figure obtained by taking five squares all the same size and fitting them together along complete edges is called a "pentomino." (As in pentagon, the prefix "penta-" means "five." The name "pentomino" is motivated by the fact that a "domino" is obtained by fitting together two squares.)

   You may have discovered in the Central section of this chapter that the Greek Cross tiles the plane. Below are some more pentominoes. Which will tile the plane?   *All will tile the plane.*

   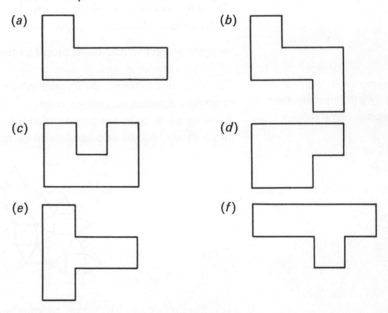

3. There are twelve different pentominoes. Seven of these were given in the preceding exercise.

Exercise 4 can be fun or frustration, depending on the student.

(a) Find the other five pentominoes.

(b) Which of these will tile the plane? *All tile the plane.*

4. Fit together the twelve different pentominoes to form

(a) a 5 by 12 rectangle (b) a 4 by 15 rectangle

(c) a 6 by 10 rectangle (d) two 5 by 6 rectangles.

5. Is it possible to fit together the twelve pentominoes to form a square (without holes in it)? Explain. *No, $\sqrt{60}$ is not an integer.*

Encourage students to consider these references.

6. There are many interesting recreational problems involving pentominoes, and the more general "polyominoes." If you are interested in learning more about this, read the following articles:

Martin Gardner, *The Scientific American Book of Mathematical Puzzles and Diversions* (New York: Simon and Schuster, 1959), chapter 13.

Martin Gardner, *New Mathematical Diversions from Scientific American* (New York: Simon and Schuster, 1966), chapter 13.

There is a comprehensive book on the subject:
S. Golomb, *Polyominoes* (New York: Charles Scribner's Sons, 1965).

A good project would be to report to the class on what you have learned.

Exercise 7 is very easy. Many interesting patterns are possible. Part (d) has an excellent reference.

7. Pleasing patterns are formed when we tile the plane with equilateral triangles, squares, or regular hexagons. It is possible to obtain more patterns by using different polygons in the same tiling. For example, here is a tiling using squares and equilateral triangles together:

(a) Find other ways to make "regular" patterns in the plane using squares and equilateral triangles together. Compare your results with those of your classmates.

(*b*) Tile the plane using regular octagons and squares together.

(*c*) Find other tiling patterns in the plane using regular polygons. Compare your results with those of your classmates.

(*d*) Read about regular tilings of the plane, called *mosaics*, in the book of D. W. Stover, *Mosaics*, Houghton Mifflin Mathematics Enrichment Series (Boston: Houghton Mifflin Co., 1966).

Tiling Space

Inexpensive cubes may be made by coating sugar cubes with liquid plastic.

8. A square will tile the plane. What is the 3-dimensional analogue of a square? Can it be used to "tile space"?   *cube, yes*

9. What is the 3-dimensional analogue of a Greek Cross? Will it tile space?   *Glue seven cubes together; it does tile space.*

Exercises 9 and 10 are excellent exercises. Encourage students to make models.

10. What is the 3-dimensional analogue of an equilateral triangle? Will it tile space? (Make some models!)   *regular tetrahedron, no*

11. There are some interesting 3-dimensional analogues to the type of problem discussed in Exercise 6 of this section. For example, the Soma Cube puzzle involves seven pieces (obtained by gluing together small cubes) which must be assembled into a single large cube. If you are interested, read about this, and related puzzles, in Martin Gardner's *Second Scientific American Book of Mathematical Puzzles and Diversions* (New York: Simon and Schuster, 1961), chapter 6.

Problems with Angles

12. When a ray of light is reflected by a mirror, the "angle of approach" equals the "angle of leaving."

$\angle 1 = \angle 2$

Imagine two *perpendicular* mirrors (forming a corner) and a ray of light reflected by the two mirrors:

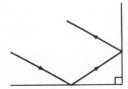

67 Projects

(*a*) The preceding picture indicates that the ray leaves the corner *parallel* to the path it enters the corner. Prove that this must always be true. [*Hint:* Use Central Exercise 21 (*b*) of this chapter.] *Label all the angles in the preceding figure.*

(*b*) Is it possible to make an angle *different* from 90° using two mirrors such that light rays leave parallel to the path of entering? *no*

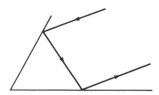

Prove that your answer is correct. Discuss with your classmates.

13. A 3-dimensional version of the last exercise is to imagine three mutually perpendicular mirrors forming a corner, with a ray of light entering the corner, being reflected once by each mirror, and then leaving the corner.

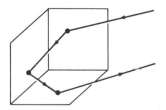

It is still true that the light always leaves the corner parallel to the path it entered! Hence if we were to shine a light into such a corner from any direction, we would always see the reflection. In 1969 the Apollo 11 astronauts left an array of such "corner reflectors" on the moon's surface. Scientists on earth are now able to determine the variation in distance from the earth to the moon with an accuracy of about 6 inches by shining a bright light (laser beam) at these moon reflectors and timing the return reflection. Read about this experiment in the article by J. E. Faller and E. J. Wampler, "The Lunar Laser Reflector," *Scientific American,* March 1970, pp. 38–49.

14. E. T. Bell writes on page 365 in *Mathematics: Queen and Servant of Science* (New York: McGraw-Hill Book Co., 1951) concerning why degrees are used as a unit in measuring angles:

Imaginative students should enjoy a lively debate.

Possibly we use degrees because the earliest shepherd-astronomers of Sumeria rated the length of the year as 360 days and passed their crude approximation to the pioneer mathematicians and astronomers of Babylon. It is known that 360 came from the Babylonians; where they got it is not known. Wherever it came from, 360 is a primitive monstrosity undeserving of mathematical survival. It should have been pitched to the undiscriminating dodo or into the backwash of Noah's Babylonian ark when the decimal system of numeration was invented, if any artificial system was to be retained.

(*a*) If you agree with E. T. Bell, what unit of measure for angles would you suggest?

(*b*) If you disagree, show why he is wrong.

15. Suppose a square is cut up into triangles, where no corner of a triangle is on a side of the square (unless it is already a corner of the square), and no three corners of triangles are collinear (i.e., in a straight line). A typical example:

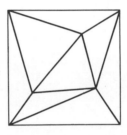

Prove that one can never cut a square into an *odd* number of triangles in this way. [*Hint:* Add together all the angles of all the triangles. Can you express this sum in terms of the number of triangles? in terms of the number of points, inside the square, where three or more triangles meet?]

The Number of Diagonals of a Convex Polygon

A polygon is *convex* if it has the property that the line segment joining two points inside the polygon is completely contained in the polygon, no matter how the two points inside are chosen. In other words, we have the following.

DEFINITION   A polygon in a plane is said to be *convex* if for each pair of points *A* and *B* inside the polygon, the segment $\overline{AB}$ is inside the polygon.

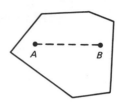

For example, any triangle is a convex polygon. In the picture below, (*i*) is convex, but (*ii*) is *not* convex:

(*i*)    (*ii*)

DEFINITION   A *diagonal* of a convex polygon is a line segment joining two vertices that are not next to each other.

For example, the pentagon below has five diagonals, shown with dotted lines.

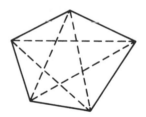

16. (*a*) Fill in the table below, giving the total number of diagonals of a convex *n*-gon.

Number of sides of n-gon	3	4	5	6	7	...	n
Total number of diagonals	0	2	5	9	14	...	$\dfrac{n(n-3)}{n}$

(If you can fill in the table correctly for *n*, you are doing great! If not, do not despair, read on.)

(*b*) For any *n*-gon, where $n \geq 3$, how many diagonals can be drawn from one vertex, for example, from vertex *A* in the following figure? $n - 3$

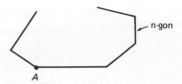

(c) How many different times can you do (b)? (How many vertices are there?)   *n*

(d) How many diagonals do you have altogether, using (b) and (c)?   *n*(*n* − 3)

(e) In (d), how many times has each diagonal been counted?   2

(f) What must you do to eliminate duplication in counting?   divide by 2

(g) Use (b)–(f) to help you find an expression for the last entry in the table in (a). $\dfrac{n(n-3)}{2}$

Algebra Review (Optional)

The area of a rectangle is the product of the length and width.

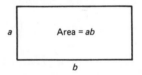

The Distributive Identity

Later in the course the formula for the area of a rectangle will be stated as an axiom. There should be no problem here because of the students' past experience.

The distributive identity is a major source of difficulty in elementary algebra.

1. Use the area of a rectangle and the figure below to explain the distributive identity:

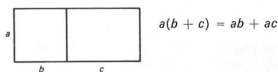

$a(b + c) = ab + ac$

The above picture of the distributive identity assumes that *a*, *b*, and *c* are positive numbers, thus the area is a positive number. However, we will assume the identity is true for all numbers on the number line, including zero and the negative numbers.

71 *Algebra Review*

2. Write each of the following as an equal expression without parentheses.

(a) $3(2x + 5)$ (b) $7(3x - 4)$ (c) $-3(2x + 5)$

(d) $-5x(2x - 3)$ (e) $-2x(-6x + 7)$ (f) $-x(-x^2 - 1)$

(g) $-1(6x - 7)$ (h) $-(9 - x^2)$

3. Exercise 1 could have been written

$$(b + c)a = ba + ca \quad \text{or} \quad ba + ca = (b + c)a.$$

Use the above to show why

(a) $7x + 2x = 9x$ (b) $7x - 2x = 5x$.

4. Simplify ("collect like terms"), if possible.

(a) $7x + 3x$ (b) $8y^2 - 6y^2$ (c) $14x^3 - 20x^3$ (d) $3x - 2x^2$

When simplifying an expression in which parentheses are within other parentheses, eliminate the innermost parentheses first. For example,

$$\begin{aligned} 2[3x^2 - 5x(x - 4)] &= 2[3x^2 - 5x^2 + 20x] \\ &= 2[-2x^2 + 20x] \\ &= -4x^2 + 40x. \end{aligned}$$

5. Simplify ("collect like terms").

(a) $5x + [x - 2(x + 4)]$ (b) $3[2x - 5(x - 2)] - 4x - 9$

(c) $\dfrac{x}{2}[2x + (x - 1)4]$ (d) $4 - (-2x + [6 - (x + 1)])$

Polynomials

In algebra, we add, subtract, multiply, and divide polynomials in a manner similar to that of arithmetic.

Comparing polynomials to integers in the fundamental operations is very helpful for most students.

Example 1. Addition

```
    207           x³        -  x +  2
   37.2                4x² + 10x - 14
 6029.07       -2x³ + 3x²        +  5
─────────      ──────────────────────
 6273.27       -x³ + 7x² +  9x -  7
```

(We line up powers of 10 in arithmetic, powers of x in algebra.)

Example 2. Subtraction (lower from upper)
(What must be added to the lower number to give the upper number?)

$$\begin{array}{r} 149 \\ 24 \\ \hline 125 \end{array} \qquad \begin{array}{r} 73 \\ 82 \\ \hline -9 \end{array} \qquad \begin{array}{r} 29 \\ -61 \\ \hline 90 \end{array} \qquad \begin{array}{r} 3x^3 - x^2 - 7 \\ 4x^2 + 5x + 2 \\ \hline 3x^3 - 5x^2 - 5x - 9 \end{array}$$

Note. You may wish to refer to preceding examples as you work the following exercises. Show all work.

6. Add.

(a) $\begin{array}{r} 7x^2 - 2x + 3 \\ 5x^2 + x - 9 \\ 6x + 4 \\ \hline \end{array}$
(b) $\begin{array}{r} 3x^3 - x + 2 \\ 7x^2 + 3x - 1 \\ -x^3 + 2x^2 + 9 \\ \hline \end{array}$
(c) $\begin{array}{r} 2x^2 - 3x + 4 \\ 7x - 5 \\ 9x^2 + 3x + 1 \\ \hline \end{array}$

(d) $x^3 - 4$, $3x^2 - 7x + 9$, and $-6x + 5$

7. Subtract the lower polynomial from the upper.

(a) $\begin{array}{r} 11x \\ -8x \\ \hline \end{array}$
(b) $\begin{array}{r} 5x - 3 \\ -2x + 4 \\ \hline \end{array}$
(c) $\begin{array}{r} 3x^2 - x + 4 \\ -4x^2 + 7 \\ \hline \end{array}$

(d) $\begin{array}{r} -2x^2 + 3x - 1 \\ 5x^3 + 4x^2 - x + 9 \\ \hline \end{array}$

8. Subtract $x^3 - 2x$ from $4x^2 - 5$.

We multiply polynomials in a manner similar to that of multiplying integers. (We line up powers of 10 in the "partial products" and then add them in arithmetic, in algebra we line up powers of x and then add them.)

Example 3. Multiplication

$$\begin{array}{r} 253 \\ 64 \\ \hline 1012 \\ 1518 \\ \hline 16192 \end{array} \qquad \begin{array}{r} 2x^2 - 4x + 3 \\ x - 5 \\ \hline -10x^2 + 20x - 15 \\ 2x^3 - 4x^2 + 3x \\ \hline 2x^3 - 14x^2 + 23x - 15 \end{array} \begin{array}{l} \\ \\ \text{(multiplying by } -5) \\ \text{(multiplying by } x) \\ \text{(adding)} \end{array}$$

9. Multiply the following polynomials.

(a) $x + 4$
$x - 2$

(b) $x - 3$
$2x + 7$

(c) $8x^2 - x + 4$
$2x^2 - 3x + 1$

(d) $x - 1$
$x^2 - 3$

(e) $x^3 - 2x + 1$
$x^2 - 1$

Similarly, the division of polynomials compares to division of integers.

Example 4. Division

$$\begin{array}{r} 376 \text{ rem. } 9 \\ 23\overline{)8657} \\ 69 \\ \hline 175 \\ 161 \\ \hline 147 \\ 138 \\ \hline 9 \end{array}$$

$$\begin{array}{r} 4x^2 + 3x + 7 \text{ rem. } 14 \\ 2x - 3\overline{)8x^3 - 6x^2 + 5x - 7} \\ 8x^3 - 12x^2 \\ \hline 6x^2 + 5x \\ 6x^2 - 9x \\ \hline 14x - 7 \\ 14x - 21 \\ \hline 14 \end{array}$$

Check:

$$(23)(376) + 9 = 8657$$

and

$$(2x - 3)(4x^2 + 3x + 7) + 14 = 8x^3 - 6x^2 + 5x - 7.$$

(We could have also written the above quotients as $376\tfrac{9}{23}$ and $4x^2 + 3x + 7 + \dfrac{14}{2x - 3}$.)

The student should realize that he can test a factor of a polynomial by division.

10. Divide (check your answer)

(a) $2x^3 - 13x^2 + 19x - 6$ by $2x - 3$

(b) $6x^3 + 29x^2 + 31x - 10$ by $2x + 5$

(c) $x^4 - 1$ by $x - 1$
[*Hint:* Write $x^4 - 1$ as $x^4 + 0 \cdot x^3 + 0 \cdot x^2 + 0 \cdot x - 1$.]

(d) $x^3 + 1$ by $x - 1$ [*Hint:* See (c).].

11. Use the figure to explain the identity:

The picture in Exercise 11 helps avoid the error in Exercise 12.

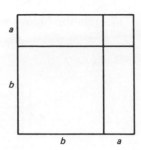

$(a + b)^2 = a^2 + 2ab + b^2$

12. Use Exercise 11 to explain why $(a + b)^2$ is *not* equal to $a^2 + b^2$.

13. Write as an expression without parentheses.

(a) $(x + 1)^2$ (b) $(x + 5)^2$ (c) $(x + 3)^2$ (d) $(2x + 3)^2$

Most students will need to consider the hint in Exercise 14. A teacher presentation may be necessary.

14. Use the figure to explain the identity:

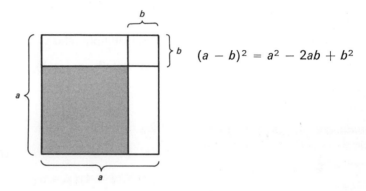

$(a - b)^2 = a^2 - 2ab + b^2$

[*Hint:*
Consider the shaded area, then $(a - b)^2 = a^2 - [2b(a - b) + b^2]$.]

15. Write as an expression without parentheses.

(a) $(x - 1)^2$ (b) $(x - 3)^2$ (c) $(x - 5)^2$ (d) $(2x - 3)^2$

Exercise 16 prepares for solving quadratic equations by completing the square.

16. Fill in the blanks.

(a) $(x + 7)^2 = $ ____ + ____ + ____

(b) $(x - 9)^2 = $ ____ + ____ + ____

(c) $(2x - 1)^2 = $ ____ + ____ + ____

(d) $(3x + 2)^2 = $ ____ + ____ + ____

(e) $(2x - 5)^2 = $ ____ + ____ + ____

A teacher presentation may be necessary in Exercise 17. Students may have difficulty visualizing the moving of the upper left rectangle.

17. Use the figures to explain the identity:

$$a^2 - b^2 = (a - b)(a + b)$$

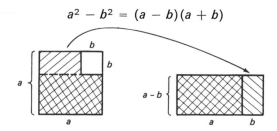

[*Hint:* Show that the shaded areas in both figures are equal.]

18. Write as an equal expression without parentheses.

(a) $(x - 1)(x + 1)$ (b) $(x - 3)(x + 3)$ (c) $(x + 5)(x - 5)$
(d) $(2x - 3)(2x + 3)$ (e) $(1 + x)(1 - x)$ (f) $(3x - 8)(3x + 8)$

19. Multiply (show all work).

(a) $\begin{array}{r} x + 3 \\ \underline{x - 2} \end{array}$ (b) $\begin{array}{r} x - 1 \\ \underline{x + 5} \end{array}$ (c) $\begin{array}{r} 2x - 3 \\ \underline{x + 4} \end{array}$ (d) $\begin{array}{r} 3x - 4 \\ \underline{x - 5} \end{array}$

Multiplying at sight is helpful for factoring in the Algebra Review in Chapter 3.

20. Multiply at sight.

(a) $(x + 3)(x - 2)$ (b) $(x - 1)(x + 5)$ (c) $(2x - 3)(x + 4)$
(d) $(3x - 4)(x - 5)$

[*Hint:* (a) See Exercise 19—there is a pattern, $(x + 3)(x - 2)$.]

21. Multiply at sight (see Exercise 20 hint).

(a) $(x + 4)(x + 4)$ (b) $(x - 6)(x - 6)$
(c) $(x + 5)(x - 3)$ (d) $(x - 7)(x + 7)$
(e) $(3x - 2)(3x - 2)$ (f) $(3x - 2)(2x + 3)$
(g) $(3x - 2)(3x + 2)$ (h) $(3x + 2)(3x + 2)$
(i) $(6x + 1)(5x - 2)$ (j) $(4x - 3)(7x + 1)$

We have not emphasized this more general identity.

22. Use the following figure to explain the identity:

$$(a + b)(c + d) = ac + ad + bc + bd$$

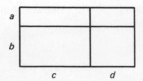

Comment. You have been asked to accept the identities in this review section based upon pictures of rectangles. We could have assumed the distributive identity in Exercise 1 and used it to show that

$$(a + b)(c + d) = (a + b)c + (a + b)d$$
$$= ac + bc + ad + bd$$
$$= ac + ad + bc + bd$$

This identity in turn could have been used to find the other identities (that we accepted from pictures).

Summary of Familiar Identities

This summary of identities may be a handy reference for some students.

The commutative property:

$$a + b = b + a \qquad \text{addition}$$
$$ab = ba \qquad \text{multiplication}$$

the associative property:

$$a + (b + c) = (a + b) + c \qquad \text{addition}$$
$$a(bc) = (ab)c \qquad \text{multiplication}$$

the distributive rule (property):

$$a(b + c) = ab + ac \qquad \text{relates addition and multiplication}$$

and

$$-a = (-1)a$$
$$a = (1)a.$$

From these it follows that

$$-a(b) = -ab$$
$$(-b)b = -b^2$$
$$-(a + b) = -a - b$$
$$-(a - b) = -a + b$$
$$a(b - c) = ab - ac$$
$$(a + b)(c + d) = ac + bc + ad + bd$$
$$(a + b)(a - b) = a^2 - b^2$$
$$(a + b)^2 = a^2 + 2ab + b^2$$
$$(a - b)^2 = a^2 - 2ab + b^2.$$

Answers

1. The area of the total figure equals the sum of the areas of the two smaller rectangles.

2. (a) $6x + 15$ (b) $21x - 28$ (c) $-6x - 15$ (d) $-10x^2 + 15x$
   (e) $12x^2 - 14x$ (f) $x^3 + x$ (g) $-6x + 7$ (h) $-9 + x^2$

3. (a) $7x + 2x = (7 + 2)x = 9x$ (b) $7x - 2x = (7 - 2)x = 5x$

4. (a) $10x$ (b) $2y^2$ (c) $-6x^3$ (d) $3x - 2x^2$

5. (a) $4x - 8$ (b) $21 - 13x$ (c) $3x^2 - 2x$ (d) $3x - 1$

6. (a) $12x^2 + 5x - 2$ (b) $2x^3 + 9x^2 + 2x + 10$ (c) $11x^2 + 7x$
   (d) $x^3 + 3x^2 - 13x + 10$

7. (a) $19x$ (b) $7x - 7$ (c) $7x^2 - x - 3$ (d) $-5x^3 - 6x^2 + 4x - 10$

8. $-x^3 + 4x^2 + 2x - 5$

9. (a) $x^2 + 2x - 8$ (b) $2x^2 + x - 21$ (c) $16x^4 - 26x^3 + 19x^2 - 13x + 4$
   (d) $x^3 - x^2 - 3x + 3$

10. (a) $x^2 - 5x + 2$ (b) $3x^2 + 7x - 2$ (c) $x^3 + x^2 + x + 1$
    (d) $x^2 + x + 1$, rem. 2

11. The area of the total figure equals the sum of the separate areas.

12. $a^2 + b^2 \neq a^2 + 2ab + b^2$

13. (a) $x^2 + 2x + 1$ (b) $x^2 + 10x + 25$ (c) $x^2 + 6x + 9$
    (d) $4x^2 + 12x + 9$

14. See hint.

15. (a) $x^2 - 2x + 1$ (b) $x^2 - 6x + 9$ (c) $x^2 - 10x + 25$
    (d) $4x^2 - 12x + 9$

16. (a) $x^2 + 14x + 49$ (b) $x^2 - 18x + 81$ (c) $4x^2 - 4x + 1$
    (d) $9x^2 + 12x + 4$ (e) $4x^2 - 20x + 25$

17. See figure.

18. (a) $x^2 - 1$ (b) $x^2 - 9$ (c) $x^2 - 25$ (d) $4x^2 - 9$
(e) $1 - x^2$ (f) $9x^2 - 64$

19. (a) $x^2 + x - 6$ (b) $x^2 + 4x - 5$ (c) $2x^2 + 5x - 12$
(d) $3x^2 - 19x + 20$

20. (a) $x^2 + x - 6$ (b) $x^2 + 4x - 5$ (c) $2x^2 + 5x - 12$
(d) $3x^2 - 19x + 20$

21. (a) $x^2 + 8x + 16$ (b) $x^2 - 12x + 36$ (c) $x^2 + 2x - 15$
(d) $x^2 - 49$ (e) $9x^2 - 12x + 4$ (f) $6x^2 + 5x - 6$ (g) $9x^2 - 4$
(h) $9x^2 + 12x + 4$ (i) $30x^2 - 7x - 2$ (j) $28x^2 - 17x - 3$

22. The area of the total figure equals the sum of the separate areas.

Chapter 3 Triangles

The opening problems A, B, C, and D are the focus of this chapter. Give the students ample time to draw these carefully. We stress that Problem A is ambiguous. After students have finished the four problems, you may wish to guide a discussion in which the students tell what they have observed. Their experience with the four problems may convince them of the validity of the various tests for congruence of triangles.

We continue the discussion of "proof" in this chapter, and Chapter 4 explores the idea even further. Our point of view is that the emphasis throughout the course should be balanced between geometric content and proving statements about that content.

By this chapter both teacher and student should be aware that our approach is a practical compromise between "the teacher tells all" and "the student discovers all." We try to provide a background of experience for the student, and then to guide him when he is equipped to make his own conjectures and reach his own conclusions.

Central

We begin this chapter with four problems involving triangles. Solve each problem by making a *careful scale drawing* (let 1 inch = 1 mile). In each problem, compare your drawing with those of other students. It is important to observe that Problem A has *two* solutions. Does each of the other problems also have more than one solution?

In Problems A, B, C, and D, students should compare and discuss their work. Make sure they understand the ambiguity of Problem A. Encourage them to look for ambiguity in Problems B, C, and D (though they will not find any).

Problem A. Dave and Ann are 4 miles apart. Dave sees Ann and he also sees a certain oak tree. The angle formed by drawing the line from Dave to Ann and drawing the line from Dave to the tree is 43°. The tree is 3 miles from Ann. How far is the tree from Dave?

Problem B. Bill and Al are 3 miles apart, but they can see each other. Each can see a statue of the authors of this book. The angle between the line from Bill to Al and the line from Bill to the statue is 73°. The angle between the line from Al to Bill and the line from Al to the statue is 51°. How far is the statue from each person?

Problem C. Mike can see an elephant 4 miles away and a donkey

3 miles away. The angle between the line from Mike to the elephant and the line from Mike to the donkey is 126°. How far apart are the two beasts?

Problem D. Jane and Mary are 2 miles apart; each can see the other. Both see a ship at sea. The ship is 1.75 miles from Jane and 3 miles from Mary. What is the angle between the lines drawn from the ship to Jane and to Mary respectively?

Each of these problems presents an idea about triangles which we will examine extensively in this chapter. We will see why Problems B, C, and D each have one solution, while Problem A has two solutions.

Problems involving models are very helpful to the student's intuition. They should not rush through these exercises.

1. (a) Using wood strips and nails, or straws with pins, or some such materials, construct a quadrilateral with sides of lengths 3, 4, 5, and 6 cm. (Or you may construct the quadrilateral by cutting four strips (or straws) with lengths 3, 4, 5, and 6 cm. and taping them to a sheet of paper.)

The ambiguity should be noticed by all students.

(b) Compare your quadrilateral with those of your classmates. Are they exactly the same shape? Can you place one over the other so they match? *not necessarily*

2. (a) By taping straws (or wood strips, or some such materials) to a piece of paper, construct a triangle with sides of lengths 5, 7, and 9 cm.

(b) Compare your triangle with those of your classmates. Can you place one triangle on the other so that they coincide (match up perfectly)? *yes*

3. Write a short explanation describing the contrast between the results of Exercises 1 and 2. *The triangles will be congruent; the quadrilaterals will not be congruent.*

Congruent Figures

When we say two figures are *congruent,* we mean that one figure is an exact copy of the other. For example, the two squares below, each having sides 3 cm. long, are congruent:

You may wish to give other examples of congruent figures.

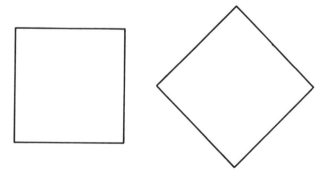

82   3: Triangles

Be sure students realize that the figures may have to be rotated or flipped over in order to coincide.

Although the squares are not in the same position, either can be moved so that it coincides with (or fits exactly over) the other.

Here is a good way to think about congruent figures. Imagine a figure drawn on a sheet of paper, and a second figure drawn on a separate thin sheet of paper (such as tracing paper), which you can see through. If you can place the thin paper on top of the first in such a way that the figures coincide, then the figures are congruent. (You may have to rotate or flip over the thin sheet of paper.)

4. (a) Recall your experiment in Exercise 1. If you know the sides of one quadrilateral have the same lengths as the sides of another quadrilateral, can you be sure the two quadrilaterals are congruent? Explain.
No, the angles may be different.
(b) Recall your experiment in Exercise 2. If you know the sides of one triangle have the same lengths as the sides of another triangle, do you think the two triangles must be congruent?   *yes*

We sometimes distort figures, not to confuse the student, but to force him to draw conclusions from the facts presented. Encourage scale drawings if the student cannot decide.

5. In each of the following problems, a pair of figures is shown, together with information about lengths of various sides and sizes of various angles. *The pictures are often distorted.* Decide which pairs are congruent (or would be congruent if accurately drawn). If you think the figures are congruent, draw them accurately to scale and explain to a classmate how to move one figure so that it coincides with the other.

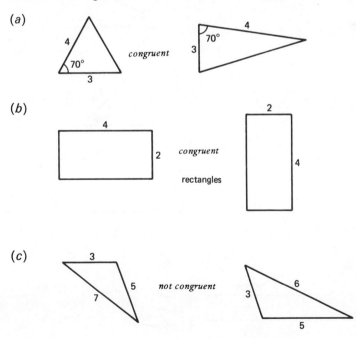

83 Central

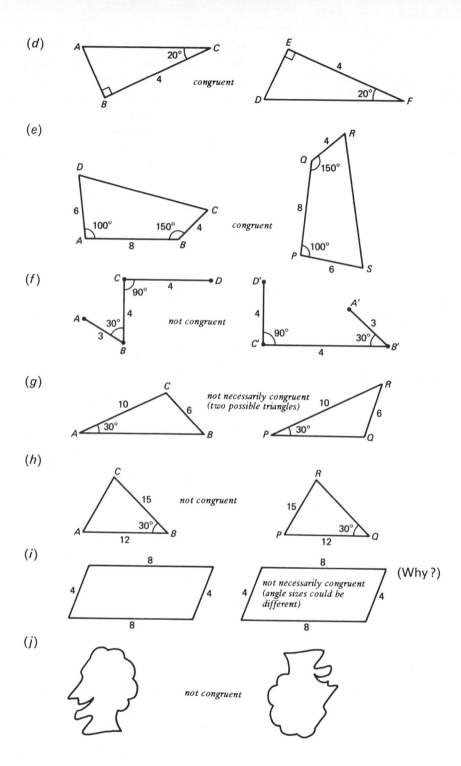

(k)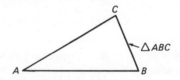

congruent

Much discussion may come from Exercise 6. Allow time for students to challenge and test conjectures. Encourage the comparison of drawings.

6. Discuss the following with your classmates (make drawings to defend your ideas). How much more information do you *think* is needed to decide if two figures are congruent, if they are both

(*a*) equilateral triangles?   (*b*) squares?   *(a), (b), (c), (d): the side lengths*

(*c*) regular hexagons?   (*d*) regular 17-gons?

(*e*) parallelograms? *one angle, two consecutive sides*   (*f*) triangles?   *SSS, SAS, or ASA*

(*g*) circles?   *radius*   (*h*) quadrilaterals?   *One possibility is 3 consecutive sides and the included angles.*

Corresponding Parts of Congruent Triangles

Since we will be discussing triangles frequently, it is convenient to have some shorthand notation for talking about them. We will often name a triangle by naming each vertex, and abbreviate the word "triangle" as "△." Thus, instead of writing "triangle *ABC*," we write "△*ABC*" for the triangle pictured below:

It is also convenient to use the symbol "≅" as a substitute for the phrase "is congruent to." For example, instead of writing "△*ABC* is congruent to △*DEF*," we write simply "△*ABC* ≅ △*DEF*."

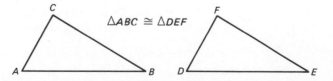

The way we label congruent triangles is very important. When we write △*ABC* ≅ △*DEF*, we mean that it is possible to place △*ABC* on top of △*DEF* in such a way that *A* is over *D*, *B* is over *E*, and *C* is over *F*.

In other words, the first letter on the left matches the first letter on the right, the second letter on the left matches the second letter on the right, and the third letter on the left matches the third letter on the right.

The arrows indicate matching letters.

The word "corresponding" is often used instead of "matching." For instance, in the picture below, $\triangle ABC \cong \triangle PQR$, and we say, "vertex A corresponds to vertex P, vertex B corresponds to vertex Q, and vertex C corresponds to vertex R."

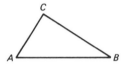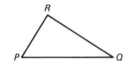

Encourage students to draw figures whenever possible.

7. Suppose you are given the following information about $\triangle ABC$: $\angle A = 65°$, $\angle B = 27°$, and $AB = 5$ cm. (By $\angle A$ we mean the angle with vertex at A; similarly for $\angle B$.) Now suppose you are told that $\triangle TOP \cong \triangle CAB$.

(a) Find $\angle T$, $\angle O$, and $\angle P$.   88° 65° 27°

(b) Find OP.   *5 cm.*

(c) Compare your answers with those of your classmates.

Listen as students discuss Exercise 8. A lecture may not be necessary. Many of us who are skilled in lecturing often neglect the art of listening to students. If we listen, much time in learning often can be saved.

8. Suppose you know that $\triangle ABC \cong \triangle DEF$.

(a) What do you think must be true about AB and DE? BC and EF? AC and DF? $\angle A$ and $\angle D$? $\angle B$ and $\angle E$? $\angle C$ and $\angle F$?   *All pairs are equal.*

(b) Discuss your answers with your classmates.

It is common to call the sides and angles of a triangle the *parts* of the triangle. Thus a triangle has six parts: its three angles and its three sides.

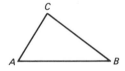

The six parts of $\triangle ABC$ are $\angle A$, $\angle B$, $\angle C$, $\overline{AB}$, $\overline{BC}$, and $\overline{CA}$. Now, suppose $\triangle ABC \cong \triangle DEF$ in the following picture.

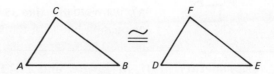

 Then we speak of their *corresponding parts:*

∡A and ∡D are corresponding angles
∡B and ∡E are corresponding angles
∡C and ∡F are corresponding angles
$\overline{AB}$ and $\overline{DE}$ are corresponding sides
$\overline{BC}$ and $\overline{EF}$ are corresponding sides
$\overline{CA}$ and $\overline{FD}$ are corresponding sides.

As you may have concluded in Exercise 8, the following important property of congruent triangles holds.

Corresponding parts of congruent triangles are equal.

In many texts, congruent triangles are defined in such a way that the boxed statement is true by definition. Our approach is to implicitly define congruence as "congruence by superposition" and take everything entailed by that as axiomatic. There is a loss in precision, but a definite gain in intuitive appeal.

This is a short way of saying that if two triangles are congruent, then their corresponding angles have the same size and their corresponding sides have the same length. Thus, if △ABC ≅ △DEF, then

$$\angle A = \angle D \qquad AB = DE$$
$$\angle B = \angle E \quad \text{and} \quad BC = EF$$
$$\angle C = \angle F \qquad AC = DF.$$

The statement that corresponding parts of congruent triangles are equal will be accepted as true, without proof, in our geometry course. In other words, this statement is an axiom in our geometry.

Conditions for Congruent Triangles

 In Exercise 2 of this section you may have discovered that when you construct a triangle with sides of given lengths, and a classmate constructs a triangle with sides of these same lengths, your triangles will be congruent. We will often use this important property of triangles, namely, *if the three sides of a triangle are equal in length to the three sides of another triangle, then the triangles are congruent.*

We will call this the SSS property of triangles (SSS stands for "side-side-side").

SSS PROPERTY

Discuss the SSS property for congruent triangles. Opening Problem D illustrated this property.

If, in △ABC and △A'B'C', we have AB = A'B', BC = B'C', and CA = C'A', then △ABC ≅ △A'B'C'.

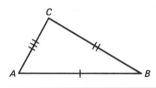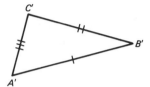

The SSS property is to be used as a test for congruence.

Note: Sides with the same number of "marks" are of equal length.

9. In each case, decide if the two triangles are congruent. Explain.

(a) SSS

(b) 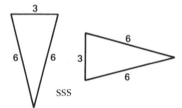 SSS

(c)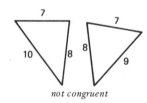
not congruent

(d) 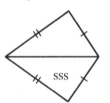 SSS

(e)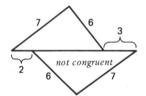
not congruent

(f) 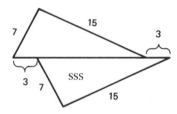 SSS

Stress the understanding of corresponding parts of congruent triangles here, and not mere notation or technical jargon.

10. Consider the two triangles below:

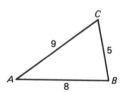 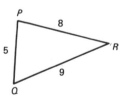

88 3: Triangles

Which of the following statements are correct and which are not? Discuss your answers with your classmates.

(a) △ABC ≅ △PQR *no* (b) △CAB ≅ △QRP *yes* (c) △BAC ≅ △PRQ *yes*

(d) △ACB ≅ △QRP *no* (e) The triangles are not congruent. *no*

Exercise 11 is an excellent way to convince students that a triangle is a rigid figure.

Exercise 11-15 are very valuable for students whose geometric intuition is not strong. Much student discussion is helpful, and a day or two in experiments is time well spent.

11. (a) A triangle is built by nailing together sticks of lengths 5″, 8″, and 10″. If a second triangle is built using sticks of the same lengths, will it necessarily be congruent to the first triangle? Explain.

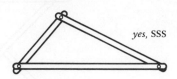

yes, SSS

(b) Build a triangle by nailing together 3 sticks, as in the figure above. Is your triangle "rigid" or is it "flexible"? Explain why you might expect this.
rigid

(c) How is this exercise related to the SSS property?   SSS *determines one and only one triangle.*

12. (a) Suppose a quadrilateral is built by nailing together 4 sticks of lengths 5″, 8″, 10″, and 7″. If a second quadrilateral is built using sticks of the same lengths, will it necessarily be congruent to the first quadrilateral?

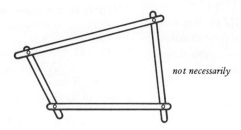

not necessarily

(b) Build a quadrilateral by nailing 4 sticks as in the figure above. Is your quadrilateral "rigid" or is it "flexible"?   *flexible*

(c) Is there an "SSSS property" for quadrilaterals? Explain.   *No, many quadrilaterals may be determined by* SSSS.

13. Which of the following frameworks are rigid? Discuss with your classmates.

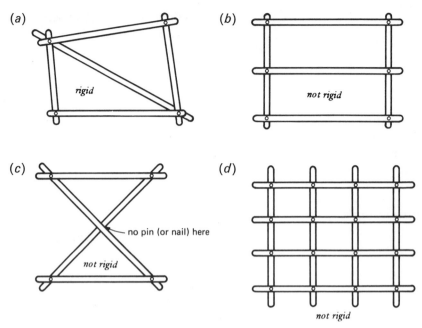

14. If any of the frameworks in the preceding exercise are not rigid, how could you make them rigid? (You may wish to make models of the frameworks.) *crossbar in (b), pin in (c), crossbar in (d)*

Exercise 15 is important. Encourage students to report subsequent observations to the class.

Exercise 16 should be drawn on the blackboard and time allowed for discussion. Opening Problem A presented this same ambiguous situation.

15. Explain why the SSS property is the basis for the triangular patterns found in constructions, such as bridges and radio towers. Observe and report to the class a use of triangles in construction that you have seen. *rigidity of the triangle*

16. (a) Draw a triangle that has one side 7 cm. long and another side 10 cm. long, with a 35° angle opposite the 7 cm. side. *Draw the 10 cm. length first, then draw the angle.*

(b) Is your triangle congruent with those of other students? *not necessarily*

(c) How many *different* triangles can you draw if you use the information in (a)? (Careful!) *two*

(d) Observe that the 35° angle is not "included" between the 7 cm. and 10 cm. sides. Does "side-side-angle" mean the triangles are always congruent? Explain. *no, since more than one triangle can be formed*

17. (a) Make an accurate drawing of $\triangle ABC$, where $AB = 5$ cm., $AC = 4$ cm., and $\angle A = 50°$.

90 *3: Triangles*

Do not rush Exercise 17, for along with Problem C, it prepares the student to accept the SAS property.

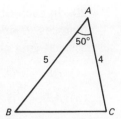

(b) Is your triangle congruent to those of other students? *yes*

18. Compare Exercises 16 and 17. What is the main difference between them? *In Ex. 17, the angle is included between the sides.*

Exercise 17 illustrates another important property of triangles, which we will call the SAS property ("side-angle-side"). This tells us that *if two sides and the included angle of a triangle are equal to two sides and the included angle of some other triangle, then the triangles are congruent.*

We are being telegraphic in our use of "equal" here. Emphasize that the angle is included between the two sides.

SAS PROPERTY

If, in $\triangle ABC$ and $\triangle A'B'C'$, we have $AB = A'B'$, $AC = A'C'$, and $\angle A = \angle A'$, then $\triangle ABC \cong \triangle A'B'C'$.

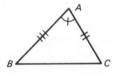 

SAS contrasts with the ambiguous SSA case of opening Problem A, where the angle was not included between the two sides.

Note. Just as with the sides, angles with the same number of marks are the same size.

19. Are the following pairs of triangles congruent? Explain.

(a)  *yes, SAS*

(b)   *no, angle not included*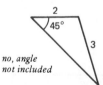

(c) 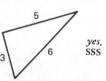 *yes, SSS*

(d)   *yes, SAS*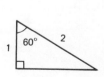

20. In each of the following cases you are given information about two triangles, $\triangle ABC$ and $\triangle PQR$. Decide if the triangles must be congruent. Discuss your answers with your classmates.

(a) $AB = PQ$, $BC = QR$, and $\angle B = \angle Q$. SAS

(b) $AB = PR$, $BC = RQ$, and $\angle B = \angle R$. SAS

(c) $AB = RQ$, $BC = QP$, and $AC = RP$. SSS

(d) $AB = PQ$, $AC = RQ$, and $\angle A = \angle P$. no

(e) $AB = PR$, $AC = RQ$, $\angle A = \angle P$, $\angle B = \angle Q$, and $\angle C = \angle R$. no

Exercise 21, along with opening Problem B, prepares the student to accept the ASA *property.*

21. (a) Make an accurate drawing of a triangle ABC such that $AB = 8$ cm., $\angle A = 25°$, and $\angle B = 56°$.

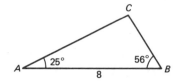

(b) Check that your triangle is congruent to that of other students. (Place your paper over that of another student; do this with the papers of several students.)

(c) Do you think ASA ("angle-side-angle") is a condition for two triangles to be congruent?

ASA PROPERTY — If, in $\triangle ABC$ and $\triangle A'B'C'$, we have $\angle A = \angle A'$, $\angle B = \angle B'$, and $AB = A'B'$, then $\triangle ABC \cong \triangle A'B'C'$.

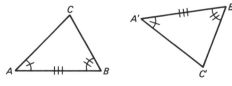

22. (a) In the figure, suppose $BC = YZ$. Are the triangles congruent?

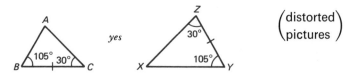

yes

(distorted pictures)

92 3: Triangles

(b) If $AB = 5$ cm., then what is XY? Explain. 5 cm.

(c) If $AC = 8$ cm., then what is XZ? Explain. 8 cm.

(d) What is the size of $\angle A$? of $\angle X$? 45°, 45°

You might want to point out that if two pairs of corresponding angles are equal, then the remaining angles are equal. Then ASA shows the triangles are congruent if a pair of corresponding sides are equal. Thus there is a valid "AAS" property.

23. (a) In the figure, suppose $AC = PQ$. Are the triangles congruent? Explain.

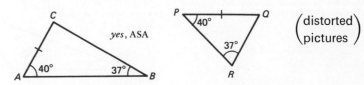

yes, ASA   (distorted pictures)

(b) What can you say about AB and PR? about BC and RQ? Explain.
equal, equal

24. In each of the following cases, use the information given in the picture to explain why the two triangles must be congruent.

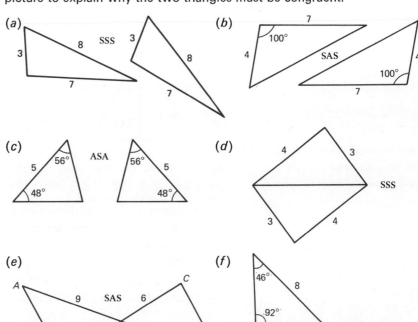

You may wish to call part (f) AAS.

E is the point where $\overline{AD}$ and $\overline{BC}$ intersect.

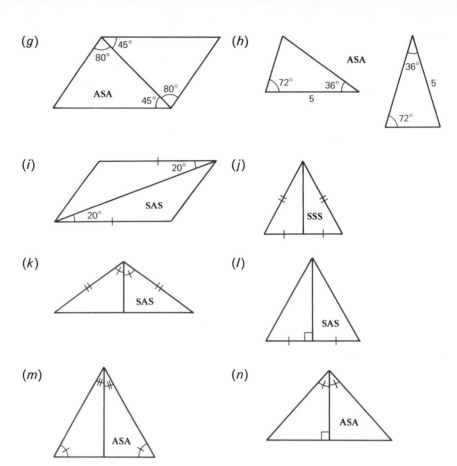

Do not rush Exercise 25, as it anticipates similarity in Chapter 8.

25. (a) Draw two triangles, △ABC and △DEF, that are *not* congruent, yet ∡A = ∡D, ∡B = ∡E, and ∡C = ∡F. *The triangles will be the same shape, but not the same size.*
(b) Suppose you know that the corresponding angles of two triangles are equal. How much more information do you need to be sure the two triangles are congruent? Discuss this with your classmates.
one corresponding side for each triangle

26. The diagram below indicates a method that might be used to find the distance between two locations *A* and *B* on opposite ends of a lake.

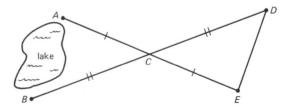

A location C is selected. Then $\overline{BC}$ is extended to D so that CD = BC, and $\overline{AC}$ is extended to E so that CE = AC. How is the distance from A to B now found? Why does the method work? $\triangle ABC \cong \triangle EDC$ by SAS, DE = AB

27. (a) What is the difficulty in opening problem A of the chapter?   no SSA *property*
(b) Discuss the principles underlying opening problems B, C, and D of this chapter.   ASA *in* B, SAS *in* C, SSS *in* D

Congruent Right Triangles

DEFINITION   The *hypotenuse* of a right triangle is the side opposite the right angle. The other sides are called the *legs* of the triangle.

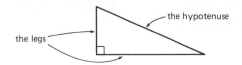

28. Suppose corresponding legs of two right triangles have equal length. Must the triangles be congruent? Explain.   *yes*, SAS

29. (a) Make an accurate drawing of a right triangle, one of whose angles is 37° and whose hypotenuse is 6 cm. long.

(b) Will your triangle be congruent to that of every other student? Explain.   *yes*, ASA

Some students may have difficulty visualizing Exercise 30.

30. During one of Napoleon's marches, his troops were forced to cross a river of unknown width. One of his officers found the width of the river by the following method. He sighted the opposite bank of the river by pulling the visor of his cap down to meet his line of vision. Remaining at the same spot, without changing the angle of his head, he turned and sighted along the bank of the river until his eyes rested on a point in line with his visor. He paced off the distance along the bank to this point and claimed that this was the width of the river. Was he correct? Explain.

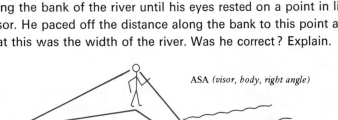

ASA *(visor, body, right angle)*

An Introduction to Proofs

We expect students to take a long time to learn how to write good proofs. We feel that often the student is placed too soon in an inflexible mold (say, a rigid double column style), in which the main requirement seems to be that he write a proof in a certain format, even though the statement he is proving is dull or obvious to him.

We present the double column format as a traditional way to write proofs in introductory geometry. However, we prefer a paragraph style because the flow of ideas is much smoother and there is less tendency toward an excess of symbols. Also, being able to give "reasons" for a string of "statements" has nothing to do with truly understanding a proof. Understanding the idea of the proof is more important, and this is given more emphasis in the paragraph style.

You will often be asked to "prove" a statement. This means you are to give a logical argument showing why the statement is true. In your proof, or argument, you will use previously established or accepted statements. When writing a proof, be sure that it is clear which statements you are assuming, and be sure that your logic is correct. Each part of the argument must be justified by stating the previously established or accepted statement upon which it is based. You will acquire a knack for writing accurate proofs only through practice.

It is very important to distinguish between *discovering* a geometric fact experimentally and *proving* that fact. To illustrate this, let us recall certain properties of the reflection of a point across a line which we considered in Chapter 1.

In the picture below, B^* is the reflection of B across line ℓ, and P is a point on ℓ:

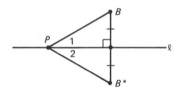

In Chapter 1 we did some experiments which indicated that

$$\angle 1 = \angle 2 \quad \text{and} \quad PB = PB^*.$$

Let us see how we can *prove* this fact about reflection, using our work on congruent triangles.

Theorem: If B^* is the reflection of point B across line ℓ, as indicated in the picture below, then $\angle 1 = \angle 2$ and $PB = PB^*$.

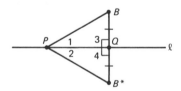

Proof: (We shall prove that $\triangle PQB \cong \triangle PQB^*$ and then use the axiom that corresponding parts of congruent triangles are equal.) Now, $\angle 3$ and $\angle 4$ are both right angles, because of the definition of reflection of a point across a line. Also $QB = QB^*$ because B^* is the reflection of B across ℓ.

Of course, $PQ = PQ$. Thus $\triangle PQB \cong \triangle PQB^*$ by the SAS property for congruent triangles. It now follows that $\angle 1 = \angle 2$ and $PB = PB^*$, since corresponding parts of congruent triangles are equal. Q.E.D.

The "Q.E.D." at the end of the preceding proof signals the end of the proof. It is an abbreviation for "quod erat demonstrandum," a Latin phrase meaning "which was to be proved."

The Perpendicular Bisector of a Line Segment

This is a good time to introduce the concept of the "perpendicular bisector" of a line segment, because it is closely related to the idea of reflection of a point across a line. First we define what we mean by the "midpoint" of a line segment.

DEFINITION The *midpoint* of a line segment $\overline{AB}$ is the point M on the segment such that $AM = MB$.

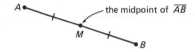

DEFINITION The *perpendicular bisector* of a line segment $\overline{AB}$ is the line passing through the midpoint of $\overline{AB}$ and perpendicular to $\overline{AB}$.

It is being tacitly assumed that $\overline{AB}$ lies in a certain plane and the perpendicular bisector lies in the same plane. You should mention this to the class if you think the ambiguity of the definition has been noticed and is causing confusion.

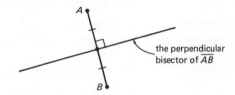

31. (*a*) Suppose ℓ is the perpendicular bisector of $\overline{AB}$. Then B is the _____ of A across ℓ.   *reflection*

(*b*) In the figure below, ℓ is the perpendicular bisector of $\overline{AB}$ and P is a point on ℓ:

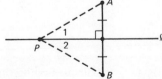

What is true about ∡1 and ∡2? about PA and PB?

equal, equal

(c) If a point P is the same distance from two other points A and B, then we say P is *equidistant* from A and B. Complete the statement below:

Each point on the perpendicular bisector of a line segment is _____ from the endpoints of the line segment.

equidistant

(d) Discuss with your classmates why parts (b) and (c) are true.

SAS

32. Complete the missing steps in the proof of the following theorem. (You may use the results of the preceding exercise.)

Theorem: If the line ℓ is the perpendicular bisector of $\overline{BD}$ in the picture below, then $\triangle ADC \cong \triangle ABC$.

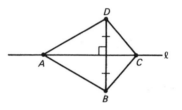

Students may say AD = AB because the upper right triangle is congruent to the lower right triangle, and similarly for CD = CB. This is certainly acceptable.

Proof: AD = AB, because ____Ex. 31(c)____. CD = CB, because ____Ex. 31(c)____. Of course AC = AC. Hence $\triangle ADC \cong \triangle ABC$ by the __SSS__ property for congruent triangles. Q.E.D.

33. Suppose in the picture below that ℓ is the perpendicular bisector of $\overline{AB}$ and m is the perpendicular bisector of $\overline{BC}$.

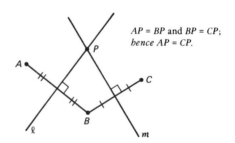

AP = BP and BP = CP; hence AP = CP.

Discuss with your classmates why AP = CP.

98   3: Triangles

34. Complete the missing statements in the proof of the following theorem.

Theorem: As shown in the picture below, suppose we are given line segments $\overline{AD}$ and $\overline{BC}$ intersecting at point E, with $AE = DE$ and $\angle A = \angle D$. Then $AB = DC$.

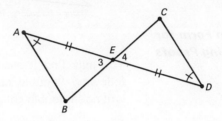

Interjecting, in a proof, the idea of the proof for purposes of clarification to the reader is acceptable form.

Proof: (The idea is to prove that $\triangle ABE \cong \triangle DCE$, and then use the fact that $\overline{AB}$ and $\overline{DC}$ are corresponding sides of two congruent triangles.) We are given that $AE = DE$ and $\angle A = \angle D$. We know that $\angle 3 = \angle 4$, because ___vertical angles___. Hence $\triangle ABE \cong \triangle DCE$ by the ___ASA___ property for congruent triangles. Hence $AB = DC$, because _____. Q.E.D.   *corresponding parts of $\cong \triangle$'s are equal*

35. Explain what is wrong in each picture. (Assume that lines that look like straight lines *are* straight lines.)

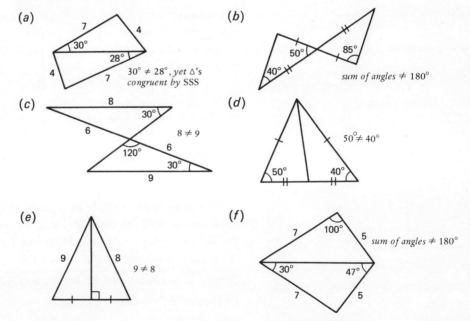

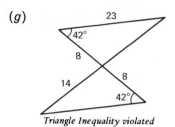

(g) Triangle Inequality violated

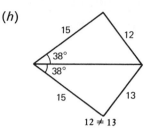

(h) $12 \neq 13$

Double Column Form for Writing Proofs

You may find it easier to keep your thoughts organized by using the following "double column" form in writing a proof. We illustrate by writing in double column form a proof of the following theorem.

Theorem: Suppose we are given the information indicated in the figure below. (That is, $\overline{AD}$ and $\overline{BC}$ are line segments intersecting at point E with $AE = DE$ and $BE = CE$.) Then we prove $\angle 3 = \angle 4$.

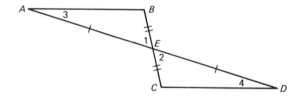

Proof:

Statement	Reason
1. $AE = DE$	1. Given
2. $BE = CE$	2. Given
3. $\angle 1 = \angle 2$	3. Vertical angles are equal.
4. Therefore $\triangle ABE \cong \triangle DCE$	4. The SAS property for congruent triangles
5. Hence $\angle 3 = \angle 4$	5. Corresponding angles of congruent triangles are equal.

Here is another example of a proof in double column form. As is often done, we present the theorem by telling what information is "given," with the aid of a picture. This is followed by a statement of what we want "to prove," followed by the proof itself.

Given: Line ℓ is parallel to line m, and $AB = CD$

To Prove: $BC = DA$

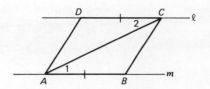

Statement	Reason
1. $AB = CD$	1. Given
2. $AC = AC$	2. Any number equals itself.
3. $\angle 1 = \angle 2$	3. Alternate interior angles formed by a line crossing two parallel lines are equal.
4. Therefore $\triangle BAC \cong \triangle DCA$	4. SAS
5. Therefore $BC = DA$	5. Corresponding sides of congruent triangles have equal length.

   The word "therefore" is used so often in proofs that it is convenient to use the symbol "∴" in its place. This is done in the proof in the next exercise.

36. Supply the reasons in the following proof.

Given: $RQ = SQ$ and $RP = SP$
To Prove: $\angle R = \angle S$

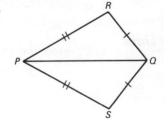

Statement	Reason	
1. $RQ = SQ$	1. Given	
2. $RP = SP$	2. Given	
3. $PQ = PQ$	3. Identity	You supply the reasons.
4. $\therefore \triangle PQR \cong \triangle PQS$	4. SSS	
5. $\therefore \angle R = \angle S$	5. Corr. parts	

37. Supply the reasons in the following proof.

Given: $AC = BC$ and $\angle 1 = \angle 2$
To Prove: $\angle A = \angle B$

Statement	Reason
1. $AC = BC$	1. Given
2. $\angle 1 = \angle 2$	2. Given
3. $CD = CD$	3. Identity
4. $\therefore \triangle ACD \cong \triangle BCD$	4. SAS
5. $\therefore \angle A = \angle B$	5. Corr. parts

As we mentioned earlier, you may find it convenient to use the double column form when writing a proof. In this book we will usually write proofs in a paragraph form (as in Exercises 32 and 34). You should practice with both forms, but use the form your teacher prefers.

Auxiliary Lines

Consider the "kite" in the picture below, with $AB = AD$ and $CB = CD$.

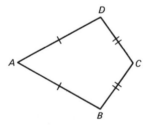

You might guess, looking at the picture, that $\angle B = \angle D$. Suppose you want to prove this is true. You could stare at the picture a long time without getting any idea about how to begin proving $\angle B = \angle D$ using the given information! But suppose you draw the line segment joining A to C, as in the following figure.

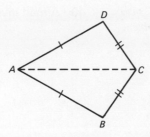

Suddenly you see two triangles, △ACD and △ACB. By the SSS property for congruent triangles, △ACD ≅ △ACB. Hence ∡B = ∡D, because they are corresponding angles of two congruent triangles.

The key to the proof was drawing the line segment $\overline{AC}$. Once that was done, it became clear how to prove the desired result. This happens frequently in a proof: By adding a certain line to a figure, you see more clearly how to prove the theorem. A line that is added to a figure to help in a proof is called an *auxiliary line*. When drawing auxiliary lines, always draw them dotted so as not to obscure the original figure.

Do not try to give a full treatment of properties of parallelograms at this stage. This chapter is supposed to be preparatory in nature. An overdose of "proof" might prove fatal.

38. Suppose in the figure below that line ℓ is parallel to line m and $PQ = RS$.

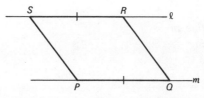

(a) Prove that $QR = SP$. [*Hint:* Draw an appropriate auxiliary line.]
Use alternate interior angles.
(b) Compare your proof with those of your classmates.

(c) Is there more than one auxiliary line you may choose to draw? Discuss this with your classmates.   yes, $\overline{PR}$ or $\overline{SQ}$

39. Suppose in the quadrilateral below we are given that $PQ = RQ$ and $PS = RS$.
Prove that ∡P = ∡R.   Draw $\overline{QS}$.
△SPQ ≅ △SRQ by SSS

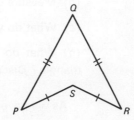

The "mad inventor problem" may suggest a student-devised problem for the Projects.

40. A mad inventor builds a pair of "super-tongs," as shown below.

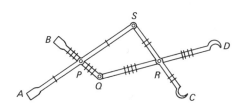

As indicated in the picture $BP = PQ$, $AP = PS$, $SR = RC$, and $QR = RD$. There are joints at P, Q, R, and S, so that as the handles A and B are moved in and out, the ends C and D move in and out. The inventor claims that the distance CD always equals the distance AB. Explain why he is right. [*Hint:* Draw $\overline{AB}$, $\overline{SQ}$, and $\overline{CD}$.] $\triangle ABP \cong \triangle SQP$ by SAS
$\triangle CDR \cong \triangle SQR$ by SAS

Isosceles Triangles

We can get more practice in writing proofs and using auxiliary lines by studying isosceles triangles, defined as follows.

DEFINITION An *isosceles* triangle is a triangle having two or more sides of equal length.

Students sometimes define an isosceles triangle as having exactly two sides equal.

$AC = BC$, so $\triangle ABC$ is isoceles.

In $\triangle ABC$, $\angle A$ and $\angle B$ are called the *base angles* of the isosceles triangle. (The base angles are the angles opposite the given equal sides.)

41. (a) Draw an isosceles triangle.

(b) Measure the angles of your triangle.

(c) What do you notice about the base angles of your triangle?   *equal*

(d) What do you think is true about the base angles of any isosceles triangle? Discuss with your classmates.   *equal*

As we have mentioned before, although careful drawings and measurements often help to convince us that certain geometric statements are true, such experiments never prove anything. The next exercise contains a proof of the theorem you observed in the preceding exercise.

42. Supply the missing steps in the proof of the following theorem.

Theorem: **If a triangle is isosceles, then its base angles are equal.**

Proof: Consider an isosceles triangle, △ABC, with AC = BC. We want to prove ∡A = ∡B. Let D be the midpoint of $\overline{AB}$, and draw the auxiliary line $\overline{CD}$, as shown below.

Now AC = BC was given, and AD = BD, because D is __*midpt. of $\overline{AB}$*__. Also, CD = CD. Hence △ADC ≅ △BDC by __SSS__. Therefore ∡A = ∡B, because __*corr. parts*__. Q.E.D.

In the preceding exercise we proved that *if two sides of a triangle are of equal length, then the angles opposite those sides are equal.* We can represent the statement in pictures as follows:

If  then

In the next exercise, we shall prove the *converse* of the above theorem. That is, we shall prove that *if the angles opposite two sides of a triangle* are equal, then those two sides are of equal length. In pictures:

If  then

43. (*a*) Write a proof of the following theorem.

Theorem: **If two angles of a triangle are equal, then the sides opposite those angles are of equal length.**

[*Hint:* Draw △*ABC*, with ∡*A* = ∡*B*. Draw an auxiliary line *bisecting* ∡*C*, as suggested in the picture below.

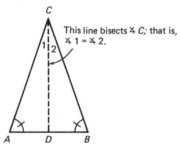

Let *D* be the point where the auxiliary line meets $\overline{AB}$. You want to prove *AC* = *BC*.]   ∡*CDA* = ∡*CDB* and *CD* = *CD*, ASA

(*b*) Check the proofs of your classmates. Do you feel they are correct?

44. Find *x*.

(a) 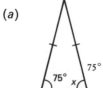 (b) 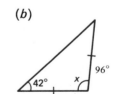 (c)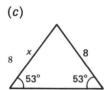

(d) 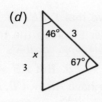 (e) 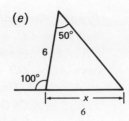 (f)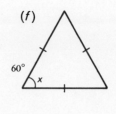

45. Supply the reasons in the following proof that $\triangle CAE \cong \triangle DBE$.

Given: $AC = BD$ and $\angle 1 = \angle 2$

To Prove: $\triangle CAE \cong \triangle DBE$

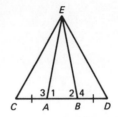

Statement	Reason
1. $AC = BD$	1. Given
2. $AE = BE$	2. Sides opposite equal angles
3. $\angle 3 = \angle 4$	3. $180° - \angle 1 = 180° - \angle 2$
4. $\therefore \triangle CAE \cong \triangle DBE$	4. SAS

Colored lines help the eye to distinguish triangles when there is overlapping.

The next exercise involves proving a certain pair of "overlapping" triangles are congruent. If you find it difficult to keep the two triangles separated in your mind, it might be helpful to make a drawing using colored pencils.

46. Consider the figure below with the given information, and fill in the missing steps in the proof.

Given: $AD = CE$ and $\angle 1 = \angle 2$

To Prove: $AE = CD$

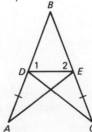

107 Central

Proof: We shall prove that $\triangle ABE \cong \triangle CBD$, from which it will follow that $AE = CD$. In order to prove the triangles congruent, first note that $DB = EB$, because *sides opp. equal ⦨'s* . Since $AD = CE$ is given, we have $AD + DB = CE + EB$. But $AB = AD + DB$ and $CB = CE + EB$, so $AB = CB$. Thus we have $AB = CB$, $EB = DB$, and $\angle B = \angle B$. Hence $\triangle ABE \cong \triangle CBD$, by SAS . Hence $AE = CD$, because *corr. parts* . Q.E.D.

47. Given: $AE = DE$ and $\angle 1 = \angle 2$

    Prove: $BE = CE$

    ⦨A and ⦨D are base angles of an isosceles △.
    $\triangle AEB \cong \triangle DEC$

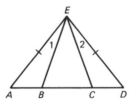

48. Given: $\angle 1 = \angle 2$

    Prove: $\triangle ABC$ is isosceles

    $AC = BC$, since sides opp. equal base angles are equal.

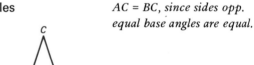

49. Given: $AC = BC$ and $\angle 1 = \angle 2$

    Prove: $\overline{CD}$ is perpendicular to $\overline{AB}$

    $\triangle CDB \cong \triangle CDA$ by SAS
    ∴ ⦨CDB = ⦨CDA
    But ⦨CDB + ⦨CDA = 180°.

50. Given: $\angle 1 = \angle 2$, $\overline{AD}$ is perpendicular to $\overline{BA}$, and $\overline{DC}$ is perpendicular to $\overline{BC}$

    Prove: $AD = DC$

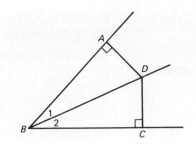

$\triangle BDA \cong \triangle BDC$
by ASA

Important Ideas in This Chapter

Summarize the chapter, perhaps using this outline.

Definitions

An *isosceles* triangle is a triangle having two or more sides of equal length.

The *base angles* of an isosceles triangle are the angles opposite sides of equal length.

The *hypotenuse* of a right triangle is the side opposite the right angle. The other two sides are called the *legs*.

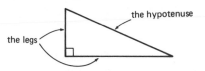

The *midpoint* of a line segment $\overline{AB}$ is that point M on $\overline{AB}$ such that $AM = MB$.

The *perpendicular bisector* of a line segment $\overline{AB}$ is the line passing through the midpoint of $\overline{AB}$ and perpendicular to $\overline{AB}$.

Axioms

The SSS property of triangles

The SAS property of triangles

The ASA property of triangles

Corresponding parts of congruent triangles are equal.

Theorems

Each point on the perpendicular bisector of a line segment is equidistant from the endpoints of the line segment.

If a triangle is isosceles, then its base angles are equal.

Conversely, if two angles of a triangle are equal, then the triangle is isosceles (the sides opposite the equal angles have equal length).

Words

congruent, corresponding parts, auxiliary lines, double column form for writing a proof, equidistant

Symbols   △*ABC* (triangle *ABC*),   ≅ (is congruent to),
Q.E.D. (which was to be proved),   ∴ (therefore)

Summary

This chapter introduced the concept of congruent figures. The SSS, SAS, and ASA congruence properties of triangles were studied and applied in various problems. We had some practice in proving theorems; the *double column* form is often used in writing proofs in geometry.

Theorem:

Given: (*Figure*)

To Prove:

Statement   Reason
_____

The *paragraph* form for writing proofs is most often used in mathematics. It has the advantage that the ideas flow more smoothly when words are mixed with symbols.

Theorem:

Proof: Paragraph (or essay) proving the theorem, include a picture, if possible.

The next chapter explores proofs further.

Historical Note

Euclid of Alexandria, who lived about 300 B.C., was the author of a number of mathematical works, the most important of these being the *Elements.* The *Elements* of Euclid is important not so much for its subject matter, namely, mathematics as known to the Greeks, but for the style of presentation. Euclid used the *axiomatic method.* He listed at the beginning of his works the fundamental concepts that would be studied and the properties that would be assumed to be true (the assumptions are called axioms). He then proceeded to systematically prove theorems based on the axioms using logical deduction. The *Elements* set a standard for mathematics that has been followed for over two thousand years.

Relatively little is known about Euclid's life. In one of the surviving stories, a certain king asks Euclid if there is not an easier way to learn geometry. Euclid replies, "There is no royal road to geometry."

Central Self Quiz

1. Prove that $\triangle PQR \cong \triangle PQS$.

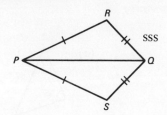
SSS

2. Prove that if $\overline{AB}$ meets $\overline{CD}$ in point E such that $CE = DE$ and $BE = AE$, then $\angle A = \angle B$.

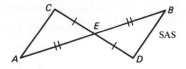
SAS

3. Prove that $\triangle ABC \cong \triangle BAD$.

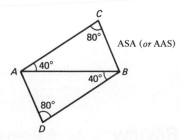
ASA (or AAS)

4. In the picture below, P and Q are points on the parade ground, and T is the top of the flagpole. Prove that $PB = QB$.

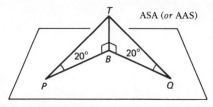
ASA (or AAS)

5. Find x. Explain your reasoning.

(a)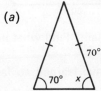

(b)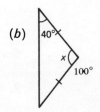

111   Central

(c)    (d)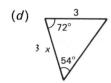

(e)    (f)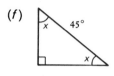

Answers

1. Use SSS.
2. $\triangle CEA \cong \triangle DEB$ by SAS (why?). Hence $\angle A = \angle B$, since corresponding angles of congruent triangles are equal.
3. The missing angles are equal (why?). Hence the triangles are congruent by ASA.
4. The right triangles have their angles at T equal (why?). Hence they are congruent by ASA. Hence $PB = QB$ (why?).
5. (a) 70° (the triangle is isosceles, hence the base angles are equal)   (b) 100°
   (c) 7 (Exercise 43(a))   (d) 3   (e) $47\frac{1}{2}°$ (f) 45°

Review

Some good arguments may occur in Review Exercise 1.

1. A surveyor's map of Loco County shows the following information about the system of highways connecting five towns A, B, C, D, E:

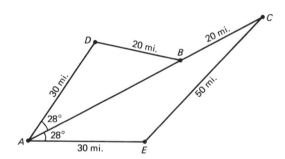

Something is wrong here! Find what it is (without measuring).

2. Use what you have learned so far to find x in each case, without

112   3: Triangles

measuring. Explain how you arrive at your answers.

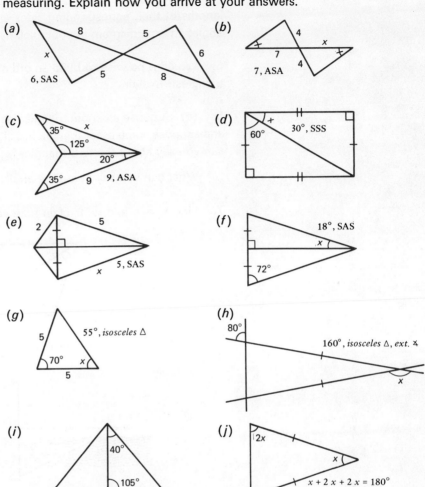

3. Suppose in the picture below $AB = PQ$, $BC = QR$, and $CD = RS$. Explain how to place one figure over the other so they coincide.

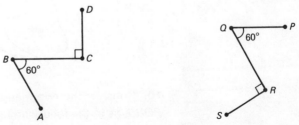

4. If two polygons are congruent, then what can you say about the lengths of their corresponding sides and the sizes of their corresponding angles? Use a diagram in your explanation.

5. Would it be wise to build an ordinary frame house without the use of triangles? Explain.

6. (a) Suppose a certain triangle has sides of lengths 7, 10, and 13 feet, and suppose some other triangle has sides of the same lengths. Then what can you say about the two triangles?

(b) What can you say about the angles of the two triangles? Explain.

7. Find x, y, z, and w. Explain how you arrive at your answers.

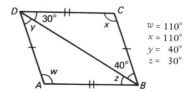

8. Use the figure below to give a method for measuring the width of a river.

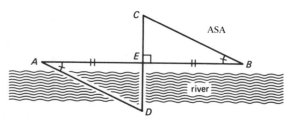

9. What is the distance from A to B? Why?

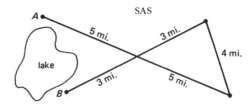

10. Consider the two triangles formed by drawing a diagonal of a square PQRS, as in the following figure:

114 3: Triangles

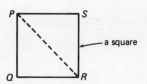
— a square

Prove that △PQR ≅ △RSP.

11. In the figure below we are given two line segments, $\overline{AB}$ and $\overline{CD}$, intersecting at a point M which is the midpoint of both.
Prove that △AMC ≅ △BMD.

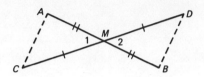

12. In the figure below, AD = BC and ∡1 = ∡2.
Prove that △ABC ≅ △BAD.

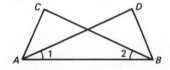

13. Given: SR = QR and ∡1 = ∡2
Prove: PS = PQ

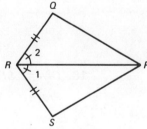

14. In the figure below AC = CE = EG, BC = CD, and DE = EF.
Prove that ∡B = ∡F and AB = EF.

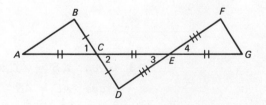

115 Review

15. *Given:* $PQ = RS$ and $PS = RQ$

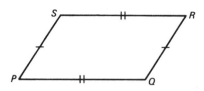

(a) Prove that $\angle P = \angle R$. (b) Prove that $\angle Q = \angle S$.

16. Using the information in the figure below, prove that $AB = CD$.

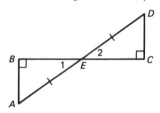

17. *Given:* $\angle 1 = \angle 2$ and $\angle 3 = \angle 4$
Prove: $AB = AD$

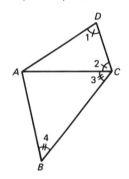

18. *Given:* $SP = SQ = SR$
Prove: $\angle 1 + \angle 4 = \angle 2 + \angle 3$

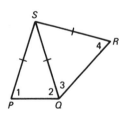

19. *Given:* $\angle B = \angle C$ and $BD = CF$
Prove: $AC = AB$

116   3: Triangles

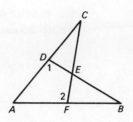

20. *Given:* ℓ is parallel to m and s is parallel to t

Prove: $a = b$ and $c = d$

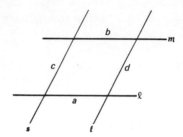

[*Hint:* Draw an auxiliary line.]

21. Using the information shown in the figure below, prove that $AB = HG$.

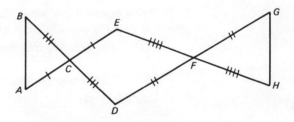

22. Given this figure, prove that $PA = PC$.

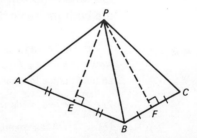

23. (*a*) Is there any isosceles triangle which has a 35° angle and a 54° angle? Explain.

(b) A certain isosceles triangle has a side of length 17 cm. and a side of length 35 cm. What is the length of the third side? Explain your answer.

24. Prove that $x + y = a + b$.

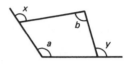

Label the other angles of the quad. c and d.
$a + b + c + d = 360°$
$x + c + y + d = 360°$

25. A ship's navigator can determine his distance to a fixed object on shore by the following procedure:

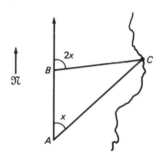

At point A, he measures the angle between point C (on shore) and the ship's heading (in our picture the ship is going north). At some point B the angle is twice what it was at A. If he knows the distance AB, then he knows the distance BC. Explain.

26. A certain triangle has two 60° angles. Prove that it must be an equilateral triangle.

27. Jim and Bill are at different places, but both are 20 feet away from the foot of a flagpole. How can you use congruent triangles to *prove* that they are both the same distance from the top of the pole?

Review Answers

1. Draw $\overline{BE}$ and observe that $\triangle ABD \cong \triangle ABE$ by SAS. Hence $BE = BD = 20$ mi. Now notice that $\triangle EBC$ has two sides of length 20 mi. and a third side of length 50 mi., so the Triangle Inequality does not hold!

2. (a) $x = 6$, since the two triangles are congruent by SAS.

(b) $x = 7$. The marked angles are equal and vertical angles are equal, hence the remaining two angles are equal. Then the triangles are congruent by ASA.

(c) $x = 9$. The missing angle in the upper triangle must be 20° (why?), and the missing angle in the lower triangle is 125° (why?). Hence the triangles are congruent by ASA.

(d) $x = 30°$, why?

(e) $x = 5$. Use congruent triangles.

(f) $x = 18°$. The triangle on the right is congruent to that on the left by SAS. Now use the fact that the sum of the acute angles in a right triangle is 90°.

(g) $x = 55°$. The unmarked angle also has size x (base angles of an isosceles triangle). Hence $2x + 70° = 180°$, so $x = 55°$.

(h) $x = 160°$. The isosceles triangle has base angles of 80°, and x is the sum of the two opposite interior angles.

(i) $x = 5$. The upper left-hand triangle has angles 35°, 105°, 40° (use the fact that the sum of the angles "around a point" is 360°), and the upper right-hand triangle has angles 35°, 105°, 40° (why?). Hence these two triangles are congruent by ASA.

(j) The unmarked angle is $2x$ (why?). Therefore $5x = 180°$, so $x = 36°$.

3. The figure on the right can be "flipped over" and placed with S on D, R on C, Q on B, and P on A.

4. equal

5. Base your discussion on the rigidity of triangles. How is this related to SSS?

6. (a) They are congruent by SSS.

(b) Corresponding angles of congruent triangles are equal.

7. $\triangle DAB \cong \triangle BCD$ by SSS. Hence $y = 40°$ and $z = 30°$, since corresponding angles of congruent triangles are equal. Also $x = 110°$ (why?), so $w = 110°$ also.

8. Using a point D on the opposite shore, obtain points A and B such that $AE = BE$ and DE is perpendicular to AB (how would you do this?). Now find $\angle A$ by sighting from A. Then sight from B in a direction such that $\angle B = \angle A$. The ray from B in this direction intersects the extended line through D and E in a point C. Then $CE = DE = $ the width of the river, because $\triangle CEB \cong \triangle DEA$ by ASA.

9. $AB = 4$ miles (why?).

10. Use SSS. *(or SAS)*

11. We are given $AM = BM$ and $CM = DM$. $\angle 1 = \angle 2$, since vertical angles are equal. Hence $\triangle AMC \cong \triangle BMD$ by SAS.

12. $AD = BC$ and $\angle 1 = \angle 2$ are given. Also $AB = AB$. Hence $\triangle ABC \cong \triangle BAD$ by SAS.

13. $\triangle RSP \cong \triangle RQP$ by SAS. Hence $PS = PQ$, since they are corresponding sides of congruent triangles.

14. $\triangle ABC \overset{SAS}{\cong} \triangle EDC$ (why?); hence $\angle B \overset{corr.\ parts}{=} \angle D$ (why?). $\triangle EDC \overset{SAS}{\cong} \triangle EFG$ (why?); hence $\angle D = \angle F$ (why?). Thus $\angle B = \angle F$. Explain why $AB = EF$.

15. (a) Draw $\overline{SQ}$. Note that $\triangle SPQ \cong \triangle QRS$ by SSS. Hence $\angle P = \angle R$, since they are corresponding angles of congruent triangles.

    (b) Draw $\overline{PR}$. Proceed as in part (a).

16. $\angle 1 = \angle 2$, since vertical angles are equal. It follows that $\angle A = \angle D$ (why?). $AE = DE$ is given. Hence $\triangle ABE \cong \triangle DCE$ by ASA. Hence $AB = CD$, since corresponding sides of congruent triangles have equal length.

17. $AD = AC$, by the theorem in Central Exercise 43. Similarly, $AC = AB$. Thus, $AD = AC = AB$.

18. $\angle 1 = \angle 2$ and $\angle 3 = \angle 4$ (base angles of an isosceles triangle are equal). Hence $\angle 1 + \angle 4 = \angle 2 + \angle 3$.

19. Separate $\triangle ABD$ and $\triangle ACF$ in your mind. In these two triangles we have $\angle B = \angle C$ and $\angle A = \angle A$. It follows that $\angle 1 = \angle 2$ (why?). Hence $\triangle ABD \cong \triangle ACF$ by ASA. Therefore $AC = AB$, since corresponding parts of congruent triangles are equal.

20. Draw a diagonal of the parallelogram, as indicated below.

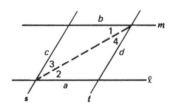

Then $\angle 1 = \angle 2$ (alternate interior angles formed by line crossing parallel lines ℓ and m) and $\angle 3 = \angle 4$ (alternate interior angles formed by line crossing parallel lines s and t). Thus the two triangles are congruent by ASA. Hence $a = b$ and $c = d$, since corresponding sides of congruent triangles have equal length.

21. Draw $\overline{ED}$. Then $\triangle BAC \cong \triangle DEC$ (why?), so $AB = ED$ (why?). Also $\triangle EDF \cong \triangle HGF$ (why?), so $ED = HG$ (why?). Hence $AB = ED = HG$.

22. You could prove this by using a pair of congruent right triangles to get $PA = PB$, and another pair to show $PB = PC$. You could also proceed as follows: $PA = PB$, since P is on the perpendicular bisector of $\overline{AB}$, and $PB = PC$, since P is on the perpendicular bisector of $\overline{BC}$. Thus $PA = PC$.

23. (*a*) Since two of the angles of the triangle would have to be equal, the triangle would have to have angles of sizes 35°, 35°, 54° or 35°, 54°, 54°. In neither case is the sum of the angles 180°, hence there does not exist such a triangle.

(*b*) The third side of the triangle is either 17 cm. long or 35 cm. long. It could not be 17 cm., since then the triangle inequality would not hold. Hence it must be 35 cm.

24. What is the sum of the interior angles of the quadrilateral?

25. Using the Exterior Angle Theorem, we see that $\angle C = x$. Hence $BC = AB$ by the theorem in Central Exercise 43.

26. The third angle must be 60°, since the sum of the angles is 180°. The result now follows by using the theorem in Central Exercise 43.

27. Use SAS.

Review Self Test

1. If the pair of triangles is congruent, prove it. If they are not congruent, explain why not. (Warning: The pictures may be distorted.)

(a)

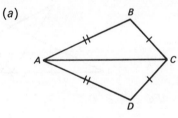

(b)

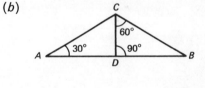

(c)

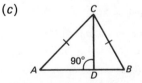

(d)

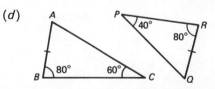

(e)

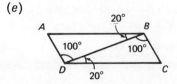

(f)

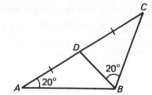
(g) (h)
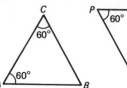

2. Prove that $\angle 1 - \angle 4 = \angle 2 - \angle 3$.

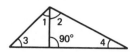

3. The dotted line k bisects $\angle B$ (that is, $\angle 1 = \angle 2$). The distance from D (on the dotted line k) to line ℓ is DC; similarly, the distance from D to line m is DA.

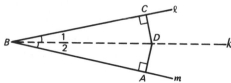

Prove: (i) $\triangle ABD \cong \triangle CBD$ (ii) $DC = DA$

4. In the following picture, assume $\angle 1 = \angle 3$, $\angle A = \angle B$, and $AF = FB$.

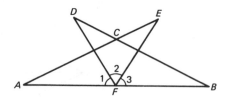

Prove $\triangle AFE \cong \triangle BFD$.

Answers

1. (a) congruent by SSS

(b) The missing angle at D is 90°; hence the missing angle at C is 60° (since the sum of the angles of $\triangle ADC$ is 180°). Therefore $\triangle ADC \cong \triangle BDC$ by ASA.

(c) $\angle A = \angle B$ (base angles of an isosceles triangle), and the missing angle at D must be 90°. Therefore $\triangle ADC \cong \triangle BDC$ by ASA.

(d) Although the two triangles have equal angles, they are not necessarily congruent! Note that ASA does *not* apply here.

(e) congruent by ASA (f) congruent by SAS
(g) not necessarily congruent (h) not necessarily congruent

2. $\angle 2 + \angle 4 = 90° = \angle 1 + \angle 3$ (why?)

3. In △ABD and △CBD, ∡1 = ∡2, BD = BD, and their angles at D are equal (why?). Hence the triangles are congruent by ASA.

4. ∡A = ∡B (given), AF = BF (given), ∡1 + ∡2 = ∡3 + ∡2 (since ∡1 = ∡3 is given). Hence △AFE ≅ △BFD by ASA.

Cumulative Review

1. Explain what is wrong in each picture. (For this problem you may assume that lines which look like straight lines actually are straight lines.)

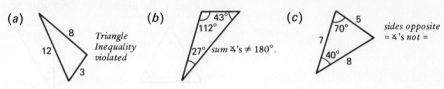

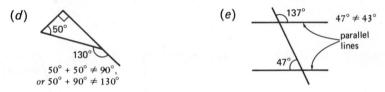

2. Here is part of a regular n-gon. What is n?   8x + x = 180°
x = 20°, 8x = 160°
160° = $\frac{(n-2) 180°}{n}$

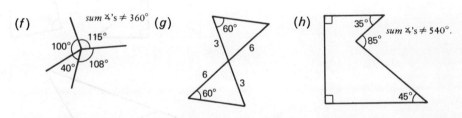

3. (a) What is the definition of a regular n-gon?   a polygon with n sides, all of whose angles and sides are equal
(b) Using a protractor and ruler, draw a 5-gon all of whose angles are equal but which is not a regular 5-gon.

(c) Draw a 5-gon all of whose sides have equal length but which is not a *regular* 5-gon.

4. Find that point P on ℓ which makes $AP + PB$ as small as possible.

•B

Construct B*.

A•

──────────────── ℓ

5. Find x. Be able to explain your answer.

(a)

(b) (c)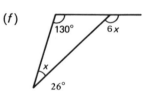

(d) (e) (f)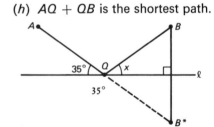

(g) (h) $AQ + QB$ is the shortest path.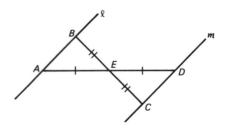

6. Prove that ℓ is parallel to m in the figure below:

124 3: Triangles

[*Hint:* You are given that E is the point where $\overline{AD}$ intersects $\overline{BC}$, $AE = DE$, and $BE = CE$. You should now be able to prove that $\triangle ABE \cong \triangle DCE$. Thus, certain angles are equal, so ℓ is parallel to m because of Central Exercise 22(a) of Chapter 2.]

7. (a) If a triangle is a right triangle, then its two acute angles sum to _____.

(b) If a triangle has two acute angles which sum to 90°, then it is a _____.

(c) What is the relationship between the statements in parts (a) and (b)?

8. (a) Give an example of a pair of triangles, $\triangle ABC$ and $\triangle PQR$, such that $AB = PQ$, $AC = PR$, $\angle B = \angle Q$, but such that the triangles are *not* congruent.

(b) Why does part (a) not contradict the SAS property for congruent triangles?   *The angle is not included between the sides.*

9. Find x.   *The triangle is isosceles, $x = 7$.*

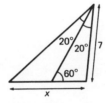

Answers

1. (a) The Triangle Inequality is violated.
   (b) The sum of the angles is not 180°.
   (c) The triangle has two equal angles (why?); hence it should be isosceles.
   (d) The exterior angle, 130°, is not equal to the sum of the two opposite interior angles.

2. $n = 18$

5. (a) $x = 60°$ (b) $x = 30°$ (c) $x = 48\frac{1}{2}°$ (g) $x = 20°$

7. (a) 90°   (b) right triangle   (c) converses

Projects

Congruent Triangles in Space

1. (a) Make a framework in the shape of a cube, using 12 straws of

equal length and common pins.

(b) Is the framework rigid? *no*

(c) What is the least number of "braces" needed in order to make the framework in part (a) a rigid figure?   *4, if you pin the major diagonals where they intersect*

2. (a) Use straws and pins to make a framework in the shape of a *regular tetrahedron* (4 faces, each an equilateral triangle).

(b) Is this framework rigid? Why would you expect this?   *yes, because it is made of triangles*

3. Imagine that the cube below is made of wire. Each letter names the midpoint of that edge.

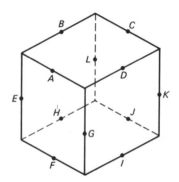

(a) *DF* is equal to which of the following?   *BJ*

   (i) *DG*   (ii) *AF*   (iii) *DH*   (iv) *BJ*   (v) none of these

(b) Which of the following is an equilateral triangle?   △*IJK*, △*AHK*

   (i) △*AEG*   (ii) △*DEF*   (iii) △*IJK*   (iv) △*AFJ*
   (v) △*AHK*   (vi) none of these

(c) Which of the following is a square?   *IFHJ, LCKJ*

   (i) *ADIF*   (ii) *EAGI*   (iii) *IFHJ*   (iv) *LCKJ*
   (v) *LDGH*   (vi) none of these

126   3: Triangles

(d) How many triangles, with vertices at the labeled points, are congruent to △IJK? 8

  (i) 2   (ii) 4   (iii) 8   (iv) 16

(e) The points F, G, D, C, L, H all lie in the same plane. Prove that they are the vertices of a regular hexagon. (You must prove that all sides have equal length and all angles are equal. Use congruent triangles.)

4. In the cube, each letter is at a vertex (or corner).

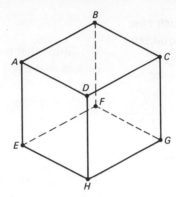

(a) Name the largest equilateral triangle that will fit inside the cube. △DEG, etc.

(b) What is the largest regular tetrahedron (see Exercise 2) that will fit inside the cube? *the tetrahedron such that each edge is a diagonal of a face of the cube*

(c) What other distances are equal to AG? FD = BH = CE = AG

Congruent Quadrilaterals

We do not introduce the notation "⪤ABC" for angles until Chapter 10. However, in this proof the notation may be handy (for example, in denoting ⪤ABD of △ABD) so you may want to introduce it to your students now.

5. Suppose we are given quadrilaterals ABCD and A'B'C'D' with AB = A'B', BC = B'C', CD = C'D', ⪤B = ⪤B', and ⪤C = ⪤C'. Prove that AD = A'D'.

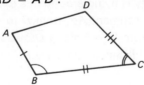 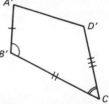

[Hint: Draw the auxiliary lines $\overline{BD}$ and $\overline{B'D'}$. Prove that △ABD ≅ △A'B'D'.]

Quick Proof That the Base Angles Are Equal

6. In this chapter we proved that the base angles of any isosceles triangle

127 Projects

Students invariably have difficulty understanding this proof because of its subtle use of the SSS axiom. It might help to draw $\triangle ABC$ and $\triangle BAC$ on the blackboard separately, and remind students to think of them as the same triangle with corresponding vertices superimposed.

are equal. In our proof, we used an auxiliary line drawn from a vertex to the midpoint of the base, and then we showed that the two smaller triangles were congruent. Find another proof which does not use any extra lines. [*Hint:* If $\triangle ABC$ is the given isosceles triangle with $AC = BC$, then think of $\triangle BAC$ as a different triangle. Note that corresponding sides of these "two" triangles are of equal length, that is, $AB = BA$, $AC = BC$, and $BC = AC$. Hence $\triangle ABC \cong \triangle BAC$. Hence the corresponding angles are equal.]

Some Exercises with the Triangle Inequality

$$PA + PB > 8$$
$$PB + PC > 7$$
$$PC + PA > 5$$
$$\overline{}$$
$$2(PA + PB + PC) > 20$$

7. Show that no matter how P is chosen inside the triangle, it will always be true that $PA + PB + PC > 10$.

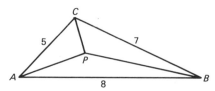

8. A man stands in a square field, 100 feet on a side. Show that no matter where he stands, the sum of the distances from him to the corners of the field will be more than 282 feet.

[*Hint:* The length of a diagonal is $100\sqrt{2}$ feet, and $\sqrt{2} > 1.41$.]

9. A fly is flying around in a room. He flies from point A to point B along the line segment AB, from B to C along $\overline{BC}$, from C to D along $\overline{CD}$, and finally from D to E along $\overline{DE}$. Assuming only the Triangle Inequality, prove that the distance he flies is at least as large as AE.

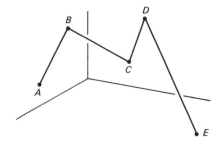

128 3: Triangles

The following exercise prepares us for Exercise 11, which is quite challenging and uses many ideas from this chapter.

10. If ∡x = 15°, prove that DE = DA.

Let y = ∡FDE.
∡D + ∡A + ∡E = 180°
= (x+y) + (x+60) + (y+60)
x+y = 30, y = 15
△DFE ≅ △DFA by SAS

11. In the figure below, ABCD is a square. Prove that DE = DA.

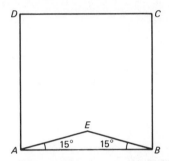

[*Hint:* Introduce an auxiliary triangle △ADF ≅ △ABE. Examine △ADE. Use the preceding exercise.]

Algebra Review (Optional)

Factoring

DEFINITION

The wording of this definition for a prime integer may be new to some students. It is a parallel wording to that of a prime polynomial.

A positive integer greater than 1 is *prime* if it is not the product of two smaller positive integers.

Observe that 2, 3, 11, and 23 are prime, but 6, 14, and 15 are not:

$$6 = 2 \cdot 3 \qquad 14 = 2 \cdot 7 \qquad 15 = 3 \cdot 5$$

1. (a) List the prime integers between 2 and 30.

   (b) List the prime integers between 70 and 100.

   (c) Is 101 prime? 201? 301?

In the polynomial
$$3x^2 - 2x + 1,$$
the *coefficient* of x^2 is 3; the coefficient of x is -2; the constant coefficient is 1. The highest exponent of x that occurs in the polynomial is the *degree* of the polynomial. The degree of $3x^2 - 2x + 1$ is 2. Some other examples:

$$5x^4 - 3x - 1 \text{ has degree 4}$$
$$x - 7 \text{ has degree 1}$$
$$x^3 + 4x^2 - 2x + 1 \text{ has degree 3}$$
$$6 \text{ has degree 0 (it is } 6x^0)$$

You may wish to give a presentation concerning coefficient *and* degree.

2. Give the degree of each polynomial.

   (a) $x + 2$ (b) $x^5 - 29x^2 + x - 4$ (c) 75

   (d) $x^3 + 64x^2 - x + 129$

DEFINITION A polynomial with integral coefficients is *prime* (with respect to integral coefficients) if the polynomial

Compare the definition of prime polynomial with that of prime integer.

(i) has degree at least one, and

(ii) is not the product of polynomials of lower degree with integral coefficients.

Thus, $x^2 + 3x + 2$ is not a prime polynomial with respect to integral coefficients, for
$$x^2 + 3x + 2 = (x + 2)(x + 1).$$
Neither is $3x^2 + 5x - 2$:
$$3x^2 + 5x - 2 = (3x - 1)(x + 2)$$
But $x^2 - 2$ is prime with respect to integral coefficients. However, it is *not* prime with respect to real coefficients:
$$x^2 - 2 = (x - \sqrt{2})(x + \sqrt{2})$$

130 3: Triangles

Emphasize that the student should check his factoring by multiplying.

3. Use the identities at the end of the Algebra Review in Chapter 2 to help you write the factors (with integral coefficients) of the following. (Check your answer by multiplying.)

   (a) $5x^2 + 5x$ (b) $x^2 - 4$ (c) $9x^2 + 6x + 1$
   (d) $x^2 - 8x + 16$ (e) $x^2 - x - 6$ (f) $25x^2 - 16$
   (g) $2x^2 - 7x$ (h) $1 - x^2$ (i) $4x^2 + 12x + 9$

4. Factor $x^3 + x^2 - 20x$. [*Hint:* First take out a common factor.]

5. (a) Can $x^2 - 1$ be factored with integral coefficients?
   (b) Can $x^2 + 1$ be factored with integral coefficients?

6. Factor into polynomials with integral coefficients (check your answer).

   (a) $x^3 + x^2 + x$ (b) $25x^2 + 40x + 16$ (c) $4x^2 - 49$
   (d) $16x^2 - 24x + 9$

7. Factor as before (check your answer).

   (a) $x^2 + 5x + 6$ (b) $x^2 - 5x + 6$ (c) $x^2 + 5x - 6$
   (d) $x^2 - 5x - 6$ (e) $2x^2 + 11x + 12$ (f) $6x^2 + 7x - 20$
   (g) $6x^2 - 7x + 2$ (h) $3x^2 - 2x - 5$

8. Factor as before (check your answer).

   (a) $2x^2 + 5x + 2$ (b) $2x^2 + 5x - 3$ (c) $6x^2 + 13x + 6$
   (d) $2x^2 - x - 6$ (e) $6x^2 - 11x - 10$ (f) $10x^2 + 3x - 4$
   (g) $2x^2 - 3x - 14$ (h) $8x^2 - 26x + 21$

Rules of Equality

Recall the rules of equality: we may add, subtract, multiply, or divide (not by zero) each side of an equality by the same number. That is, if $a = b$, then each of the following is true:

$$a + c = b + c \qquad a - c = b - c \qquad ac = bc \qquad \frac{a}{c} = \frac{b}{c}, c \neq 0$$

9. Solve for x (show all work and check your answer).

131 Algebra Review

This short section on solving easy linear equations is needed to solve quadratic equations by factoring in Exercises 11-14. In a later Algebra Review the student solves more complicated linear equations.

(a) $x - 3 = 7$

(b) $x + 9 = 2$

(c) $-2x = 11$

(d) $\dfrac{x}{4} = 8$

(e) $x + a = b$

(f) $x - c = d$

(g) $-ax = b, a \neq 0$

(h) $\dfrac{x}{a} = b, a \neq 0$

(i) $2x - 3 = 0$

10. Solve for x (show all work and check your answer).

(a) $7x - 9 - 2x = 15 + x - 8$

(b) $3(2x - 4) + 3x = 5(x - 4)$

(c) $5x + [x - 2(x + 4)] = 0$

(d) $2[(x - 1) + (x - 2) - (x - 3)] = 0$

Exercise 9 will be useful in solving *quadratic* (degree 2) equations by factoring. First, we recall a very important idea.

The Zero Product Rule

If $ab = 0$, then $a = 0$ or $b = 0$ (or both may be 0).

Emphasize that the Zero Product Rule is the key step in solving quadratics by factoring.

Example 1. Solve by factoring:

$$6x^2 + 11x = 10$$

In order to use the zero product rule we write

$$6x^2 + 11x - 10 = 0$$

and factor:

$$(2x + 5)(3x - 2) = 0$$

Thus, by the zero product rule,

$$2x + 5 = 0 \quad \text{or} \quad 3x - 2 = 0$$
$$2x = -5 \qquad\qquad\quad 3x = 2$$
$$x = -\dfrac{5}{2} \qquad\qquad\quad x = \dfrac{2}{3}$$

The solutions are $-\dfrac{5}{2}$ and $\dfrac{2}{3}$.

Check:

$$6\left(-\dfrac{5}{2}\right)^2 + 11\left(-\dfrac{5}{2}\right) \stackrel{?}{=} 10 \qquad\qquad 6\left(\dfrac{2}{3}\right)^2 + 11\left(\dfrac{2}{3}\right) \stackrel{?}{=} 10$$

132   3: Triangles

$$6\left(\frac{25}{4}\right) + 11\left(-\frac{5}{2}\right) \stackrel{?}{=} 10 \qquad\qquad 6\left(\frac{4}{9}\right) + 11\left(\frac{2}{3}\right) \stackrel{?}{=} 10$$

$$3\left(\frac{25}{2}\right) + 11\left(-\frac{5}{2}\right) \stackrel{?}{=} 10 \qquad\qquad 2\left(\frac{4}{3}\right) + 11\left(\frac{2}{3}\right) \stackrel{?}{=} 10$$

$$\frac{75}{2} - \frac{55}{2} \stackrel{?}{=} 10 \qquad\qquad \frac{8}{3} + \frac{22}{3} \stackrel{?}{=} 10$$

$$\text{yes} \qquad\qquad \frac{30}{3} \stackrel{?}{=} 10$$

$$\text{yes}$$

Comment. Observe when you use the zero product rule to solve a quadratic equation that you must put the equation in the form:

polynomial in $x = 0$

11. Solve for x by factoring (show all work and check your answer).

(a) $4x^2 - 9 = 0$ (b) $2x^2 + x - 6 = 0$

The students must be able to detect the error in Exercise 12.

12. (a) Is the following solution correct? Explain.

Solve for x: $\qquad\qquad 10x^2 + 3x = 4$

We factor,

$$x(10x + 3) = 4.$$

Then

$$x = 2 \qquad \text{or} \qquad \begin{aligned} 10x + 3 &= 2 \\ 10x &= 2 - 3 \\ 10x &= -1 \\ x &= -\frac{1}{10} \end{aligned}$$

The solutions are 2 and $-\frac{1}{10}$.

(b) Check the above solutions in the original equation.

13. Solve for x by factoring (show all work).

(a) $2x^2 - 5x = 0$ (b) $2x^2 - 3x - 14 = 0$

(c) $3x^2 - 5x = 2$ (careful!) (d) $x^2 + 35 = 12x$

(e) $18 - 5x - 2x^2 = 0$ (f) $2x^2 + 5x - 12 = 0$

133 Algebra Review

Make sure students see that there are three solutions for each equation in Exercises 14-15.

14. Solve for x.

(a) $(x - 2)(x + 3)(x - 5) = 0$

(b) $(x^2 - 1)(x + 2) = 0$

(c) $(x^2 - 3x + 2)(x - 3) = 0$

15. Solve for x: $2x^3 + 5x^2 + 2x = 0$ [*Hint:* Find all the factors.]

16. Are the following statements correct? Explain.

(a) If $xy = 0$, then $x = 0$.

(b) If $xy = 0$, then $y = 0$.

(c) If $xy = 0$, then $x = 0$ or $y = 0$.

(d) If $xy = 0$ and $x = 3$, then $y = 0$.

Answers

1. (a) 2, 3, 5, 7, 11, 13, 17, 19, 23, 29

(b) 71, 73, 79, 83, 89, 97

(c) yes, $201 = 3 \cdot 67$, $301 = 7 \cdot 43$

2. (a) 1   (b) 5   (c) 0   (d) 3

3. (a) $5x(x + 1)$ (b) $(x + 2)(x - 2)$ (c) $(3x + 1)^2$

(d) $(x - 4)^2$ (e) $(x - 3)(x + 2)$ (f) $(5x - 4)(5x + 4)$

(g) $x(2x - 7)$ (h) $(1 - x)(1 + x)$ (i) $(2x + 3)^2$

4. $x(x + 5)(x - 4)$

5. (a) $(x - 1)(x + 1)$ (b) no

6. (a) $x(x^2 + x + 1)$ (b) $(5x + 4)^2$

(c) $(2x - 7)(2x + 7)$ (d) $(4x - 3)^2$

7. (a) $(x + 3)(x + 2)$ (b) $(x - 2)(x - 3)$ (c) $(x + 6)(x - 1)$

(d) $(x - 6)(x + 1)$ (e) $(2x + 3)(x + 4)$ (f) $(2x + 5)(3x - 4)$

(g) $(3x - 2)(2x - 1)$ (h) $(3x - 5)(x + 1)$

8. (a) $(2x + 1)(x + 2)$ (b) $(x + 3)(2x - 1)$ (c) $(3x + 2)(2x + 3)$

(d) $(x-2)(2x+3)$ (e) $(3x+2)(2x-5)$ (f) $(5x+4)(2x-1)$
(g) $(2x-7)(x+2)$ (h) $(2x-3)(4x-7)$

9. (a) 10   (b) -7 (c) $-\frac{11}{2}$ (d) 32   (e) $b-a$
(f) $c+d$ (g) $-\frac{b}{a}$ (h) ab (i) $\frac{3}{2}$

10. (a) 4   (b) -2 (c) 2   (d) 0

11. (a) $\frac{3}{2}, -\frac{3}{2}$ (b) $-2, \frac{3}{2}$

12. (a) No, one should first write $10x^2 + 3x - 4 = 0$, in order to use the zero product rule.
(b) The solutions are *not* 2 and $-\frac{1}{10}$. (The correct solutions are $\frac{1}{2}$ and $-\frac{4}{5}$.)

13. (a) $0, \frac{5}{2}$ (b) $\frac{7}{2}, -2$ (c) $-\frac{1}{3}, 2$ (d) 7, 5   (e) $-\frac{9}{2}, 2$
(f) $\frac{3}{2}, -4$

14. (a) $2, -3, 5$ (b) $1, -1, -2$ (c) $1, 2, 3$

15. $0, -\frac{1}{2}, -2$

16. (a) not necessarily, y could be 0   (b) not necessarily, x could be 0
(c) yes   (d) yes

135 *Algebra Review*

Chapter 4 What is a Proof?

This chapter continues the study of proof, but do not expect students to be master "provers" by the end of it. They will have many occasions throughout the remaining chapters to develop their skill at writing proofs.

Central

At this stage, let them guess about the solution of the carpenter problem.

A carpenter has just completed the construction of a floor for a room. The floor is supposed to be a rectangle 24 feet long and 15 feet wide. The only measuring device he has is a steel tape measure. How can he measure to be sure the floor he has built is a rectangle?

This chapter provides a theorem that the carpenter can apply to check his floor. However, the main purpose of this chapter is to offer an opportunity for you to develop more skill in proofs, while learning some useful properties of triangles and quadrilaterals.

"If . . . , Then . . ." Statements

Emphasize the key word, "can."

Recall that a *theorem* is a statement that can be proved, based on statements which have been previously proved or accepted as true. Here is an example of a theorem:

"The base angles of any isosceles triangle are equal."

We proved this theorem in Chapter 3, Central Exercise 42. You will notice that many theorems can be phrased as "If . . . , then . . ." statements. For example, the isosceles triangle theorem just quoted can be restated:

137

Students who are weak in English generally have trouble with a theorem stated as a simple sentence. The "if..., then..." form helps those students. At this point they must know the role of the if and then clauses in a proof.

"If a triangle is isosceles, then its base angles are equal."

The "if ..." part tells us what information is given, while the "then ..." part tells us what we have to prove on the basis of the given information. It is often convenient to restate a theorem in the "if ..., then ..." form before trying to prove it (if it is not already in that form). For example, the statement,

"The angles of any equilateral triangle are equal,"

may be rephrased as,

"If a triangle is equilateral, then its angles are equal."

As another example, the statement,

"The sum of the angles of any triangle is 180°,"

is equivalent to the statement,

"If a polygon is a triangle, then the sum of its angles is 180°."

As a final example, the statement,

"A point on the perpendicular bisector of a line segment is equidistant from the endpoints,"

is equivalent to the statement,

"If a point is on the perpendicular bisector of a line segment, then it is equidistant from the endpoints."

By putting a theorem in the form of an "if ..., then ..." statement, we can see more easily what is assumed (the "if ..." part) and what is to be proved (the "then ..." part).

When considering a simple sentence, the complete subject is the "given" and the complete predicate is the "to prove."

1. In each case, write an "if ..., then ..." statement equivalent to the given statement. Compare your answers with those of your classmates.

(*a*) Any triangle with two equal angles is isosceles.

(*b*) Alternate interior angles formed by a line crossing two parallel lines are equal.

(*c*) The sum of the acute angles of a right triangle is 90°.

(*d*) Every triangle has at least one acute angle.

(*e*) The sum of the angles of any quadrilateral is 360°.

(*f*) Any good soap has suds.

(*g*) Any man living in Paris also lives in France.

Properties of Parallelograms

Exercises 2–12 concern properties of a parallelogram.

There should be careful discussion of these early proofs — student-student, and teacher-student where necessary. Resist any temptation to rush through these exercises.

Remind students that we use "previously established" (or proved) statements as well as previously accepted statements. Emphasize that in each of the following exercises they may use the results of previous exercises.

The following exercises, dealing with properties of parallelograms, will give you more practice in devising and writing proofs. *In each exercise, you may use the results of previous exercises in your proof.* It will be very instructive for you to check your proofs against those of your classmates.

For our first theorem about parallelograms, the next exercise proves that a parallelogram is cut into two congruent triangles by a diagonal (line segment joining opposite vertices).

2. Fill in the missing steps in the proof of the following theorem.

Theorem: If *ABCD* is a parallelogram with diagonal $\overline{AC}$, then $\triangle ABC \cong \triangle CDA$.

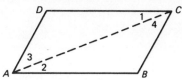

Emphasize to the class that a parallelogram is defined to be a quadrilateral whose opposite sides are parallel.

Proof: By the definition of a parallelogram, $\overline{DC}$ is parallel to $\overline{AB}$. Hence $\angle 1 = \angle 2$, because ___alt. int. angles___ formed by a line crossing two parallel lines are equal (see Chapter 2, Central Exercise 10(*d*)). Similarly $\overline{AD}$ is parallel to $\overline{BC}$, so $\angle 3 = \angle 4$. Side $\overline{AC}$ is common to both triangles. Hence $\triangle ABC \cong \triangle CDA$ by ___ASA___. Q.E.D.

We have felt that little is gained by a proliferation of symbols; hence the symbol "∥" for "is parallel to" has not been introduced. However, you may find it convenient to introduce the class to the use of the symbol.

3. *Prove:* If a quadrilateral is a parallelogram, then each pair of opposite sides have equal length.

[*Hint:* Given a parallelogram *ABCD*, draw a diagonal $\overline{AC}$ as shown below. Why is *AB* = *CD*? *AD* = *CB*?] $\triangle ABC \cong \triangle CDA$ by Exercise 2.

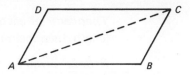

4. *Prove:* Opposite angles of a parallelogram are equal.

[*Hint:* Consider a parallelogram *ABCD* as shown below. What auxiliary

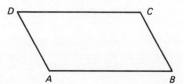

line will help you to prove ∡D = ∡B? to prove ∡A = ∡C?]

5. The next theorem says that the diagonals of a parallelogram "bisect each other." This means that the point of intersection of the diagonals is the midpoint of each. Supply the missing statements in the proof.

Theorem: The diagonals of a parallelogram bisect each other.

Proof: Given a parallelogram ABCD, with diagonals $\overline{AC}$ and $\overline{BD}$ intersecting at E, we want to prove that DE = BE and AE = CE.

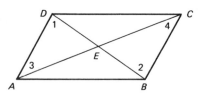

We shall prove that △AED ≅ △CEB, from which it will follow that AE = CE and DE = BE. Now, ∡1 = ∡2, because _alt. int. angles_. Also, ∡3 = ∡4, because _alt. int. angles_. AD = CB, because _opp. sides of a parallelogram_ (see Exercise 3!). Therefore △AED ≅ △CEB by _ASA_. Hence AE = EC and DE = BE, because _corr. parts of congruent triangles_. Q.E.D.

We have used the term converse *a few times already. Students are not expected to really understand what it means at this point. An explicit treatment of converses is given later in this chapter.*

It was proved in Exercise 3 that *if a quadrilateral is a parallelogram, then each pair of opposite sides have equal length.* The following exercise proves the converse statement.

6. Supply the missing statements in the proof of the following theorem.

Theorem: If each pair of opposite sides of a quadrilateral are of equal length, then the quadrilateral is a parallelogram.

Proof: Given a quadrilateral ABCD, with AB = CD and AD = CB, we want to prove it is a parallelogram (that is, we want to prove opposite sides are parallel).

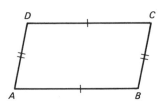

140   4: What is a Proof?

We must prove that $\overline{AB}$ is parallel to $\overline{DC}$ and $\overline{AD}$ is parallel to $\overline{BC}$. In order to do this, first draw diagonal $\overline{AC}$:

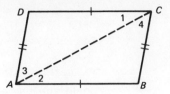

We are given that $AB = CD$ and $BC = DA$. Of course, $AC = AC$. Hence $\triangle ABC \cong \triangle CDA$ by ___SSS___. Therefore $\angle 1 = \angle 2$, because ___corr. parts___. Hence $\overline{AB}$ is parallel to $\overline{DC}$, because the line containing $\overline{AC}$ crosses $\overline{AB}$ and $\overline{DC}$ in such a way that alternate interior angles are equal. (In Chapter 2, Central Exercise 22, we proved that this implies $\overline{AB}$ is parallel to $\overline{DC}$.) Next, $\angle 3 = \angle 4$, because ___corr. parts___. Therefore $\overline{AD}$ is parallel to $\overline{BC}$, because ___alt. int. angles___. Thus both pairs of opposite sides of $ABCD$ are parallel, so it is a parallelogram. Q.E.D.

For some students, this model will give a "feeling" for parallelograms that they did not have before.

7. In the days of sailing ships, parallel rulers were used to measure parallel lines on navigation charts. They were "walked" across the chart.

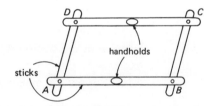

At A, B, C, and D are pins through wooden sticks. Opposite sticks have equal length.

(*a*) Make a set of parallel rulers (optional exercise).

(*b*) Demonstrate in class, explaining how and *why* they work (that is, why is $\overline{AB}$ always parallel to $\overline{CD}$?). *Opposite sides remain equal; hence the quadrilateral is always a parallelogram (Ex. 6).*

The next exercise proves the converse of the statement proved in Exercise 5.

8. *Prove:* If the diagonals of a quadrilateral bisect each other, then the quadrilateral is a parallelogram.

[*Hint:* Consider quadrilateral $ABCD$, with $AE = CE$ and $BE = DE$, in the following picture:

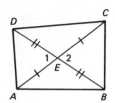

Prove that $\triangle AED \cong \triangle CEB$. Why does it follow that $AD = CB$? How could you prove that $AB = CD$? After you have proved that $AD = CB$ and $AB = CD$, why can you conclude that $ABCD$ is a parallelogram?]
corr. parts, $\triangle EDC \cong \triangle EBA$, opposite sides equal

9. *Prove:* The diagonals of a rectangle have the same length.

[*Hint:* Consider a rectangle $ABCD$:

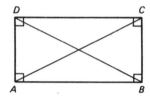

Prove that $\triangle ABC \cong \triangle BAD$.]   *opposite sides equal, right angles equal*

The next exercise is the key to the solution of the problem of the carpenter at the beginning of this chapter.

10. Supply the missing statements in the proof of the following theorem.

Theorem: If the diagonals of a parallelogram have equal length, then the parallelogram is a rectangle.

Proof: Given a parallelogram $ABCD$ with $AC = BD$, we must prove that each angle of the parallelogram is a right angle.

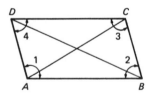

$AD = BC$, since opposite sides of a parallelogram have equal length (Exercise 3). $AC = BD$ is given, and $AB = AB$. Therefore $\triangle BAD \cong \triangle ABC$ by ___SSS___. Hence $\angle 1 = \angle 2$, because ___*corr. parts*___.
Next, $AB = CD$, since ___*opposite sides of a parallelogram*___. Again $AC = BD$ is given,

142   4: What is a Proof?

and $BC = BC$. Hence $\triangle ABC \cong \triangle DCB$ by ___SSS___. Therefore $\angle 2 = \angle 3$, because ___corr. parts___. But $\angle 1 = \angle 3$ and $\angle 4 = \angle 2$, because ___opp. angles___ (see Exercise 4). Thus we have all four angles, $\angle 1$, $\angle 2$, $\angle 3$, and $\angle 4$, equal. But $\angle 1 + \angle 2 + \angle 3 + \angle 4 = $ ___$360°$___, because the sum of the angles in any quadrilateral is ___$360°$___ (see Chapter 2, Central Exercise 28). Hence each angle must be a right angle because ___the angles are equal and their sum is $360°$___ and $ABCD$ is a rectangle. Q.E.D.

Assume that the carpenter can check by "sighting" that the floor lies in one plane.

11. This chapter began with a story about a carpenter building a floor. When he is finished, the floor has the shape of some quadrilateral. He wants to check, by measuring with a steel tape measure, that this quadrilateral is actually a rectangle. Discuss with your classmates how he can check, using the theorems in Exercises 6 and 10. *Measure both pairs of opposite sides and the diagonals.*

Exercises 6 and 8 give conditions on a quadrilateral which guarantee that it is a parallelogram. The next exercise gives another such condition.

12. *Prove:* If two sides of a quadrilateral are parallel and of equal length, then the quadrilateral is a parallelogram.

[*Hint:* Given a quadrilateral $ABCD$ with $AB = CD$ and $\overline{AB}$ parallel to $\overline{CD}$, draw the diagonal $\overline{AC}$:

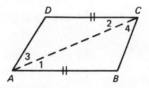

Why is $\angle 1 = \angle 2$? Why is $\triangle ABC \cong \triangle CDA$? Why is $\angle 3 = \angle 4$? Why is $\overline{AD}$ parallel to $\overline{BC}$? Why is $ABCD$ a parallelogram?] *alt. int. angles, SAS, corr. parts, $\angle 3 = \angle 4$ implies $\overline{AD} \parallel \overline{BC}$.*

13. Suppose ℓ and m are parallel lines, and a line segment $\overline{AB}$ is drawn, with A on ℓ, B on m, and $\overline{AB}$ perpendicular to both lines.

You may want to mention to the class that the result of Exercise 13 is often stated as "parallel lines are everywhere equidistant."

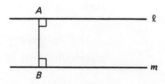

Next, another line segment $\overline{CD}$ is drawn, with C on ℓ, D on m, and $\overline{CD}$ perpendicular to both lines, as in the following figure.

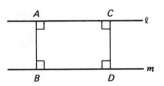

(a) Prove: $AB = CD$. $\overline{AB} \parallel \overline{CD}$, hence ABDC is a parallelogram.

(b) Compare your proof with those of your classmates.

The Converse of a Statement

It is important to learn to distinguish between a statement and the converse of that statement. The converse of a statement is defined as follows:

DEFINITION The *converse* of an "if ... , then ..." statement is obtained by reversing the "if" and "then" parts of the statement.

You may wish to involve the entire class in a presentation and discussion of converses.

Here are some examples of statements and their converses:

Statement	Converse
If a triangle is equilateral, then all the angles of the triangle are equal.	If all the angles of a triangle are equal, then the triangle is equilateral.
If a quadrilateral is a parallelogram, then each pair of opposite sides of the quadrilateral have equal length.	If each pair of opposite sides of a quadrilateral have equal length, then the quadrilateral is a parallelogram.
If a parallelogram is a rectangle, then the diagonals of the parallelogram have equal length.	If the diagonals of a parallelogram have equal length, then the parallelogram is a rectangle.
If a soap is fine, then the soap has suds.	If a soap has suds, then the soap is fine.

The following exercises concern the common error of confusing a statement with its converse.

14. In Exercise 8, you were asked to prove:

If the diagonals of a quadrilateral bisect each other, then the quadrilateral is a parallelogram.

144 4: *What is a Proof?*

Alex Smart says, "This is simple. I can use Exercise 5."

Alex's "Proof": Suppose we are given that the diagonals of quadrilateral *ABCD* bisect each other.

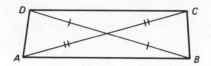

The diagonals of any parallelogram bisect each other (proved in Exercise 5). The diagonals of *ABCD* bisect each other. Therefore *ABCD* is a parallelogram.

Is there anything wrong with Alex's "reasoning"? Discuss with your classmates. *He confused the statement with its converse.*

Lead a discussion on this section. A single counterexample is sufficient to prove a statement false.

15. An advertiser says:

Any fine soap has suds. New Muddo has suds. Therefore New Muddo is a fine soap.

(*a*) Is there anything wrong with the advertiser's logic? *obviously, yes*

(*b*) Is there a similarity between the advertiser's "proof" that New Muddo is a fine soap and Alex's "proof" in the preceding exercise?
The "reasoning," such as it is, is the same.

16. (*a*) Complete the statement correctly:

Any cowardly crocodile has sharp fangs. This crocodile has sharp fangs. Therefore *this crocodile may or may not be cowardly*.

(*b*) Discuss your answer to part (*a*) with your classmates.
no conclusion possible

Here we appeal to the student's common sense. The "proof" in Exercise 15 is ridiculous. In Exercise 16 it would be silly (and possibly fatal) to conclude that this crocodile is cowardly. Yet Alex's "reasoning" in Exercise 14 is precisely of this nature. Thus, common sense should show that his "reasoning" is no reasoning at all.

The logical error made by Alex in Exercise 14 of this chapter involved the *converse* of a statement. Alex wanted to prove that:

If the diagonals of a quadrilateral bisect each other, then the quadrilateral is a parallelogram.

His "proof" was essentially the following:

We know that if a quadrilateral is a parallelogram, then the diagonals bisect each other. The diagonals of our quadrilateral bisect each other, therefore our quadrilateral is a parallelogram.

"If a crocodile is cowardly, then it has sharp fangs." Would Alex believe that "if a crocodile has sharp fangs, then it is cowardly?"

Alex's basic mistake was in thinking that if a statement is true, then the converse of that statement is also necessarily true. But we shall see in the following exercises that often a statement is true, but the converse is false. It can also happen that a statement is false, while the converse is true.

17. In each of the following cases, write the converse of the given statement, and state if the converse is true or false. Discuss your answers with your classmates.

The converse is true.   (a) If a triangle has a pair of equal angles, then the triangle is isosceles.

false   (b) If a quadrilateral is a square, then the quadrilateral has four right angles.

false   (c) If $x = 3$, then $x^2 = 9$.

false   (d) If $x = 2$, then $x^2 = x + 2$.

false   (e) If $a = 0$, then $ab = 0$.

false   (f) If $a > 0$ and $b > 0$, then $ab > 0$.

false   (g) If a man lives in Paris, then the man lives in France.

false   (h) If a soap is good, then the soap has suds.

false   (i) If a crocodile is cowardly, then it has sharp fangs.

true   (j) If there is smoke, there is fire.

Student examples are more valuable than the ones in the text.

18. Give an example of an "if . . . , then . . ." statement such that

(a) the statement is false, but its converse is true;

(b) the statement is true, but its converse is false;

(c) the statement and its converse are both true;

(d) the statement and its converse are both false.

Compare your examples with those of your classmates.

19. A politician gives a speech:

It is my sincere belief that if a politician truly wishes to serve the people, then he will never steal any public money. I pledge now that I will never steal one penny of public money. Therefore . . .

(a) What does this man want you to conclude?

(b) What can you conclude?

(c) What does this have to do with our study of converses?

Midpoints of the Sides of a Triangle

We can use our work on parallelograms to prove a certain useful property of triangles.

Exercise 20 prepares the student for the important theorem in Exercise 21.

20. (a) Draw a triangle, $\triangle ABC$, with $AB = 6$ cm. Find the midpoint D of $\overline{AC}$ and the midpoint E of $\overline{BC}$. Draw $\overline{DE}$:

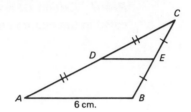

3 cm.
(b) Measure DE, and compare your result with those of your classmates.

(c) Do parts (a) and (b), drawing $\triangle ABC$ with $AB = 8$ cm. *DE = 4 cm.*

(d) Guess a relationship between the lengths DE and AB, for any $\triangle ABC$. Discuss this with your classmates. $DE = \frac{1}{2}AB$

(e) Draw the lines formed by extending $\overline{DE}$ and $\overline{AB}$. Do you notice anything special about the lines? Discuss this with your classmates.
They are parallel.

Your experiments in the preceding exercise may have convinced you of the truth of the following statement:

The line segment joining the midpoints of two sides of a triangle has half the length of the third side and is parallel to the third side.

The next exercise proves that this is really true.

The students should know that this is one of the key theorems in the course. It will be used repeatedly.

21. Supply the missing statements in the proof of the following theorem.

Theorem: The line segment joining the midpoints of two sides of a triangle has half the length of the third side and is parallel to the third side.
(In the next figure, we want to prove $DE = \frac{1}{2}AB$ and $\overline{DE}$ is parallel to $\overline{AB}$.)

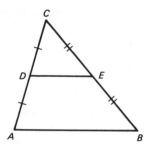

Proof: Extend $\overline{DE}$ to a point F so that $DE = EF$, and draw $\overline{BF}$, as indicated in the picture below:

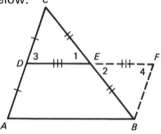

Then $\angle 1 = \angle 2$, since ___vertical angles___. $CE = BE$, because ___E is midpoint___, and $DE = FE$, because ___construction___. Hence $\triangle BFE \cong \triangle CDE$ by ___SAS___. Hence $BF = CD$, because ___corr. parts___. But $CD = AD$, since D is the midpoint of $\overline{AC}$, therefore $BF = AD$. $\angle 3 = \angle 4$, because ___corr. parts___. Therefore $\overline{AC}$ is parallel to $\overline{BF}$, because the alternate interior angles they form with $\overline{DF}$ are equal (Chapter 2, Central Exercise 22). Thus we see that the sides $\overline{AD}$ and $\overline{BF}$ of quadrilateral $ABFD$ are parallel and equal in length. Hence $ABFD$ is a parallelogram because ___two sides parallel and equal___. Therefore $DF = AB$, because ___opp. sides___. Hence $DE = \frac{1}{2} AB$, because ___2(DE) = AB___. Finally, $\overline{DE}$ is parallel to $\overline{AB}$, because ___opp. sides of parallelogram___. Q.E.D.

Encourage student discussion in Exercise 22.

22. What is wrong in each picture?

(a)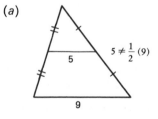

$5 \neq \frac{1}{2}(9)$

(b)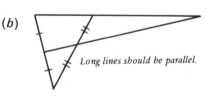

Long lines should be parallel.

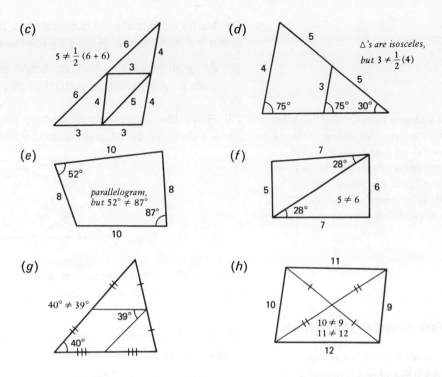

23. *Prove:* The four small triangles formed by joining the midpoints of the sides of any triangle are congruent.

Use Exercise 21 and SSS.

Compare your proof with those of your classmates.

An Unexpected Property of Quadrilaterals

24. (*a*) Make a careful drawing of a quadrilateral, using your ruler, and join the *midpoints* of the sides, as in the picture below.

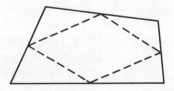

149 Central

Do this for a few different quadrilaterals. Do you observe anything special about the quadrilateral formed by joining the midpoints? *parallelogram*

(b) State an interesting theorem based on your observations in part (a). Compare your observation with those of your classmates.

An alternate proof to the "opposite sides are equal" argument: Observe that PQ = SR and that PQ ∥ SR (both being parallel to the diagonal), hence PQRS is a parallelogram. This latter proof generalizes to skew quadrilaterals.

25. *Prove* the theorem suggested by the preceding experiment! [*Hint:* Draw a diagonal of the original quadrilateral:

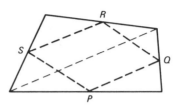

Compare PQ with the length of the diagonal, using the theorem in Exercise 21. Why is PQ = SR? Can you prove PS = QR?]

More About Right Triangles

If a student insists on seeing a proof that ℓ ∥ m and m ∥ n implies ℓ ∥ n in Exercise 25, note that ℓ could not intersect n in a point P, since two distinct lines through P parallel to m contradicts the Parallel Postulate.

26. In the figure below, △ABC is a right triangle, E and F are the midpoints of the legs, and D is the midpoint of the hypotenuse.

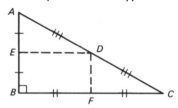

Prove that BFDE is a rectangle. [*Hint:* Use the theorem about the midpoints of the sides of a triangle proved in Exercise 21.]

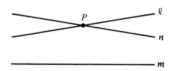

27. *Prove:*

In any right triangle, the segment joining the vertex of the right angle to the midpoint of the hypotenuse is half as long as the hypotenuse.

[*Hint:* We want to prove that $BD = \frac{1}{2}AC$ in the figure on the next page. Consider the figure in the preceding exercise, and recall that the diagonals of any rectangle have equal length.] $BD = FE$, but $FE = \frac{1}{2}AC$

150   4: What is a Proof?

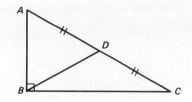

The 30°–60° Right Triangle

This is Timaeus speaking in the Dialogue Timaeus *[B. Jowett, trans., Dialogues of Plato, vol. 3, 3rd ed. (New York: Oxford University Press, 1892)]. Taken in context one sees that he is considering only the class of non-isosceles right triangles. As to why the 30°-60° is most beautiful: "The reason would be long to tell. He who disproves what we are saying, and shows that we are mistaken, may claim a friendly victory."*

In the Dialogue *Timaeus,* Plato said, "The triangle which we maintain to be the most beautiful of the many right triangles is that which with a duplicate forms a triangle which is equilateral." Plato was referring to a right triangle with acute angles of 30° and 60° (that is, a "30°–60° right triangle").

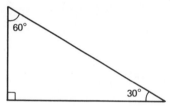

Whether you agree with Plato or not, you will find that this triangle occurs often in geometry and its applications.

The next exercise is related to Plato's statement and brings out an important property of 30°–60° right triangles.

Emphasize that students will use the 30°-60° triangle repeatedly during the course.

28. (*a*) Show how two congruent copies of any 30°–60° right triangle can be placed together to form an equilateral triangle.

Students will find it valuable to memorize this result.

(*b*) Use part (*a*) to show that:

In any 30°–60° right triangle, the side opposite the 30° angle has half the length of the hypotenuse.

---

29. Give a proof of the theorem in Exercise 28(*b*) using the theorem in Exercise 27. [*Hint:* Consider a 30°–60° right triangle, △*ABC*, and let *D* be the midpoint of the hypotenuse $\overline{AC}$, as in the picture on the next page. What kind of triangle must △*ABD* be? Why?]   △*ABD is equilateral.*

151 Central

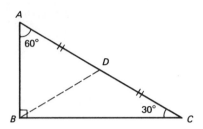

30. A ladder leans against a house. The foot of the ladder is 4 feet from the house and the ladder makes a 60° angle with the ground. How long is the ladder?

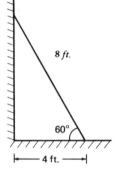

Some students have difficulty seeing all the 30°-60° triangles in Exercise 31.

31. If in the figure below $AC = 20$, find AD:

$AD = \frac{1}{2}AB = \frac{1}{2}(\frac{1}{2}AC)$

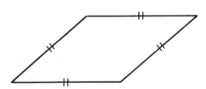

Types of Quadrilaterals

This is a good time to become familiar with some quadrilaterals that have special names.

DEFINITION A *rhombus* is a parallelogram all of whose sides have equal length.

152   4: What is a Proof?

32. Prove that a diagonal of a rhombus bisects the angle at the vertex from which it is drawn. That is, prove that in the figure below we must have ∡1 = ∡2.   △ABD ≅ △CBD by SSS

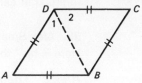

33. (*a*) Prove that the diagonals of a rhombus are perpendicular to each other.

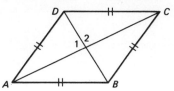

[*Hint:* Show that ∡1 = ∡2. Why will this prove that $\overline{AC}$ is perpendicular to $\overline{BD}$?]

(*b*) Prove that each diagonal of a rhombus is the perpendicular bisector of the other. [*Hint:* Use part (*a*) above and Exercise 5 of this section.]

DEFINITION A quadrilateral with two parallel sides is called a *trapezoid*.

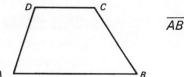

$\overline{AB}$ is parallel to $\overline{CD}$.

34. Suppose you are given trapezoid *ABCD*, with $\overline{AB}$ parallel to $\overline{CD}$, and suppose you know that ∡*A* = ∡*B*. Prove then that *AD* = *BC*. [*Hint:* Draw perpendicular segments from *D* and *C* to $\overline{AB}$. What triangles are congruent? Why?]   Use Exercise 13(*a*).

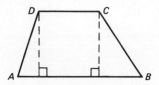

Wrong "Proofs"

These examples of incorrect proofs are good total class discussion topics.

We have already considered the mistake of confusing a statement with its converse in making a proof. Here are some amusing examples of other things that can go wrong in "proofs."

35. Alex Smart claims he can prove *every* triangle is an isosceles triangle! Here is his "theorem" and the "proof."

Alex's "Theorem": If △ABC is any triangle, then AC = BC.

Alex's "Proof": Consider any △ABC. As in the figure below, choose a point D on $\overline{AB}$ such that D is the midpoint of $\overline{AB}$ and ∡1 = ∡2.

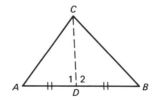

Now, in △ADC and △BDC we have AD = BD, CD = CD, and ∡1 = ∡2. Hence △ADC ≅ △BDC, by SAS. Hence AC = BC, because corresponding sides of congruent triangles have the same length. Q.E.D.

(*a*) Compare Alex's "proof" with the proof (Chapter 3, Central Exercise 42) that if a triangle is isosceles then its base angles are equal. *In this exercise, CD is overdefined.*
(*b*) Alex has "proved" something ridiculous, so surely there is some mistake in his "proof." What is it? Discuss this with your classmates. [*Hint:* Is it true that in every triangle, △ABC, we can draw a line $\overline{CD}$ as above with AD = BD and ∡1 = ∡2?] *The two conditions on $\overline{CD}$ define distinct lines in non-isosceles triangles.*
Alex's error shows that we must be careful how we use auxiliary lines. In particular, we must be sure that any auxiliary lines in a proof actually *exist*.

36. Alex Smart has a "proof" that an obtuse angle can be the same size as a right angle.

"Proof": Consider a quadrilateral ABCD, with ∡A = 90°, ∡B an obtuse angle, and AD = BC, as in the picture below.

Let P be the point where the perpendicular bisector of $\overline{AB}$ intersects the perpendicular bisector of $\overline{CD}$.

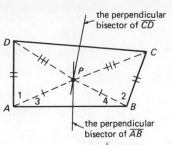

The perpendicular bisectors do not meet within the quadrilateral if $\angle B$ is obtuse.

We have $PC = PD$, since P is a point on the perpendicular bisector of $\overline{CD}$, and any such point is equidistant from the endpoints C and D (Chapter 3, Central Exercise 31). Similarly $PA = PB$, since P is a point on the perpendicular bisector of $\overline{AB}$. Hence $\triangle APD \cong \triangle BPC$ by SSS. Hence $\angle 1 = \angle 2$, since corresponding angles of congruent triangles are equal. Also, $\angle 3 = \angle 4$, since they are base angles of the isosceles triangle $\triangle APB$. Thus $\angle 1 + \angle 3 = \angle 2 + \angle 4$. That is, $\angle A = \angle B$, as we claimed.

Alex has done it again—he has proved something ridiculous, so there must be a mistake in his "proof." Find it. [*Hint:* Draw the picture very accurately.]

Alex's last "proof" shows we must be careful how we use pictures. Although it is a good idea to draw a picture before writing a proof, do not be deceived by your picture. For example, never assume something is true just because it *looks* true in your picture. In the next exercise, Bill makes this mistake. The theorem is correct, but his proof is not.

37. Here is a theorem and its proof by Bill.

Theorem: In any triangle, an exterior angle is always larger than each opposite interior angle.

"Proof": Consider any triangle with exterior angle, $\angle 1$, and opposite interior angles, $\angle 2$ and $\angle 3$. We want to prove that $\angle 1$ is larger than $\angle 2$ and that $\angle 1$ is larger than $\angle 3$.

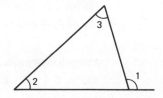

Now, ∡2 and ∡3 are acute angles, and ∡1 is an obtuse angle. But an obtuse angle is larger than an acute angle. Hence ∡1 is larger than ∡2 and is also larger than ∡3, as we wanted to prove.

(*a*) What is wrong with Bill's proof? Discuss with your classmates. [*Hint:* Is it really true that in every triangle ∡2 and ∡3 will be acute angles?]

(*b*) Give a *correct* proof of the theorem. Compare your proof with those of your classmates. [*Hint:* ∡2 + ∡3 = ∡1   .]

Axioms

You may wish to give a total class presentation on this section.

We have now had quite a bit of practice in proving geometric theorems. Correct proofs are arguments based on statements that have been previously proved or accepted as true. In Chapter 2 we mentioned that those statements which we have accepted as true are traditionally called "axioms" or "postulates," and we formally define them now.

DEFINITION   An *axiom* (sometimes called a *postulate* or *assumption*) is a statement which we accept as true, without proof, and use to prove other statements.

For example, in Chapter 2 we considered a certain statement about parallel lines, namely:

If a line *t* crosses two parallel lines ℓ and m, then corresponding angles are equal.

We have never proved this statement, although we have often used it in our proofs. This statement will not be proved in our geometry course. It will remain one of our axioms, or assumptions. Could we prove it, using other assumptions? Yes, but we would have to base our proof on still other axioms. A little thought shows that we cannot prove *everything*. We must make *some* assumptions in order to have a starting point for our arguments.

Another example of an axiom we are using is the Parallel Postulate (Chapter 2, Central section). Namely,

Given any line ℓ and a point *P* not on ℓ, there is one and only one line through *P* and parallel to ℓ.

Why do we *assume* this is true? Cannot such a simple statement be proved using the statements we have already either assumed or proved? If you try to prove the parallel postulate, using only the axioms we have so far assumed or the theorems we have proved, you will inevitably fail. You will probably find yourself reasoning in circles. We will try briefly to indicate why. If you could *prove* the Parallel Postulate, this would mean that any "geometry" which satisfies the other assumptions we have made so far also satisfies the Parallel Postulate. But there are "geometries" *different* from the geometry we are studying in this course. For example, if instead of assuming the Parallel Postulate, we were to assume that through a given point not on a given line there passes *more* than one line parallel to the given line, we would obtain a geometry quite different from the one we are studying now. The geometry obtained this way is called *Non-Euclidean Geometry.* The geometry discussed in this course is called *Euclidean Geometry,* in honor of Euclid, a mathematician at the University in Alexandria about 300 B.C. Euclid is credited as the first to write down a short list of axioms for geometry (one of the axioms being equivalent to the Parallel Postulate) and systematically prove theorems on the basis of those axioms. Thus Euclid was the first to apply the *axiomatic method* in mathematics. (See the Historical Notes in Chapter 3 and in this chapter.)

Important Ideas in This Chapter

Definitions

The *converse* of an "if . . . , then . . ." statement is obtained by reversing the "if" and "then" parts of the statement.

An *axiom,* or *postulate,* is a statement that is accepted as true without proof.

A *proof* is a logically correct argument that uses only statements either previously proved or accepted as true.

A *rhombus* is a parallelogram whose sides all have the same length.

A *trapezoid* is a quadrilateral with two parallel sides.

Theorems

Any parallelogram has the following properties:
    A diagonal cuts it into two congruent triangles.
    Opposite sides have equal length.
    Opposite angles are equal.
    The diagonals bisect each other.

A quadrilateral is a parallelogram if any of the following hold:
    Each pair of opposite sides are of equal length.

Two of its sides are parallel and of equal length.
Its diagonals bisect each other.

The diagonals of a rectangle have the same length.

If the diagonals of a parallelogram have the same length, then it is a rectangle.

If ℓ and m are parallel lines, then any two line segments drawn with endpoints on ℓ and m respectively, and perpendicular to both ℓ and m, have the same length.

The line segment joining the midpoints of two sides of a triangle is parallel to the third side and has half the length of the third side.

In any right triangle, the line segment joining the vertex of the right angle to the midpoint of the hypotenuse has half the length of the hypotenuse.

In any 30°–60° right triangle, the side opposite the 30° angle has length half the hypotenuse.

Summary

This chapter was devoted to proofs and some common pitfalls in devising proofs. Along the way, some useful properties of quadrilaterals, especially parallelograms, were developed. These ideas will be needed in the next chapter, in which figures of various kinds are constructed using compass and ruler.

Historical Note

An excellent account of Lobatchevsky's life, and a brief description of Non-Euclidean geometry, is found in E.T. Bell's Men of Mathematics *(New York: Simon and Schuster, 1937).*

For many centuries after Euclid it was believed that the Parallel Postulate could be proved on the basis of the other axioms of Euclidean Geometry. Girolamo Saccheri published a book in 1733 (the year of his death), purporting to prove the Parallel Postulate. He was not successful in his attempt; however, he unwittingly had proved many important theorems of what was later called Non-Euclidean Geometry. Nicolai Ivanovitch Lobatchevsky in 1826 realized that the Parallel Postulate could not be proved on the basis of the other axioms of Euclid and recognized the existence of Non-Euclidean Geometry. He has been hailed as the "Copernicus of geometry," because the revolution in thought caused by his discovery was comparable to that following the introduction of the Copernican theory in astronomy. The Hungarian mathematician Janos Bolyai discovered Non-Euclidean Geometry at about the same time, without knowledge of Lobatchevsky's work. In order to obtain the opinion of the world's greatest living mathematician, Bolyai communicated his thoughts to Carl Friedrich Gauss. He was dismayed and angered to find that Gauss had already made essentially the same discoveries but had not published his work for fear of public misunderstanding and ridicule.

Central Self Quiz

1. Write the converse of each statement and state whether the converse is true or false.

(a) If the diagonals of a quadrilateral bisect each other, then the quadrilateral is a parallelogram.

(b) If $x > 0$, then $x^2 > 0$.

(c) If a point is equidistant from the endpoints of a line segment, then the point is on the perpendicular bisector of the line segment.

(d) If a triangle has a 30° angle and a 60° angle, then it is a right triangle.

(e) If the diagonals of a parallelogram are perpendicular to each other, then the parallelogram is a rhombus.

2. Suppose $PM = 5$ in the right triangle below. Find RQ.

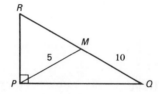

3. Find x, and explain your answer.

(a)

(b)

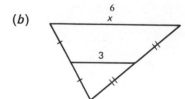

(c)

(d)

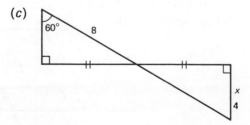

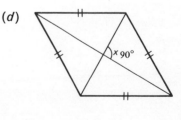

Answers

1. (a) If a quadrilateral is a parallelogram, then its diagonals bisect each other. *true*

   (b) If $x^2 > 0$, then $x > 0$. *false* (Notice that $(-1)^2 > 0$, but $-1 < 0$!)

   (c) If a point is on the perpendicular bisector of a line segment, then it is equidistant from the endpoints of the line segment. *true*

   (d) If a triangle is a right triangle, then it has a 30° angle and a 60° angle. *false*

   (e) If a parallelogram is a rhombus, then its diagonals are perpendicular to each other. *true*

2. $RQ = 10$ (See Central Exercise 27.)

3. (a) Opposite sides of the quadrilateral are equal in length, hence the quadrilateral is a parallelogram (Central Exercise 6). Therefore $x = 110°$ (why?).

   (b) $x = 6$ (See Central Exercise 21.)

   (c) $x = 4$. The two triangles are congruent by ASA, and x is one-half the length of the hypotenuse (Central Exercise 28(b)).

   (d) $x = 90°$. The quadrilateral is a rhombus (why?). Use Central Exercise 33.

Review

1. In the figure below, ℓ and m are two lines intersecting in point O. Points M, N, P, and Q have been chosen so that $ON = OP$ and $OM = OQ$. Prove that $MN = QP$.

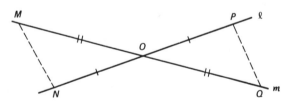

2. In the figure below, line ℓ is parallel to line m, and $\angle 1 = \angle 2$. Prove that $AB = CD$.

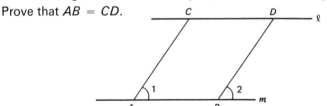

3. Given the figure below, with CD = FD, BD = ED, AC = GF, and AB = GE. Prove that ∡A = ∡G.

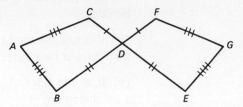

[*Hint:* Draw $\overline{BC}$ and $\overline{EF}$.]

4. The figure below shows a triangle, △ABC. Line ℓ is the perpendicular bisector of $\overline{AC}$, line m is the perpendicular bisector of $\overline{BC}$, and P is the point where they intersect. Prove that P is equidistant from A, B, and C.

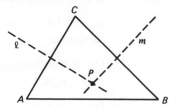

[*Hint:* Recall that each point on the perpendicular bisector of a line segment is equidistant from the endpoints.]

5. Given the figure below, prove that ADEF is a parallelogram.

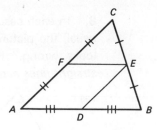

6. Why must CD = AD in the figure below?

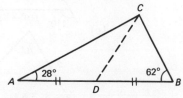

161 Review

7. Answer *true* or *false*.

(a) If a quadrilateral is a parallelogram, then its opposite sides have equal length.   *Central Exercise 3*

(b) If a quadrilateral has one pair of opposite sides of the same length, then it must be a parallelogram.

(c) If all the sides of a quadrilateral are the same length, then it must be a parallelogram.   *Central Ex. 6*

(d) If all the sides of a quadrilateral are the same length, then it must be a rhombus.

(e) If all the sides of a quadrilateral are the same length, then it must be a square.

(f) The diagonals of any rectangle have equal length.   *Central Ex. 9*

(g) If the diagonals of a parallelogram have equal length, then it is a rectangle.   *Central Ex. 10*

(h) If the diagonals of a quadrilateral have equal length, then it must be a rectangle.

(i) The diagonals of any rhombus are perpendicular.   *Central Ex. 33(a)*

(j) The diagonals of any rhombus have equal length.

(k) If the diagonals of a parallelogram are perpendicular and have equal length, then the parallelogram must be a square.

8. In each case, explain what is wrong with the information given. (As usual, the picture might be distorted, so you are not allowed to say "it looks wrong." However, *you may assume that lines which look like straight lines actually are.* Find a geometric fact which is violated.)

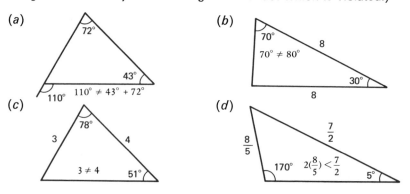

162   *4: What is a Proof?*

Ex. 8(i): See Central Exercise 23.

(e)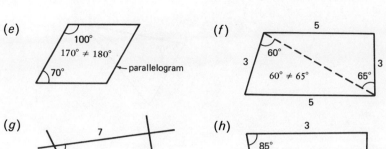

(f)

(g)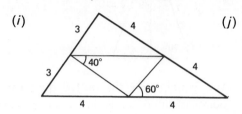

(h)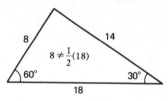

(i)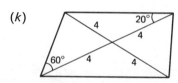

(j)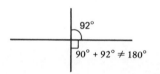

(k)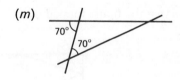

(l)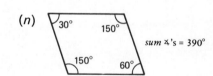

(m)

(n)

9. Find x. Be able to *prove* that your answer is correct.

(a)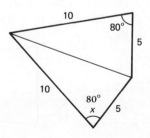

(b)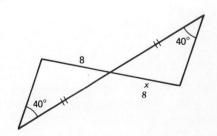

163 Review

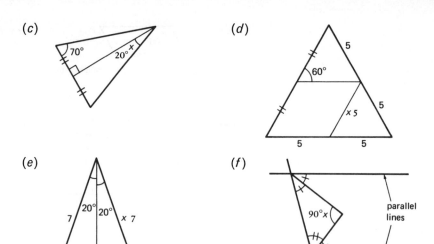

10. Explain why the dotted quadrilateral must be a parallelogram. (All points of the figure lie in the same plane.)

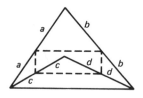

11. Prove that the figure formed by joining the midpoints of a rectangle is a rhombus.

Use Central Ex. 24 and the congruence of the right triangles.

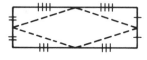

12. Alex Smart says he can prove that in every quadrilateral opposite angles are supplementary. That is, if *ABCD* is any quadrilateral, then $\angle A + \angle C = 180°$ and $\angle B + \angle D = 180°$.

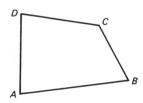

When working Review Exercise 11 in Chapter 10, you may want to refer back to this exercise. It forms the basis for a proof, different from the usual, that opposite angles of a quadrilateral inscribed in a circle (cyclic quadrilateral) are supplementary.

Alex's "Proof": Choose a point P inside the quadrilateral, with P equidistant from the vertices. That is, such that $PA = PB = PC = PD$.

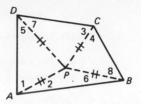

We have $\angle 1 = \angle 5$, since $\triangle APD$ is isosceles. Similarly, $\angle 2 = \angle 6$, $\angle 3 = \angle 7$, and $\angle 4 = \angle 8$. Thus $\angle 1 + \angle 2 + \angle 3 + \angle 4 = \angle 5 + \angle 6 + \angle 7 + \angle 8$. Hence $\angle A + \angle C = (\angle 1 + \angle 2) + (\angle 3 + \angle 4) = \angle 5 + \angle 6 + \angle 7 + \angle 8 = (\angle 6 + \angle 8) + (\angle 5 + \angle 7) = \angle B + \angle D$. Now we have shown that $\angle A + \angle C = \angle B + \angle D$. But $(\angle A + \angle C) + (\angle B + \angle D) = 360°$, because the sum of the angles in any quadrilateral is $360°$. Since $\angle A + \angle C$ and $\angle B + \angle D$ are equal and their sum is $360°$, it follows that $\angle A + \angle C = 180°$ and $\angle B + \angle D = 180°$, as we wanted to prove.

Do you see anything wrong with this "proof"? [*Hint:* Is there always some point inside a quadrilateral which is equidistant from all the vertices?]

Review Answers

1. $\triangle MON \cong \triangle QOP$ by SAS. Hence $MN = QP$ since corresponding sides of congruent triangles have equal length.

2. $\overline{AC}$ is parallel to $\overline{BD}$ (why?), and ℓ is given parallel to m. Hence $ABCD$ is a parallelogram. But opposite sides of a parallelogram have equal length (Central Exercise 3), so $AB = CD$.

3. Prove $\triangle CDB \cong \triangle FDE$, then prove $\triangle CAB \cong \triangle FGE$.

4. $PA = PC$, since P is on the perpendicular bisector of $\overline{AC}$. $PB = PC$, since P is on the perpendicular bisector of $\overline{BC}$. Thus $PA = PC = PB$.

5. $FE = AD$ and $DE = AF$ (Central Exercise 21), hence $ADEF$ is a parallelogram (Central Exercise 6).

6. $\triangle ABC$ is a right triangle (why?) and $\overline{CD}$ is the line segment joining the vertex of the right angle to the midpoint of the hypotenuse.

7. (a) true (b) false (c) true (d) true (e) false (f) true (g) true (h) false (i) true (j) false (k) true

8. (a) The Exterior Angle Theorem does not hold.

(b) Note that the triangle is isosceles.

(c) The missing angle must be 51° (why?), but 3 ≠ 4.

(d) The third side has length $\frac{8}{5}$ (why?). What is wrong?

(f) The quadrilateral must be a parallelogram (why?). We have a line crossing two parallel lines, but alternate interior angles are not equal.

(g) The four lines must form a parallelogram (why?), but the opposite sides are not of equal length.

(h) The quadrilateral has a pair of parallel sides of equal length (why are the horizontal sides parallel?). Hence it is a parallelogram. What is wrong?

(j) The triangle is a 30°–60° right triangle. But the side opposite the 30° angle is not half the length of the hypotenuse.

(k) The quadrilateral has four 80° angles (why?). What is wrong?

(n) The sum of the angles is 390°.

10. The horizontal sides of the dotted quadrilateral are both parallel to and half the length of the bottom side of the triangle.

Review Self Test

1. If $AB = BC$ and $AD = CD$, then prove that

(a) $\overline{BD}$ is perpendicular to $\overline{AC}$

(b) $AX = XC$.

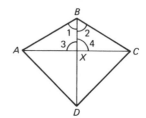

2. Prove:

(a) If the diagonals of a parallelogram are perpendicular, then the parallelogram is a rhombus.

(b) If a diagonal of a parallelogram bisects the angles, the parallelogram is a rhombus.

Answers

1. (a) $\triangle ABD \cong \triangle CBD$ by SSS. It follows that $\angle 1 = \angle 2$, for these angles are corresponding parts of the above congruent triangles. Thus $\triangle ABX \cong \triangle CBX$ by SAS and $\angle 3 = \angle 4$, for they are corresponding parts of these congruent triangles.

(b) *AX* = *XC*, for they are corresponding parts of the congruent triangles △*ABX* and △*CBX*.

2. (a) Consider parallelogram *ABCD*, with diagonal $\overline{AC}$ perpendicular to diagonal $\overline{BD}$:

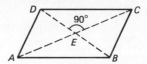

We want to prove that *ABCD* is a rhombus. All we need to show is that *DC* = *BC* (why?). But △*DEC* ≅ △*BEC* by SAS (prove it!); hence *DC* = *BC*, because corresponding sides of congruent triangles have equal length. This completes the proof.

(b) Consider parallelogram *ABCD* with ∡1 = ∡2 and ∡3 = ∡4:

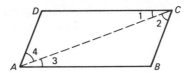

We want to prove that *ABCD* is a rhombus. As in part (b), we see that all we have to show is that *DC* = *BC*, since we already know *ABCD* is a parallelogram. But △*ADC* ≅ △*ABC* by ASA (why?); hence *DC* = *BC* (why?).

Projects

1. A "plumb level" consists of a wooden framework in the shape of an isosceles triangle, with a string and weight hung as below.

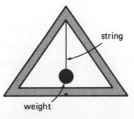

(a) Explain to a classmate how the plumb level can be used to determine whether an object is level. *When the base of the plumb level is level, the weight hangs over the midpoint of the base.*

(b) Explain *why* the method is valid, in terms of known geometrical facts.

2. Suppose *ABCD* is any *skew quadrilateral* (the four vertices are not all in the same plane). Then if *E, F, G,* and *H* are the *midpoints* of the sides, prove that *EFGH* is a parallelogram.

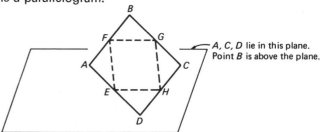

A, C, D lie in this plane. Point *B* is above the plane.

If you have difficulty visualizing this figure, then make a model by cutting a quadrilateral out of paper and folding it along one diagonal. The edges will form a skew quadrilateral.

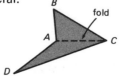

[*Hint:* In order to prove *EFGH* is a parallelogram, draw $\overline{AC}$.]

3. Prove that in the box below $AX = BY$.

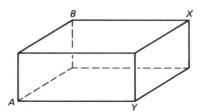

(The box is "rectangular," that is, the six faces of the box are rectangles.)

4. The figure below shows a trapezoid with parallel sides of lengths *a* and *b* respectively. The points *E* and *F* are the midpoints of the respective sides on which they lie.

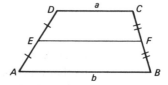

Prove that $EF = \dfrac{a+b}{2}$. [*Hint:* Let *G* be the point where the line through *D* and *F* intersects the line through *A* and *B*:

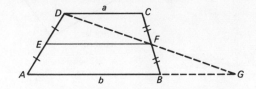

Prove that $\triangle DCF \cong \triangle GBF$. Then it follows that $BG = a$ (why?), and $EF = \frac{1}{2} AG$ (why?).]

Polyhedra

These projects will enhance a student's spatial intuition.

A *polyhedron* is a 3-dimensional solid whose surface is made up of polygons. These polygons are called the *faces* of the polyhedron. For example, the figure below shows a *triangular prism.* This prism has 5 faces altogether, 3 of which are parallelograms and 2 of which are triangles.

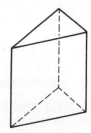

You could build a model of this prism by gluing together 3 parallelograms and 2 triangles along their edges in an appropriate fashion.

The box in Exercise 3 on the previous page is another example of a polyhedron (pl. *polyhedra*).

DEFINITION

Note that the number of faces *meeting at a vertex of a polyhedron is the same as the number of* edges *meeting at that vertex, an edge being defined as a line segment along which two faces meet.*

A *regular polyhedron* is a polyhedron such that

(*i*) all the faces are congruent *regular* polygons, and

(*ii*) the number of faces which meet at each vertex (corner) is the same for all vertices of the polyhedron.

For example, a *cube* is a regular polyhedron, since (*i*) all the faces are regular 4-gons, and (*ii*) there are 3 faces which meet at each vertex. The ancient Greek geometers discovered five regular polyhedra, called the *Platonic solids:*

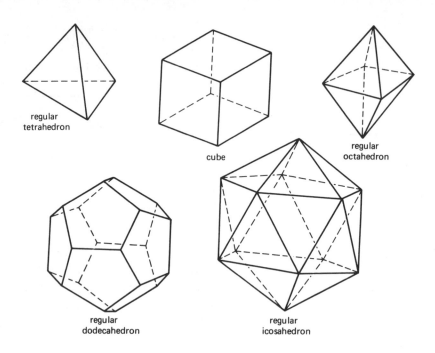

5. The regular tetrahedron has 4 faces, each an equilateral triangle, with 3 faces meeting at each vertex. The regular octahedron has 8 faces, each an equilateral triangle, with 4 faces meeting at each vertex.

(a) The regular dodecahedron has __12__ faces, each face a regular __pentagon__, and __3__ faces meeting at each vertex.

(b) The regular icosahedron has __20__ faces, each face an __equilateral triangle__, and __5__ faces meeting at each vertex.

Encourage students to do Exercise 6.

6. Build models of the five Platonic solids.

7. Is there a regular polyhedron all of whose faces are *regular hexagons*? If you think so, try building a model. *no*

8. Suppose A, B, C, A', B', C' are the midpoints of the six edges of a *regular tetrahedron*, as below.

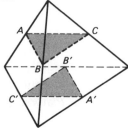

170   4: *What is a Proof?*

The lengths $AB, BC, CA, A'B'$, $B'C', C'A', AC', BC', BA'$ CA', CB', and AB' are all equal and are one-half the length of an edge of the regular tetrahedron.

(a) Explain why the 8 triangles, $\triangle ABC$, $\triangle A'B'C'$, $\triangle ABC'$, $\triangle A'B'C$, $\triangle BCA'$, $\triangle B'C'A$, $\triangle A'C'B$, and $\triangle ACB'$, are all congruent equilateral triangles.

(b) The 8 triangles in part (a) are the faces of a certain regular polyhedron. What is it? *regular octahedron*

(c) The midpoints of the edges of a regular tetrahedron are the vertices of a regular _____. *octahedron*

9. Using wire and colored thread, build a model illustrating the result in Projects Exercise 8(c).

10. The midpoints of the edges of a cube are the vertices of a polyhedron called the *cuboctahedron*.

(a) How many faces does the cuboctahedron have? What kinds of faces?
14, squares and equilateral triangles

(b) Explain why the cuboctahedron is *not* a regular polyhedron. *The faces are not all the same regular polygon.*

(c) Build a model of a cuboctahedron.

11. The midpoints of the edges of a regular octahedron are the vertices of some polyhedron. Describe it. *cuboctahedron*

12. Show how to tile space using cuboctahedra and regular octahedra together. [*Hint:* Recall Exercise 10. Begin by tiling space with cubes, then imagine cutting the corners from the cubes.]

13. (a) Which regular polygons tile the plane? *triangles, squares, hexagons*

(b) Which regular polyhedra tile space? Check your answer using models.
cubes

Algebra Review

Fractions

We add, subtract, multiply, and divide fractions in algebra just as we do in arithmetic, as the following examples show.

The review of algebraic fractions is based on arithmetic fractions.

Example 1. Addition

$$\frac{2}{3} + \frac{4}{7} = \frac{2}{3} \cdot \frac{7}{7} + \frac{4}{7} \cdot \frac{3}{3}$$

$$= \frac{(2)(7) + (4)(3)}{(3)(7)}$$

$$= \frac{14 + 12}{21}$$

$$= \frac{26}{21}$$

$$\frac{a}{b} + \frac{c}{d} = \frac{a}{b} \cdot \frac{d}{d} + \frac{c}{d} \cdot \frac{b}{b}$$

$$= \frac{ad + cb}{bd}$$

Example 2. Subtraction

$$\frac{4}{9} - \frac{3}{7} = \frac{4}{9} \cdot \frac{7}{7} - \frac{3}{7} \cdot \frac{9}{9}$$

$$= \frac{(4)(7) - (3)(9)}{(9)(7)}$$

$$= \frac{28 - 27}{63}$$

$$= \frac{1}{63}$$

$$\frac{2m}{n} - \frac{p}{3q} = \frac{2m}{n} \cdot \frac{3q}{3q} - \frac{p}{3q} \cdot \frac{n}{n}$$

$$= \frac{6mq}{3nq} - \frac{np}{3nq}$$

$$= \frac{6mq - np}{3nq}$$

1. Write the sum (or difference) as a single fraction (show all work).

(a) $\dfrac{6}{7} - \dfrac{4}{11}$ (b) $\dfrac{c}{d} + \dfrac{e}{f}$ (c) $\dfrac{2}{x} + \dfrac{x}{3}$ (d) $\dfrac{2x}{y} - \dfrac{x}{7}$

(e) $\dfrac{x+2}{2} + \dfrac{x+6}{4}$ (f) $\dfrac{3}{xy} - \dfrac{2}{xz}$ (g) $\dfrac{x}{y} + \dfrac{y}{z}$

(h) $\dfrac{2x-y}{x^2y} + \dfrac{x-y}{xy^2}$

2. Multiply at sight.

(a) $(2x + 1)(x - 2)$ (b) $(1 - x)(1 + x)$

(c) $(3x + 5)(2x - 3)$ (d) $(x + y)(z + w)$

(e) $(x - 2)(x + y)$ (f) $(3x - 1)(x + 2)$

3. Factor into prime polynomials with integral coefficients.

(a) $x^2 - y^2$ (b) $x^2 + 4x - 4$ (c) $3x^2 + 13x - 30$

(d) $x - 9x^2$ (e) $5x^2 + 13x - 6$ (f) $9x^2 - 12x + 4$

(g) $x^3 - x^2 + x$ (h) $x^4 - 1$ [*Hint:* $x^2 + 1$ is one factor.]

Example 3. Multiplication

$$\frac{3}{8} \cdot \frac{2}{9} = \frac{\cancel{3}}{(\cancel{2})(4)} \cdot \frac{\cancel{2}}{(\cancel{3})(3)} \qquad \frac{3}{x^2 - x} \cdot \frac{x}{3} = \frac{\cancel{3}}{\cancel{x}(x-1)} \cdot \frac{\cancel{x}}{\cancel{3}}$$

$$= \frac{1}{4} \cdot \frac{1}{3} \qquad\qquad\qquad\qquad = \frac{1}{x-1}$$

$$= \frac{1}{12}$$

Example 4. Division

$$\frac{6}{7} \div \frac{5}{4} = \frac{6}{7} \cdot \frac{4}{5} \qquad\qquad \frac{a}{b} \div \frac{c}{d} = \frac{a}{b} \cdot \frac{d}{c}$$

$$= \frac{24}{35} \qquad\qquad\qquad\qquad = \frac{ad}{bc}$$

4. Write the product (or quotient) as a single fraction.

(a) $\left(3\dfrac{1}{5}\right)\left(1\dfrac{7}{8}\right)$ (b) $\left(1 - \dfrac{1}{x}\right)\left(1 + \dfrac{2}{x}\right)$

[*Hint:* (b) Write $1 - \dfrac{1}{x}$ as a single fraction.]

(c) $\dfrac{x^2 - 9}{x^2 - 2x - 3} \cdot \dfrac{x + 1}{x + 3}$ (d) $\dfrac{4x + 12}{3x - 18} \cdot \dfrac{9x - 54}{2x + 6}$

(e) $\dfrac{x^4 - y^4}{x - y} \cdot \dfrac{x + y}{x^2 + y^2}$

In a numerator that has two or more terms, place each term in parentheses to avoid error:

Example 5.

$$\frac{4x}{4x^2 - 1} - \frac{1}{2x - 1} = \frac{4x}{(2x + 1)(2x - 1)} - \frac{1}{2x - 1}$$

$$= \frac{4x}{(2x+1)(2x-1)} - \frac{(1)(2x+1)}{(2x-1)(2x+1)}$$

$$= \frac{4x - (1)(2x+1)}{(2x-1)(2x+1)}$$

$$= \frac{4x - 2x - 1}{(2x-1)(2x+1)}$$

$$= \frac{(2x-1)}{(2x-1)(2x+1)}$$

$$= \frac{1}{2x+1}$$

5. Write the sum (or difference) as a single fraction.

(a) $\frac{2}{3} + \frac{3}{5}$ (b) $\frac{2}{x+3} + \frac{3}{x-5}$ (c) $\frac{x+2}{4} - \frac{x-3}{6}$

(d) $\frac{2}{x^2 + 4x + 3} - \frac{x+2}{x+3}$ (e) $\frac{x-y}{x+y} - \frac{x+y}{x-y}$

6. Write as a single fraction.

(a) $\frac{3}{x-1} - \frac{2}{x-1}$ (b) $\frac{a}{x+y} + \frac{b}{x+y}$

(c) $\frac{x^2}{x+4} - \frac{16}{x+4}$ (d) $x + \frac{1}{x^2}$ (e) $\frac{1}{x-y} - \frac{3}{x+y}$

(f) $\frac{x}{2} + \frac{x}{3} + \frac{x}{4}$ (g) $\frac{1}{a} + \frac{1}{b}$ (h) $\frac{4x}{4x^2 - 1} - \frac{3}{6x - 3}$

(i) $\frac{1}{a} + \frac{1}{b} - \frac{1}{c}$

7. Write the reciprocal of each of the following.

(a) $\frac{2}{7}$ (b) $3\frac{2}{7}$ (c) $\frac{a}{b}$ (d) $x + \frac{y}{z}$

[Hint: (d) Write $x \cdot \frac{z}{z} + \frac{y}{z}$, then continue.]

8. Write as a single fraction.

(a) $\dfrac{\frac{2}{7}}{\frac{3}{4}}$ (b) $\dfrac{\frac{a}{b}}{\frac{c}{d}}$ (c) $\dfrac{3\frac{3}{4}}{1\frac{5}{8}}$ (d) $\dfrac{\frac{1}{x}+1}{\frac{1}{x}-1}$

[Hint: (d) What is the reciprocal of $\frac{1}{x} - 1$?]

9. Write as a single fraction.

(a) $\dfrac{3\frac{1}{4}}{4\frac{1}{3}}$ (b) $\dfrac{x+1}{1+\frac{1}{x}}$ (c) $\dfrac{\frac{1}{x}+\frac{1}{y}}{\frac{1}{x}-\frac{1}{y}}$ (d) $\dfrac{3x - \frac{1}{3x}}{\frac{3x+1}{2x}}$

10. Write as a single fraction:

$$\frac{b^2}{4a^2} - \frac{c}{a}$$

11. Solve for x. [Hint: Clear the equation of fractions.]

(a) $\dfrac{x}{2} = \dfrac{4}{7}$ (b) $\dfrac{5}{x} = \dfrac{7}{9}$ (c) $2x = 0$ (d) $\dfrac{x}{3} - \dfrac{2}{7} = 4$

(e) $\dfrac{1}{x} + \dfrac{1}{b} = \dfrac{1}{a}$ (f) $\dfrac{1}{x} + a = 0$

[Hint: (e) Clear of fractions, multiply each side by xab.]

12. Solve for x.

(a) $\dfrac{3x}{4} + \dfrac{7x}{8} = 26$ (b) $\dfrac{14}{x-3} = 2$ (c) $\dfrac{5}{x} - 3 = \dfrac{9-x}{4x}$

(d) $\dfrac{x}{6} = \dfrac{12-x}{3}$ (e) $\dfrac{3}{x-2} = \dfrac{-5}{x+1}$

(f) $\dfrac{8}{x^2-9} - \dfrac{6}{x+3} = \dfrac{2}{x-3}$

13. Solve for x.

(a) $\dfrac{x}{9} = 0$
(b) $2 = \dfrac{6}{x}$
(c) $0 = 1 - \dfrac{x}{3}$

(d) $\dfrac{x}{2} + x = 3$
(e) $\dfrac{3x - 4}{10} = 5$
(f) $\dfrac{2x + 8}{x + 7} = 5$

(g) $\dfrac{x}{2} + \dfrac{x}{3} = \dfrac{5}{6}$
(h) $\dfrac{1}{x} = \dfrac{1}{36} - \dfrac{1}{60}$
(i) $\dfrac{4}{x} - \dfrac{3}{x} + \dfrac{5}{x} = \dfrac{2}{3}$

(j) $\dfrac{2}{x - 2} = \dfrac{3}{x + 1}$
(k) $\dfrac{x + 5}{2} + \dfrac{x + 1}{4} = 5$

14. Solve for x in terms of the other letters.

(a) $\dfrac{x}{a} = 2$
(b) $\dfrac{1}{x} = \dfrac{1}{a}$
(c) $\dfrac{a}{x} = \dfrac{b}{c}$

(d) $\dfrac{x}{b} - a = 0$
(e) $\dfrac{x}{a} + b = c$
(f) $\dfrac{a}{x} + b = c$

(g) $\dfrac{x}{3} - \dfrac{a}{3} = a$
(h) $\dfrac{1}{x} = \dfrac{1}{a} + \dfrac{1}{b}$
(i) $\dfrac{x - b}{x + b} = \dfrac{1}{2}$

15. Solve for x.

(a) $\dfrac{4}{x} + \dfrac{2}{x} = 3$
(b) $\dfrac{x + 3}{2} = \dfrac{x}{5}$
(c) $\dfrac{x - 3}{3x + 2} = \dfrac{1}{5}$

(d) $\dfrac{x + 1}{x - 1} = \dfrac{2x - 7}{2x + 3}$
(e) $1 - \dfrac{2}{2x + 1} = -\dfrac{x}{1 - x}$

16. Solve for x. [*Hint:* Clear the equation of fractions, then write in the form: polynomial $= 0$.]

(a) $x + 2 + \dfrac{1}{x} = 0$
(b) $x - 3 + \dfrac{2}{x} = 0$
(c) $x = 1 + \dfrac{2}{x}$

(d) $x + \dfrac{6}{x} = 5$
(e) $6x + 1 - \dfrac{1}{x} = 0$
(f) $2x + 1 = \dfrac{1}{x}$

Answers

1. (a) $\dfrac{38}{77}$
(b) $\dfrac{cf + de}{df}$
(c) $\dfrac{6 + x^2}{3x}$
(d) $\dfrac{14x - xy}{7y}$

(e) $\dfrac{3x + 10}{4}$
(f) $\dfrac{3z - 2y}{xyz}$
(g) $\dfrac{xz + y^2}{yz}$
(h) $\dfrac{x^2 + xy - y^2}{x^2y^2}$

2. (a) $2x^2 - 3x - 2$
(b) $1 - x^2$
(c) $6x^2 + x - 15$
(d) $xz + yz + xw + yw$
(e) $x^2 - 2x + xy - 2y$
(f) $3x^2 + 5x - 2$

3. (a) $(x-y)(x+y)$ (b) already prime (c) $(3x-5)(x+6)$
   (d) $x(1-9x)$ (e) $(5x-2)(x+3)$ (f) $(3x-2)^2$
   (g) $x(x^2-x+1)$ (h) $(x^2+1)(x-1)(x+1)$

4. (a) 6 (b) $\dfrac{x^2+x-2}{x^2}$ (c) 1 (d) 6 (e) $(x+y)^2$

5. (a) $\dfrac{19}{15}$ (b) $\dfrac{5x-1}{(x+3)(x-5)}$ (c) $\dfrac{x+12}{12}$ (d) $\dfrac{-x}{x+1}$
   (e) $\dfrac{-4xy}{(x+y)(x-y)}$

6. (a) $\dfrac{1}{x-1}$ (b) $\dfrac{a+b}{x+y}$ (c) $x-4$ (d) $\dfrac{x^3+1}{x^2}$
   (e) $\dfrac{-2x+4y}{(x-y)(x+y)}$ (f) $\dfrac{13x}{12}$ (g) $\dfrac{a+b}{ab}$ (h) $\dfrac{1}{2x+1}$
   (i) $\dfrac{bc+ac-ab}{abc}$

7. (a) $\dfrac{7}{2}$ (b) $\dfrac{7}{23}$ (c) $\dfrac{b}{a}$ (d) $\dfrac{z}{xz+y}$

8. (a) $\dfrac{8}{21}$ (b) $\dfrac{ad}{bc}$ (c) $\dfrac{30}{13}$ (d) $\dfrac{1+x}{1-x}$

9. (a) $\dfrac{3}{4}$ (b) x (c) $\dfrac{y+x}{y-x}$ (d) $\dfrac{2(3x-1)}{3}$

10. $\dfrac{b^2-4ac}{4a^2}$

11. (a) $x=\dfrac{8}{7}$ (b) $x=\dfrac{45}{7}$ (c) $x=0$ (d) $x=\dfrac{90}{7}$
    (e) $x=\dfrac{ab}{b-a}$ (f) $x=-\dfrac{1}{a}$

12. (a) $x=16$ (b) $x=10$ (c) $x=1$ (d) $x=8$
    (e) $x=\dfrac{7}{8}$ (f) $x=\dfrac{5}{2}$

13. (a) $x=0$ (b) $x=3$ (c) $x=3$ (d) $x=2$
    (e) $x=18$ (f) $x=-9$ (g) $x=1$ (h) $x=90$
    (i) $x=9$ (j) $x=8$ (k) $x=3$

14. (a) $x=2a$ (b) $x=a$ (c) $x=\dfrac{ac}{b}$ (d) $x=ab$
    (e) $x=a(c-b)$ (f) $x=\dfrac{a}{c-b}$ (g) $x=4a$ (h) $x=\dfrac{ab}{a+b}$
    (i) $x=3b$

15. (a) $x=2$ (b) $x=-5$ (c) $x=\dfrac{17}{2}$ (d) $x=\dfrac{2}{7}$ (e) $x=\dfrac{1}{4}$

16. (a) $x=-1$ (b) $x=1,2$ (c) $x=2,-1$ (d) $x=3,2$
    (e) $x=-\dfrac{1}{2},\dfrac{1}{3}$ (f) $x=\dfrac{1}{2},-1$

Chapter 5 Constructions with Straightedge and Compass

Central

Students generally enjoy this chapter a great deal. With each construction they add to a valuable reservoir of geometric experiences. This chapter also shows the importance of proofs and introduces several important geometric concepts.

*If the only straightedge available is a ruler, tell the students **never** to use the marks on it for construction purposes.*

Consider this line segment $\overline{AB}$:

A •⎯⎯⎯⎯⎯⎯⎯⎯⎯⎯⎯• B

Suppose you want to find the *midpoint* of $\overline{AB}$ (that is, you want to find the point M on $\overline{AB}$ such that $AM = MB$). You could certainly do this by measuring with your ruler. But suppose all the markings on your ruler had disappeared, so that you could not use it to measure distances; that is, your "ruler" is useful only for drawing straight lines. Then could you *find the midpoint of the line segment using only your unmarked ruler and a compass?*   yes

The ancient Greek geometers were curious to discover what geometric figures they could draw using only an unmarked ruler and a compass as drawing tools. Here is a specific example of this type of problem:

Can you draw a square using only an unmarked ruler and a compass?   yes

We call an unmarked ruler a *straightedge.* Then the preceding problem can be restated:

179

Can you draw a square using only a straightedge and a compass?

Another example of the kind of problem which intrigued the Greek geometers concerns a circle:

Can you find the center of this circle using only a straightedge and a compass?

The method for drawing a figure with straightedge and compass, as well as actually drawing the figure, is called a *construction.* In this chapter, you will learn some basic constructions and you will also learn that there are some geometric figures which cannot be drawn with only straightedge and compass. Construction problems will help you understand basic geometric facts and improve your ability to think about geometric figures. Some constructions are of practical use in mechanical drawing, but constructing geometric figures using only straightedge and compass is also simply a lot of fun. Perhaps this is the main reason it interested the Greek geometers.

The first two exercises will familiarize us with circles. Since we will work with circles throughout the course, it is important that we know exactly what we mean by "circle."

DEFINITION

Let P be a point in a plane and r some positive number. Then the *circle* of radius r with center P consists of all those points in the plane that are at a distance r from P.

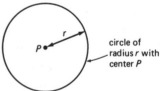

Note: The plural of radius is *radii.*

1. The following figure shows a triangle, and a circle with center P. The circle intersects the triangle in four points A, B, C, and D.

This should help students grasp Exercise 2 more readily.

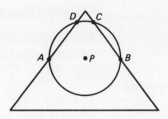

(a) Explain to a classmate why $AP = BP = CP = DP$. *equal radii*

(b) Are there any other points on the triangle that are the same distance from P as A, B, C, and D? Discuss with your classmates. *no*

Exercise 2 prepares for the construction of the perpendicular bisector of a line segment. Remind students of the definition of a rhombus.

2. Mark a point A on a piece of paper and use your compass to draw a circle with center A. Next, using the same compass setting, draw another circle *of the same radius* with center B and intersecting the first circle in two points P and Q.

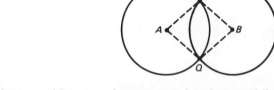

Discuss with your classmates why the quadrilateral $AQBP$ must be a rhombus. *It is a parallelogram, since opposite sides are equal. But since all sides are equal, it is in fact a rhombus.*

Constructing the Perpendicular Bisector of a Line Segment

The next exercise shows how to find the perpendicular bisector of any line segment, using only a straightedge and compass.

3. (a) Mark two points A and B and draw the line segment with endpoints A and B. Using your compass, draw a circle with center A and a circle *of the same radius* with center B. (Make sure the radius you choose is large enough so that the two circles intersect each other.) Let P and Q be the two points of intersection of the circles. Draw the line segment $\overline{PQ}$, and let M be the point where it intersects $\overline{AB}$.

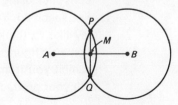

(b) What kind of quadrilateral is *AQBP*?   rhombus

(c) Discuss with your classmates why the line through *P* and *Q* is the perpendicular bisector of $\overline{AB}$ (Chapter 4, Central Exercise 33(b)).
diagonals of a rhombus

Bisecting a Line Segment

To *bisect* a line segment means to find its midpoint. The preceding exercise showed how to bisect any line segment using only straightedge and compass. Given a line segment $\overline{AB}$, all you have to do is draw a circle centered at *A* and a circle *of the same radius* centered at *B*, such that the circles intersect in two points *P* and *Q*. Then the point *M*, where $\overline{PQ}$ intersects $\overline{AB}$, is the midpoint of $\overline{AB}$. (Note that you don't really have to draw the entire circles to find the points *P* and *Q*. As shown in the picture below, you only need to draw part of each circle in order to locate *P* and *Q*.)

Demonstrate this construction to the entire class.

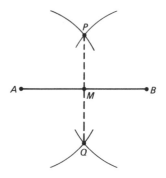

4. Draw a line segment $\overline{AB}$ on a sheet of paper.

(a) Using only straightedge and compass, bisect your line segment $\overline{AB}$.

(b) Using only straightedge and compass, divide your line segment $\overline{AB}$ into 4 segments all having the same length.   *Bisect $\overline{AB}$ at C, bisect $\overline{AC}$ and $\overline{CB}$.*

(c) Do parts (a) and (b) using a line segment of a different length.

Some Comments About Constructions

In the following exercises, when the word "construct" is used, it will mean "construct, using only a straightedge and compass as tools," unless otherwise specified. You may not use your ruler or protractor to measure. You may like to think of this as a game, whose objective is to draw certain geometric figures and to find certain lines and points. The rules of the game prohibit you from measuring lengths with your ruler or measuring angles with your protractor. Let us list these rules of construction:

> **RULES OF CONSTRUCTION**
>
> (*i*) Your straightedge is to be used only for drawing straight lines, *not* for measuring lengths.
>
> (*ii*) Your compass is to be used to draw circles or parts of circles.
>
> (*iii*) Your protractor will *not* be used.
>
> (*iv*) You should *not* use "trial and error" methods in any construction. You are supposed to use a definite method consisting of a few steps. Anyone should be able to use your method to make the construction in the same number of steps.

Review the Rules of Construction. Emphasize the value of a freehand sketch before they use construction tools; many teachers prefer that they show this freehand sketch near the actual construction.

For constructions, pencil points should be sharp; some students may prefer a hard drawing pencil.

A freehand sketch before starting a construction will help you see how to proceed with the actual construction. (Of course, it is helpful to make a freehand sketch before solving *any* geometry problem!) In order to help others see how your construction method works, you should "leave a trail" as you make the construction, a trail which can be followed by others and which you yourself can retrace in remembering the construction at a later time. This means

1. You should *not* erase any marks used in the construction.
2. You should label the figure you have constructed in such a way that all the steps of the construction and the order in which they were done can be described if you are called upon to do so.

For example, consider the construction of the perpendicular bisector of a line segment $\overline{AB}$ shown below:

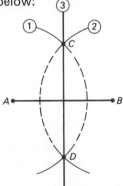

For text demonstration purposes only, we use numbers to show the sequence of steps in the construction. Students need not *be required to be so articulate for their work.*

For text purposes only, we use numbers to describe the following steps:

① Using compass, draw part of a circle with center *A*.

183 Central

② Using compass *with the same setting,* draw part of a circle with center B to obtain intersection points C and D.

③ Using straightedge, draw the straight line through C and D. This line is the perpendicular bisector of $\overline{AB}$.

The Perpendicular Bisectors of the Sides of a Triangle

In Exercise 3 you found a method for constructing the perpendicular bisector of any segment, using only straightedge and compass. The next exercise presents an interesting property of the perpendicular bisectors of the sides of any triangle.

5. (a) Using a straightedge and compass, construct the perpendicular bisectors of each of the three sides of this triangle:

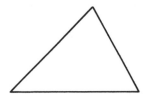

(b) What do you observe about the perpendicular bisectors of the sides of the triangle? (Find the points where pairs of perpendicular bisectors intersect.) Compare your observation with those of your classmates.

At this stage, let them enjoy making the construction. A proof comes later in the chapter.

(c) Draw three triangles of different shapes (such as those below) and the perpendicular bisectors of their sides. Does your observation in part (b) hold for these triangles?   yes

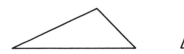

In your experiments in the last exercise you observed the following:

The perpendicular bisectors of the sides of a triangle all meet in a single _____.   *point*

We have observed something experimentally. But is this true for all triangles? A later exercise in this section proves that it is.

Constructing a Perpendicular Through a Point on a Line

6. Draw a line ℓ and mark a point P on ℓ. Construct a line m which passes

184   5: *Constructions with Straightedge and Compass*

Demonstrate how to construct a perpendicular to a line, through a point on that line.

through *P* and is perpendicular to ℓ. [*Hint:* Use your compass to draw a circle with center *P*, and let *A* and *B* be the points where this circle intersects ℓ. Note that *P* is the midpoint of $\overline{AB}$. The line *m* we are trying to construct is the perpendicular bisector of $\overline{AB}$.]

There should be no guessing of right angles here. They must be constructed.

7. Mark a point *P*, draw a circle centered at *P*, and draw a line segment $\overline{AB}$ as below:

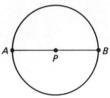

(a) Construct a square which has $\overline{AB}$ as one of its *diagonals*. *Construct the perpendicular bisector of $\overline{AB}$.*
(b) Construct a square which has $\overline{AB}$ as one of its *sides*. *Construct m and ℓ perpendicular to $\overline{AB}$ at A and B.*

8. Construct a rectangle such that

(a) one side is twice as long as another side

(b) one side is three times as long as another side.

Remember. Use only your straightedge and compass—no measuring with a ruler!

Constructing Regular Polygons

In Exercise 7 you constructed a square using only straightedge and compass. Since a square is a regular 4-gon, it is natural to ask how to construct any regular *n*-gon. The next exercise deals with the case *n* = 3. Later exercises will deal with larger values of *n*, but we will see that some regular *n*-gons *cannot* be constructed using only straightedge and compass.

9. (a) Mark a point *P* and use your compass to draw a circle with center *P*. Mark any point *Q* on this circle and draw a circle *of the same radius* with center *Q*. Let *R* be one of the two points where the circles intersect. What kind of triangle is $\triangle PQR$? Explain. *equilateral*

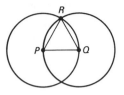

(b) Construct a regular 3-gon (equilateral triangle) one of whose sides has the same length as the segment given below:

10. (a) Construct a 30°–60° right triangle (that is, a right triangle whose acute angles are 30° and 60°). [*Hint:* Start by constructing an equilateral triangle.]   *Construct the perpendicular bisector of one of the sides.*

(b) Compare your method of construction with those of your classmates. Did they follow the rules of construction?

11. Construct a right triangle that has two 45° angles. Compare your method with those of your classmates.   *Construct a square, draw a diagonal.*

The next exercise tells how to construct a regular hexagon (regular 6-gon), using straightedge and compass. The case of the regular 5-gon is more difficult and will be considered later in the course.

Make sure students understand why this construction works.

12. (a) Use your compass to draw a circle. Now, *without changing the compass setting,* mark off successive points around the circle, as indicated in the picture.

If you are careful, you will end up exactly where you started. Explain why. [*Hint:* Draw a triangle with one vertex at the center of the circle and whose two other vertices are successive points that you drew on the circle. What kind of triangle is it? If you draw two line segments from the center to successive points on the circle, what size angle is formed? How many such angles "fit around" the center of the circle?]   *equilateral, 60°, 6*

(b) Construct a regular hexagon. *Connect the points in (a).*

(c) Prove that your construction is correct. (In other words, prove that the polygon you obtain by your method will always be a regular hexagon —that is, a hexagon all of whose angles are equal and all of whose sides have equal length.) *The sides are equal radii, the angles are each 120°.*

So far you have constructed regular 3-gons, 4-gons, and 6-gons using only straightedge and compass. You obtained a regular 6-gon by joining 6 equally spaced points on a circle. If you could construct n equally spaced points on a circle, then you would obtain a regular n-gon by joining successive points by line segments (we will discuss this further in the Projects section of this chapter). But is it possible to construct n equally spaced points on a circle using only a straightedge and compass, if n is *any* positive integer?

Before considering this question, we define what we mean by an *arc* of a circle.

DEFINITION An *arc* of a circle is a part of a circle lying between two points on the circle. If you draw a circle and mark two points on it, then you will have divided the circle into two arcs.

These definitions may be familiar to some of the students. For this reason, you must be sure that all understand them; otherwise, some will quickly become lost and confused.

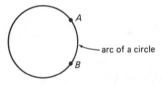

In the above figure, the arrow points to the smaller of the two arcs.

13. Mark a point P and, using your compass, draw a circle with center P. Using your straightedge, construct two points, A and B, on the circle which divide the circle into two *congruent* arcs. *Draw a line through P.*

DEFINITION A line segment joining two points on a circle is called a *chord* of the circle.

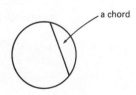

187 Central

DEFINITION   A *diameter* of a circle is any chord that passes through the center of the circle.

Exercises 14–18 show how to construct regular *n*-gons for special values of *n*.

14. (*a*) Using only straightedge and compass, draw a circle and find 4 points on the circle that divide it into 4 congruent arcs. [*Hint:* Construct a diameter and its perpendicular bisector.]

(*b*) Use part (*a*) to construct a square (regular 4-gon).

15. Using your compass, draw a circle.

(*a*) Using only straightedge and compass, find 3 points on the circle that divide it into 3 congruent arcs.   *Choose alternate points on the circle in Exercise 12(a).*

(*b*) Construct an equilateral triangle (regular 3-gon) with its vertices on the circle.

     The picture below shows how to "bisect" a given arc of a circle with center *P*.

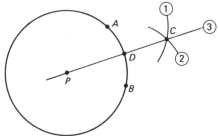

Later we prove why this construction method works. For now, they accept it as a technique on faith.

① Using compass, draw an arc of a circle with center *A*.

② Using compass *with the same setting,* draw an arc of a circle with center *B* to obtain intersection point *C*.

③ Using straightedge, draw the straight line through *P* and *C*. The point *D* where this line intersects the arc divides the given arc into two smaller congruent arcs.

You will see later why this construction method works. For now, use it in doing the following exercises.

16. (*a*) Construct a regular 8-gon. [*Hint:* Start by dividing a circle into 4 congruent arcs. Then bisect each arc.]

(*b*) Construct a regular 12-gon.

17. Explain how you could construct a regular

Bisect the center angles of an

(*a*) 16-gon (*b*) 32-gon (*c*) 64-gon (*d*) 2^n-gon.
  8-*gon*   16-*gon*   32-*gon* 2^{n-1}-*gon*

18. Explain how you could construct a regular

Bisect the center angles of a

(*a*) 24-gon (*b*) 48-gon (*c*) 96-gon (*d*) $3 \cdot 2^n$-gon.
  12-*gon*   24-*gon*   48-*gon* $3 \cdot 2^{n-1}$-*gon*

Impossible Constructions

An excellent reference on the construction of regular n-gons is Heinrich Tietze, Famous Problems of Mathematics (New York: Graylock Press, 1965, authorized translation from the Second (1959) Revised German Edition, ed. B.K. Hofstadter and H. Komm). Chapter 9 deals with construction of n-gons. Other chapters deal with angle trisection and squaring the circle.

So far you have learned how to construct, with only straightedge and compass, regular n-gons for many different values of n; for example, $n = 3, 4, 6, 8, 12, 16, 24$. The ancient Greek geometers knew these constructions and many more. For example, they knew how to construct a regular pentagon (regular 5-gon), which we will learn in a later chapter, after we have studied the Pythagorean Theorem. Thus the Greeks knew how to construct regular 3-gons, 4-gons, 5-gons, and 6-gons. What about the regular 7-gon? You may be surprised to learn that they never found a method for constructing a regular 7-gon using only a straightedge and compass! In fact, this was one of the famous unanswered questions of antiquity:

Is there a method for constructing a regular 7-gon, using only a straightedge and compass?

You may wish to refer interested students to the Projects at this stage.

Two thousand years later, the mathematicians C. F. Gauss and P. L. Wantzel proved a surprising theorem: there exists *no* method for constructing a regular 7-gon, using only a straightedge and compass! Gauss and Wantzel found all regular polygons which can be constructed with straightedge and compass and also all regular polygons which cannot be so constructed. They discovered, for example, that a regular 9-gon *cannot* be constructed using only straightedge and compass, but a regular 17-gon *can* be. In the Projects section of this chapter you will find some exercises dealing with the work of Gauss and Wantzel.

Bisecting an Angle

DEFINITION

A ray is said to *bisect* an angle if it divides the angle into two equal angles, as in the picture below, where ∡1 = ∡2.

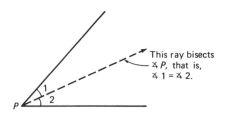

This ray bisects ∡P, that is, ∡1 = ∡2.

The ray is then called the *bisector* of the angle.

Demonstrate the bisection of an angle.

We can bisect any segment using only straightedge and compass. The following picture shows how to bisect any angle using only a straightedge and compass.

In the picture, we are given an angle with vertex P.

① Draw an arc of a circle with center P to obtain points A and B.

② Draw an arc of a circle with center A.

③ Draw an arc of a circle, *with the same compass setting as in* ②, with center B, to obtain point C.

④ Draw the ray through C with endpoint at P. This ray *bisects the angle* (in other words, ∡a = ∡b).

19. Prove that the above method for bisecting an angle is valid. In other words, prove that ∡a must equal ∡b. [*Hint:* Consider △PAC and △PBC.]   △PAC ≅ △PBC by SSS

20. (a) Using your protractor, draw a right angle. Using only straightedge and compass, bisect your right angle.

190   5: *Constructions with Straightedge and Compass*

(b) Using your protractor, check that the two smaller angles you obtained in part (a) are each 45°.

21. (a) Using your protractor, draw a 120° angle.

(b) Use a straightedge and compass to bisect your angle.

(c) Use your protractor to check that you have obtained two 60° angles.

22. (a) Draw a 180° angle.

(b) Use the method you have just learned to bisect this angle.

(c) Compare this with what you did in Exercise 6 of this chapter. Are the two constructions the same? *yes*

23. (a) Use straightedge and compass to draw lines bisecting each angle of this triangle:

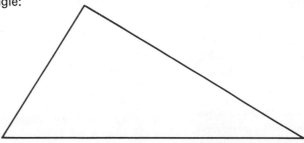

(b) What do you observe about the angle bisectors of the triangle? Compare your observation with those of your classmates.

(c) Draw a right triangle and a triangle having an obtuse angle. In each case draw the angle bisectors. Does your observation of part (b) hold for these triangles? *yes*

This theorem is proved in Central Exercise 33, Chapter 10.

In your experiments in the last exercise you observed the following fact, which will be proved in a later chapter.

The angle bisectors of any triangle all meet in a single *point* .

24. Using only straightedge and compass, construct an angle of size

(a) 120°   (b) 60°   (c) 30°   (d) 15°.

Some Basic Constructions

Check that they know how to construct a line segment equal in length to a given line segment.

Demonstrate how to construct an angle equal in size to a given angle. Some students may not realize that they must start with a line and a point on that line.

25. (*a*) Describe a method for dividing a given line segment into *n* segments of equal length, using only straightedge and compass, if *n* equals

(*i*) 2   (*ii*) 4   (*iii*) 8   (*iv*) 16.

(We will learn shortly how to do this for other values of *n*.)

(*b*) What is a similar thing that one could do with *angles*?

The next exercise deals with copying any given line segment, using only straightedge and compass.

26. (*a*) On a sheet of paper draw a line ℓ and mark a point *P* on ℓ. Now, using your compass, find a point *Q* on ℓ such that $PQ = AB$, where *AB* is the length of the segment drawn below: There are two such points.

(*b*) Compare your method with those of your classmates.

The next exercise shows how to copy any given angle, using only straightedge and compass.

27. (*a*) Draw any angle with vertex *P*. Also draw any line ℓ and mark any point *A* on ℓ. Now, use the method shown in the picture below to construct a line passing through *A* and making with ℓ an angle of the same size as the angle at *P*.

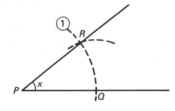 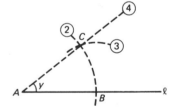

① Draw an arc of a circle with center *P* to obtain points *R* and *Q*.

② With the same compass setting as in ①, draw an arc of a circle with center *A*, to obtain point *B* on ℓ.

③ Draw an arc of a circle with center *B* and radius *RQ* to obtain point *C*.

④ Draw the line passing through *A* and *C*.

(*b*) Prove that $\angle y = \angle x$. $\triangle PQR \cong \triangle ABC$ by SSS

192   5: Constructions with Straightedge and Compass

28. (*a*) Using your protractor, draw an angle of size 37°.

(*b*) Using only your straightedge and compass, copy the angle in part (*a*) on a separate sheet of paper. Check the accuracy of your work using a protractor.

(*c*) Do parts (*a*) and (*b*) for an angle of size 146°.

Let us consider now a few more basic constructions. In Exercise 27 we learned how to copy any angle using only straightedge and compass. The next exercise shows how to copy any given triangle.

29. (*a*) Draw any triangle *PQR*. Also draw any line ℓ and mark a point *A* on ℓ. Now, using your compass, find a point *B* on ℓ such that *AB* = *PQ*, and also find a point *C* such that *AC* = *PR* and *BC* = *QR*.

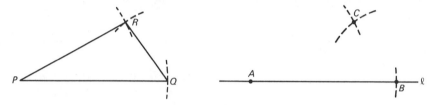

(*b*) Draw △*ABC*. Explain why △*ABC* ≅ △*PQR*. SSS

30. Draw line segments of lengths 5, 7, and 10 cm. Now, using *only* straightedge and compass, construct a triangle having sides of these lengths.

This construction gives further intuitive acceptance of SAS.

31. Using only straightedge and compass, construct a triangle, given two sides and the angle included between the two sides:

[*Hint:* We want to construct a triangle, two sides of which are the same lengths as the given line segments, and such that the angle included between those sides is the same size as the angle given above. Start by copying the given angle on a sheet of paper with straightedge and compass. Then copy the given segments on the sides of the angle.]

32. Construct a triangle with straightedge and compass, two of whose sides have lengths equal to the segments given below and the angle between the two sides the same size as the angle given below:

This construction gives further intuitive acceptance of ASA.

33. Using only straightedge and compass, construct a triangle, given two angles and the included side:

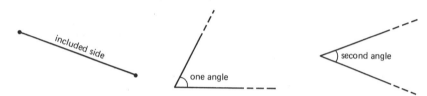

[*Hint:* Copy the given segment on a sheet of paper. Then copy one angle with the vertex at one endpoint of the segment and the second angle with the vertex at the other endpoint.]

34. Do the preceding exercise using the segment and angles given below:

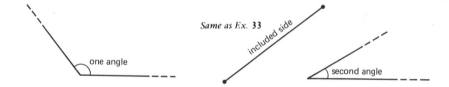

Constructing a Line Through a Point and Parallel to a Given Line

In construction problems it is often useful to know how to construct a line passing through a given point and parallel to a given line. The next exercise shows how to do this.

35. Suppose you are given a line ℓ and a point P not on ℓ, as below.

194 5: *Constructions with Straightedge and Compass*

After Exercise 35 is completed, you may wish to point out that the construction can be done using alternate interior angles instead of corresponding angles.

(a) Draw any line through P intersecting ℓ in a point Q. This line will make some angle with ℓ ($\angle 1$ in the picture below). Now construct a line m through P so that $\angle 2 = \angle 1$. *Copy $\angle 1$ at P.*

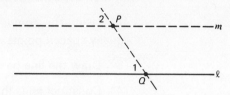

(b) Why must m be parallel to ℓ? *corr. angles*

36. On a sheet of paper, draw a line ℓ and a point P positioned as below. Construct a line passing through P and parallel to ℓ. *Use Ex. 35.*

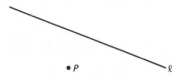

Dividing a Given Line Segment into Equal Segments

Demonstrate how to divide a line segment into equal parts, following the construction steps. For some students this may be a difficult construction.
Exercise 38 shows why the method works.

Having found a straightedge and compass construction for *bisecting* any line segment, it is natural to ask for a construction for *trisecting* any line segment (dividing it into 3 segments of equal length), or more generally, dividing a given line segment into n segments of equal length, where n is any fixed positive integer.

The following figure and description give a method for dividing any line segment into 3 segments of equal length, using only straightedge and compass.

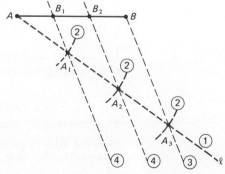

195 Central

Suppose we are given a line segment $\overline{AB}$.

① Draw any line ℓ through A (the angle at A should not be too small for then the construction becomes awkward).

② Using the compass at any fixed setting, start at A and mark off 3 equally spaced points A_1, A_2, A_3 along ℓ.

③ Draw the line passing through A_3 and B.

④ Construct lines through A_1 and A_2 parallel to $\overline{BA_3}$ (Exercise 35).

The points B_1 and B_2, where the lines drawn in ④ intersect segment $\overline{AB}$, divide the segment $\overline{AB}$ into three segments of equal length.

37. Draw a line segment $\overline{AB}$ on a sheet of paper and use the above method to trisect the segment.

The next exercise proves that the above construction for trisecting a line segment is valid.

Strive for a general understanding of this proof, otherwise students get lost in meticulous detail. You may need to demonstrate the proof at the board.

38. (a) In the figure below, the segments $\overline{A_3B}$, $\overline{A_2B_2}$, and $\overline{A_1B_1}$, are all parallel, and $AA_1 = A_1A_2 = A_2A_3$. Why must $AB_1 = B_1B_2 = B_2B$?

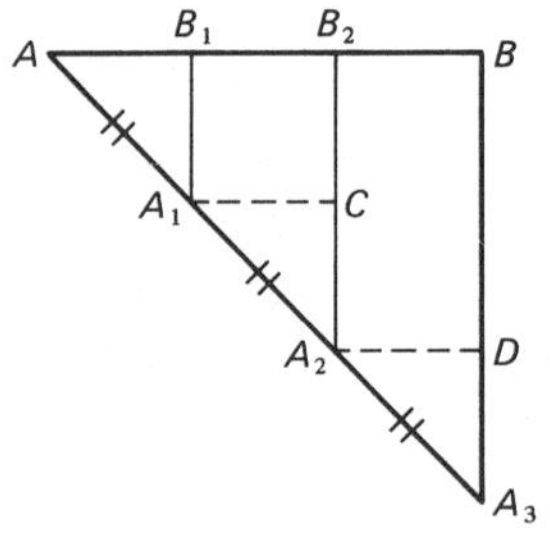

[Hint: Draw $\overline{A_1C}$ and $\overline{A_2D}$ parallel to the segment AB, and look for triangles congruent by ASA. What do you know about opposite sides of a parallelogram?]   *equal*

(b) Discuss with your classmates why part (a) shows that the construction given for trisecting line segment $\overline{AB}$ is valid.

39. (a) Find a method for dividing a given line segment into 5 segments of equal length, using only straightedge and compass. [Hint: Start by marking off 5 equally spaced points along a line drawn through one endpoint of the line segment. Proceed as you did in trisecting a line segment.]

(b) Compare your method with those of your classmates.

40. Using your ruler, draw a line segment 10 cm. long. Now, using only your straightedge and compass, construct

(a) a line segment of length 5 cm. (simply bisect the given segment!)

(b) a line segment of length 2 cm. (Exercise 39!)

(c) a line segment of length 3 cm.

41. Discuss with your classmates a method for dividing any line segment into n segments of equal length (n any fixed positive integer) using only straightedge and compass. *Use Ex. 39(a), substituting n for 5.*

42. On a sheet of paper, draw a line segment $\overline{AB}$ that is about 3 inches in length.

(a) Place a sheet of lined binder paper *under* the above sheet. (The lines should show through the paper.)

This method is quick and practical. (b) Move the lined paper in (a) so that the parallel lines help you to divide $\overline{AB}$ into 2 segments of equal length.

(c) Use the method in (b) to divide another line segment into 3 segments of equal length.

(d) Use the method in (b) to divide yet another segment into 5 segments of equal length.

(e) Discuss with your classmates the connection between this exercise and the method used in Exercises 37–41 to divide a line segment into smaller equal segments. *Both use parallel lines to divide segments.*

Impossibility of Trisecting the Angle

You have learned to bisect any given line segment using only straightedge and compass. In fact, in Exercise 41 you found a method for dividing any given line segment into n line segments of the same length, for *any* positive integer n. It is natural to ask if the same sort of thing can be done with angles.

You know how to bisect any angle using only straightedge and compass. Can you perhaps trisect any given angle using only straightedge and compass? As with the case of constructing a regular 7-gon, the Greek geometers could not find such a method. This problem was another of the unanswered questions of antiquity—trisecting any angle:

Is there a method which can be used to divide any angle into 3 equal angles, using only a straightedge and compass? no

It is easy to see that certain angles *can* be trisected using only straightedge and compass, as the next exercise shows.

43. Draw a 90° angle. Now, using only straightedge and compass, trisect your angle. That is, construct two lines (dotted in the figure below) which cut the given 90° angle into 3 equal angles (so each smaller angle would be 30°). [*Hint:* Construct an appropriate 60° angle.]

44. Draw a 45° angle. Now, using only straightedge and compass, trisect your angle. [*Hint:* You really want to construct certain 15° angles. Do you know how to construct a 15° angle? It might also be helpful to keep in mind that 45 − 30 = 15.]   *Bisect the $30°$ angle from Ex. 43.*

45. How can you trisect a 180° angle using only straightedge and compass?   *Construct a $60°$ angle.*

The reason the ancient Greeks failed in their attempts to find a general method for trisecting *every* angle with straightedge and compass became clear only in the 19th century when, after centuries of fruitless effort on the part of many people to find such a method, it was finally proved (by P. L. Wantzel in 1837) that *there does not exist such a method!* In fact, it is not even possible to trisect a 60° angle using only a straightedge and compass.

46. (*a*) Draw a circle and construct a regular hexagon having its vertices on the circle. As in the figure below, draw the six radii from the center of the circle to the vertices of the hexagon.   *See Ex. 12.*

(b) Consider the six 60° angles around the center of the circle. Suppose we *could* trisect each of these angles using only straightedge and compass. What kind of polygon would we get using the points where the trisecting lines intersect the circle, together with the six points already on the circle? How many sides would it have?   *regular 18-gon*

(c) If we actually *could* trisect each of the 60° angles in part (b) with only a straightedge and compass, explain how we could then construct a regular 9-gon.   *Connect every other point in (b).*

(d) Gauss and Wantzel proved that it is *not* possible to construct a regular 9-gon using only a straightedge and compass. Explain why this also proves that it is *not* possible to *trisect* a 60° angle using only straightedge and compass.

47. Alex Smart says, "I have discovered a method for trisecting any angle using only straightedge and compass!" Here is Alex's method.

Alex's "Trisection": Given any angle with vertex P, use a compass to mark points A and B on the sides of the angle, with PA = PB. Then draw $\overline{AB}$, using straightedge.

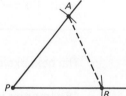

Next trisect $\overline{AB}$ using straightedge and compass, and then draw the rays from P through the trisection points.

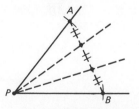

Alex claims that the given angle has now been divided into three equal angles.

This method is popular in attempting to trisect an angle.

(a) Use Alex's method on a 90° angle. Do you think the 90° angle has been trisected? Measure carefully with a protractor and see if the three smaller angles are all equal.   *no*

At this point we are only interested in having the student reject the method as a result of experiment.

(b) Try Alex's method on a 60° angle. (If you have been careful in your construction, you will find that the "center" angle is larger than the two "outside" angles. In fact, it can be proved that if Alex's method is applied to any angle, the "center" angle will always be larger than the two "outside" angles.)

The Projects section of this chapter goes further into these intriguing questions about constructions.

Some Properties of Perpendicular Bisectors

Stress the importance of this theorem and its converse.

In Chapter 3, Central Exercise 31, we proved the following fundamental property of the perpendicular bisector of a line segment:

If P is a point on the perpendicular bisector of a line segment, then P is equidistant from the endpoints.

Is the converse of this statement true? That is, is it true that

---

If P is equidistant from the endpoints of a line segment, then P is on the perpendicular bisector of the segment.

The next exercise proves that this indeed *is* true.

48. In order to prove the statement above, we must prove that if a point P is equidistant from the endpoints of a line segment $\overline{AB}$, then P is on the perpendicular bisector of $\overline{AB}$. In other words, we must prove that *if $PA = PB$ in the figure below, then P is on the perpendicular bisector of $\overline{AB}$.*

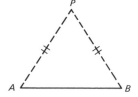

Give a proof of the statement above. [*Hint:* Draw the line segment joining P to the *midpoint M* of $\overline{AB}$, as on the next page. Why is $\triangle AMP \cong \triangle BMP$? Why is $\measuredangle 1 = \measuredangle 2$? Why then are $\measuredangle 1$ and $\measuredangle 2$ right angles? Why is the line through P and M the perpendicular bisector of $\overline{AB}$?].
  SSS, corr. parts, $\measuredangle 1 + \measuredangle 2 = 180°$, $\measuredangle 1 = 90°$ and M the midpoint of $\overline{AB}$

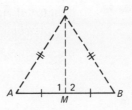

49. Suppose $\overline{AB}$ is a chord of a circle with center P.

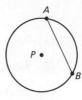

(*a*) Explain why the perpendicular bisector of $\overline{AB}$ passes through P.
$PA = PB$, Ex. 48

(*b*) The picture below shows two chords of a circle. Where must the perpendicular bisectors of the two chords intersect?   *center of the circle*

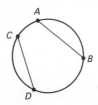

  (*c*) The perpendicular bisector of a chord passes through the _____ of the circle.   *center*

50. Use the result in the preceding exercise to solve the problem at the beginning of this chapter of finding the center of a circle using only straightedge and compass. Discuss your solution with your classmates.
the intersection of the perpendicular bisectors of two chords

51. An archaeologist digs up what appears to be a fragment of a saucer. He would like to know the radius of the original circular saucer. How can he measure it? Discuss this with your classmates.

52. Consider $\triangle ABC$ on the following page. Line ℓ is the perpendicular

201   *Central*

Exercise 52 prepares for Exercise 53.

bisector of $\overline{AC}$ and line m is the perpendicular bisector of $\overline{BC}$. Point P is the intersection of ℓ and m.

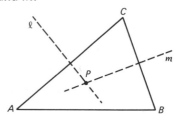

(a) Why is $PA = PC$? Why is $PB = PC$? *P is on the perpendicular bisector of $\overline{AC}$, $\overline{BC}$.*

(b) From part (a), deduce that $PA = PB$.

(c) From part (b), deduce that the perpendicular bisector of AB also passes through P. *Ex. 48*

(d) Discuss why we have proved the following statement.

> **The perpendicular bisectors of the sides of any triangle all meet in a single point.**

The Circumscribed Circle of a Triangle

For some students, it is a help to do this exercise more than once, using triangles of varying shapes.

53. Draw a large triangle, $\triangle ABC$, on a sheet of paper.

(a) Find the point P where the perpendicular bisectors of the sides of your triangle intersect.

(b) Why is $PA = PB = PC$?

(c) Draw the circle with center P and passing through A.

(d) If your constructions were done carefully, the circle you constructed in part (c) should also pass through B and C. Why? *PA = PB = PC = radius*

(e) Discuss with your classmates why the following statement is true.

> **Given any triangle, there is a circle passing through the vertices of the triangle. The center of this circle is the point of intersection of the perpendicular bisectors of the sides of the triangle.**

Do not put too much emphasis on part (f).

(f) Given △ABC, could there be more than one circle passing through A, B, and C? Discuss with your classmates. *no*

DEFINITION The circle passing through the vertices of a triangle is called the *circumscribed circle* of the triangle. The center of this circle is called the *circumcenter* of the triangle.

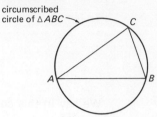

54. (a) Copy each of the following triangles on a sheet of paper.

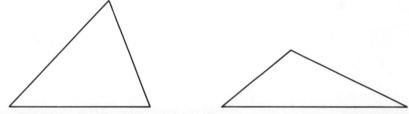

(b) Construct the circumscribed circle and circumcenter of the triangles you drew in part (a). *Use Ex. 53.*

Constructing a Perpendicular from a Point to a Line

Exercise 6 showed how to construct a line perpendicular to a given line and passing through a given point *on* that line. The next exercise shows how to construct a line perpendicular to a given line and passing through a given point *not* on the line.

Students often confuse this construction with that of constructing a perpendicular from a point on *a line.*

55. Draw a line ℓ and mark a point P *not* on ℓ. Construct a line through P perpendicular to ℓ. [*Hint:* Draw an arc of a circle centered at P intersecting ℓ in points A and B.

203 Central

Now construct the perpendicular bisector of $\overline{AB}$. Why must it pass through P?] $PA = PB$

Altitudes of a Triangle

In the figure below, the line segment $\overline{AD}$ is perpendicular to the side $\overline{BC}$ of $\triangle ABC$.

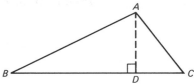

Some students have difficulty with altitudes, especially the case where a side must be extended.

We say in this case that $\overline{AD}$ is an *altitude* of $\triangle ABC$. In the following figure, $\overline{BD}$ is also called an altitude of $\triangle ABC$:

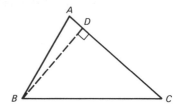

In the next figure, $\overline{AD}$ is also called an altitude of $\triangle ABC$:

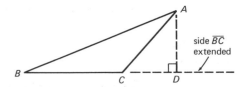

In other words, we have the following definition.

DEFINITION   An *altitude* of a triangle is a line segment joining a vertex of the triangle to a point on the opposite side (possibly extended) and which is perpendicular to that side.

Note that every triangle has three altitudes, one from each vertex to the opposite side (possibly extended).

56. (*a*) On a sheet of paper draw a large triangle, all of whose angles are acute, like the one on the next page.

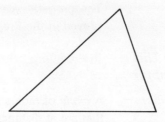

Using only a straightedge and compass, construct the three altitudes of your triangle.

(b) Do you observe anything special about the altitudes of the triangle? Compare your observation with those of your classmates.

(c) Draw a triangle having an obtuse angle and construct its altitudes. Extend the altitudes so each two intersect. If your construction is done carefully, you should observe the same thing you saw in part (b):

The altitudes of any triangle (possibly extended) all meet in a <u>single point</u>.

In the Projects section of this chapter we will prove that the special property of altitudes you discovered in the preceding exercise holds for all triangles. Let us now consider another remarkable property of certain lines in a triangle.

Medians of a Triangle

DEFINITION

See Projects Exercise 14.

A line segment joining a vertex of a triangle to the midpoint of the opposite side is called a *median* of the triangle.

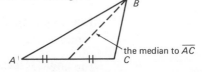

This is a good point to review altitude, angle bisector, perpendicular bisector (of the side of a triangle), and median.

57. Draw a triangle and carefully draw its three medians. Do this with several different triangles.

The medians of any triangle all meet in a <u>single point</u>.

The property you observed experimentally in the preceding exercise is proved in the Projects section of this chapter.

Important Ideas in This Chapter

Definitions

Give a short lecture summarizing the chapter.

A *circle* of radius r with center P consists of all those points in the plane that are at distance r from P.

An *arc* of a circle is a part of a circle lying between two points on the circle.

A *chord* of a circle is a line segment joining two points on the circle.

A *diameter* of a circle is a chord passing through the center of the circle.

A *bisector* of an angle is a ray through the vertex of the angle dividing it into two equal angles.

An *altitude* of a triangle is a line segment drawn from a vertex perpendicular to the opposite side.

A *median* of a triangle is a line segment joining a vertex to the midpoint of the opposite side.

The *circumscribed circle* of a triangle is the circle passing through the vertices of the triangle.

The *circumcenter* of a triangle is the center of its circumscribed circle.

Theorems

(Only the first three were proved in the Central section)

If a point is equidistant from the endpoints of a line segment, then it is on the perpendicular bisector of the segment.

The perpendicular bisectors of a triangle all meet in a point. This point is the circumcenter of the triangle.

The perpendicular bisector of any chord of a circle passes through the center of that circle.

The angle bisectors of a triangle all meet in a point.

The extended altitudes of a triangle meet in a point.

The medians of a triangle meet in a point.

Basic Constructions with Straightedge and Compass

bisecting any segment

You may wish to point out that, using these basic techniques, more difficult constructions can be mastered.

Some students may like to devise their own construction problems. They may wish to consider constructions as a continuing activity Project for the remainder of the course.

bisecting any angle

constructing the perpendicular bisector of any line segment

copying any angle

copying any triangle

constructing a line perpendicular to a given line and passing through a given point on the given line

constructing a line perpendicular to a given line and passing through a given point not on the given line

constructing a line parallel to a given line and passing through a given point

dividing any line segment into n segments of equal length, where n is any positive integer

Summary

This chapter presented some of the constructions which are possible by straightedge and compass alone. Many ideas in previous chapters were used in proving that these constructions are valid.

Gauss and Wantzel proved that some constructions are impossible. For example, it is not possible to construct a regular 7-gon using only a straightedge and compass; or, equivalently, it is impossible to divide a circle into 7 congruent arcs using only a straightedge and compass. There are many other regular polygons which cannot be so constructed. For example, a regular 9-gon cannot be constructed with straightedge and compass alone. (Of course, we learned that certain regular polygons *can* be so constructed—for example, regular 3-gons, 4-gons, and 6-gons.) Wantzel also proved that it is not possible to trisect a 60° angle using only straightedge and compass.

The following properties of triangles were observed in the course of our studies:

(*i*) The perpendicular bisectors of the sides of any triangle intersect in one point.

(*ii*) The angle bisectors of any triangle intersect in one point.

(*iii*) The extended altitudes of any triangle intersect in one point.

(*iv*) The medians of any triangle intersect in one point.

The proof of (*i*) was given in the Central section, and the proofs of (*iii*) and (*iv*) will be found in the Projects section of this chapter. We delay the proof of (*ii*) until a later chapter.

Historical Note

The following is another famous story of Gauss's precocity:

One day, when Gauss was three years old, his father was making out the payroll to pay the workers in the masonry firm where he was foreman. At the end of a long computation he was startled to hear his son say, "Father, you have made a mistake in your reckoning," and then name another figure. A check of the account showed the three-year-old Gauss was correct.

Carl Friedrich Gauss (1777–1855) is ranked, with Archimedes and Isaac Newton, among the three greatest mathematicians who ever lived and has been called the "Prince of Mathematicians." His talents showed at an early age. At ten years of age Gauss was the youngest in a certain class where the students were asked to add up all the numbers from one to a hundred. The teacher had hardly finished stating the problem when Gauss laid down his slate and said, "There it lies." Unbelieving, his teacher did not bother looking at Gauss's slate until the others had finished their work. He was then amazed to see that Gauss had written down only one number and was the only student with the correct answer. Many of Gauss's greatest mathematical discoveries were made before he was twenty years old. For example, when he was nineteen he was the first person in history to prove that a regular 17-gon can be constructed with straightedge and compass. This result so delighted him that later in life he expressed the desire to have a regular 17-gon engraved on his tombstone. This is in a classical tradition, since Archimedes wanted engraved on his tombstone a cylinder with inscribed sphere to commemorate one of his favorite discoveries.

Central Self Quiz

1. Given $\triangle ABC$.

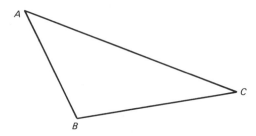

   (*a*) Construct the perpendicular bisector of $\overline{AC}$.   *Central Ex. 3(a)*

   (*b*) Construct the bisector of $\angle B$.   *Central Ex. 23*

   (*c*) Construct the median to $\overline{AB}$.   *Central Ex. 4(a)*

   (*d*) Construct the altitude from A to side $\overline{BC}$ extended.   *Central Ex. 56(c)*

2. On a separate sheet of paper, construct an angle equal in size to $\angle 1$.
    Central Ex. 27(a)

3. Construct a triangle congruent to △ABC on a separate sheet of paper.

Central Ex. 29

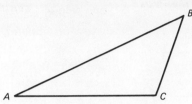

4. Construct a line through M perpendicular to line t. *Central Ex. 55*

• M

t

5. Construct a line perpendicular to line m at point A. *Central Ex. 6*

6. Divide $\overline{AB}$ into three equal parts. *Central Ex. 37*

A ●———————————————● B

Consider spending a day or so on Projects at this point.

7. Construct a line through P that is parallel to m. *Central Ex. 35(a)*

• P

————————————— m

Review

1. (a) Using a ruler, compass, and *protractor,* draw a regular 9-gon.

   (b) Can you construct a regular 9-gon using only a straightedge and compass (no protractor!)? *no*

2. Is it possible to construct, using only straightedge and compass, a regular n-gon if $n = 3$? 4? 5? 6? 7? *no* 8? 9? *no* 10? 12? 14? *no* 16? 27? *no* 384?

Encourage students to construct other patterns based on the circle as a Project.

3. (a) Using only your compass, construct a 6-leaved pattern like the one below. *First construct a regular hexagon.*

(b) Could you construct a 7-leaved pattern (with evenly spaced leaves) using only your compass? *No, since you cannot construct a regular 7-gon.*

4. Draw a line segment of length 5 cm. Using only a straightedge and compass, construct a regular *n*-gon having this segment as one of its sides, if

(a) $n = 3$ (b) $n = 4$ (c) $n = 6$.
Central Ex. 9(b)   *Central Ex. 7(b)*   *Central Ex. 12(b)*

5. Using only straightedge and compass, construct an isosceles triangle having

(a) two 75° angles   (b) exactly one 75° angle

(c) some angle twice the size of another of its angles. (There is more than one possibility. Choose one you can do.)   *See Central Ex. 11.*
45°-45°-90°, 72°-72°-36°

6. Draw a line segment $\overline{AB}$ of length 4 cm. Using only a straightedge and compass, construct

(a) an isosceles triangle, $\triangle ABC$, having $\overline{AB}$ as one of its sides and $AC = BC = 8$ cm.

(b) a triangle, $\triangle ABC$, having $\overline{AB}$ as one of its sides, $AC = 3$ cm., and $BC = 6$ cm. (no extra measurements allowed!). *Divide AB into four segments.*

7. (a) Using a straightedge and compass, construct a triangle two of whose sides are the same lengths as the given segments below and one of whose angles is the same size as the angle given below.

(b) Construct two more triangles which satisfy the conditions in part (a), but which are *not* congruent to each other and *not* congruent to the triangle you constructed in part (*a*).

(*c*) Why doesn't part (*b*) contradict the SAS congruence property of triangles?

8. (*a*) Using your ruler, draw line segments of lengths 3, 4, and 5 cm. Then, using only straightedge and compass, construct a triangle having sides of these lengths.

(*b*) Try part (*a*) using line segments of lengths 3, 5, and 9. *impossible*

(*c*) What geometric fact is behind the difficulty in part (*b*)? *Triangle Inequality*

9. Could you divide any angle into *n* smaller equal angles, using only straightedge and compass, if *n* = 2?   3?   4?   16?   1024?

10. There is no general method for trisecting every given angle with only straightedge and compass. Is there a straightedge and compass construction for *tripling* every angle? Explain.

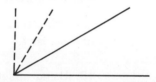

11. Describe two different methods for dividing any given line segment into 8 segments of equal length using only straightedge and compass. *successive bisections or the method described in Central Ex. 41*

12. The line segment $\overline{AB}$ below is $1\frac{1}{2}$ inches long. Using only your straightedge and compass, find all those points in the plane which are 2 inches from *A* and 1 inch from *B*.

13. Using straightedge and compass, construct a square which has the given line segment $\overline{AB}$ as one of its diagonals. *Central Ex. 7(a)*

14. Using straightedge and compass, construct a rectangle such that one side and one diagonal are the same lengths as the segments given below:

15. Using straightedge and compass, construct a circle passing through the 3 points below.   *Draw $\triangle ABC$ and use Central Ex. 54(b)*

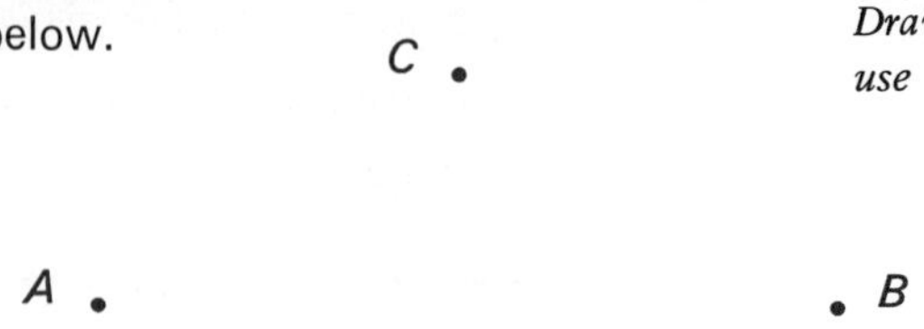

C •

A •   • B

16. Somebody leads you to a curved section of railroad track, an arc of a circle. He asks you to find the center of that circle. How might you do it?

17. In the figure below you are given a line ℓ and a point P not on ℓ.

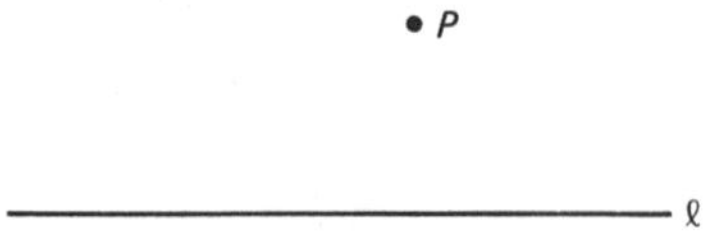

• P

———————————————— ℓ

(a) Using only a straightedge and compass, construct that point on ℓ which is closest to P.

(b) Using only a straightedge and compass, construct the reflection P^* of P across ℓ.

18. Using straightedge and compass, construct point P on the line ℓ below such that $AP + PB$ is as small as possible.

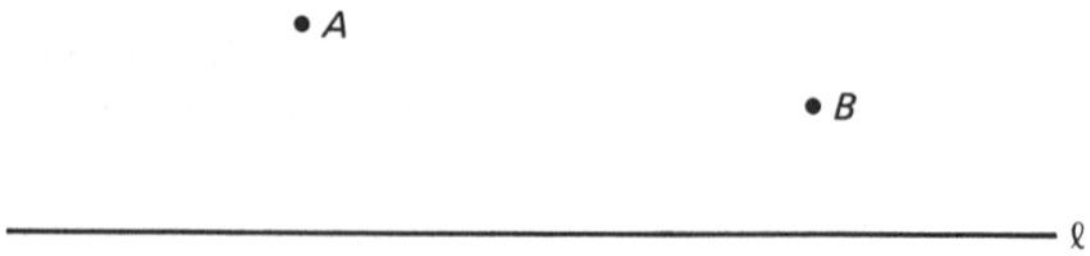

• A

• B

———————————————— ℓ

19. Using straightedge and compass, construct a point P on the line ℓ below such that $AP = BP$.   • A

• B

———————————————— ℓ

20. (a) In the plane, what geometric figure is formed by all points that are 2 inches from a given fixed point A?   *circle*

(b) Do part (a), but change "the plane" to "space."   *sphere*

21. (a) In the plane, what geometric figure is formed by all points which are equidistant (same distance) from two given fixed points A and B. *line*

(b) Do part (a), but change "the plane" to "space." *plane*

22. Prove that if A, B, C, and D are the midpoints of the sides of any quadrilateral (such as in the picture below), then $\overline{AC}$ and $\overline{BD}$ bisect each other.

[Hint: Recall what you learned in Central Exercise 25 of Chapter 4.]

23. Suppose you are given $\triangle ABC$, and through each vertex a line is drawn parallel to the opposite side, forming a new triangle, $\triangle DEF$.

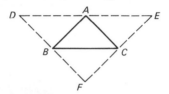

Prove that A, B, and C are the midpoints of the sides of $\triangle DEF$. [Hint: We know that $\overline{DE}$ is parallel to $\overline{BC}$, $\overline{EF}$ is parallel to $\overline{AB}$, and $\overline{FD}$ is parallel to $\overline{CA}$. We want to prove that DA = AE, EC = CF, and FB = BD. Note that ADBC is a parallelogram (why?). Hence DA = BC (why?).]

24. Mark 3 points A, B, and C (not all on the same straight line) on your paper. Now construct a triangle such that A, B, and C are the midpoints of the sides of your triangle. *Draw $\triangle ABC$. Construct lines through the vertices, parallel to the opposite sides.*

25. On your paper draw a line ℓ and mark a point P that is 2 inches from ℓ. Construct all those points that are 3 inches from ℓ and 1 inch from P.

Encourage discussion as they work Exercise 26, for some parts may seem difficult.

26. In each of the following, find x. (You should be able to prove your answer is correct if asked. The pictures are distorted!)

(a)    (b)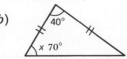

213 Review

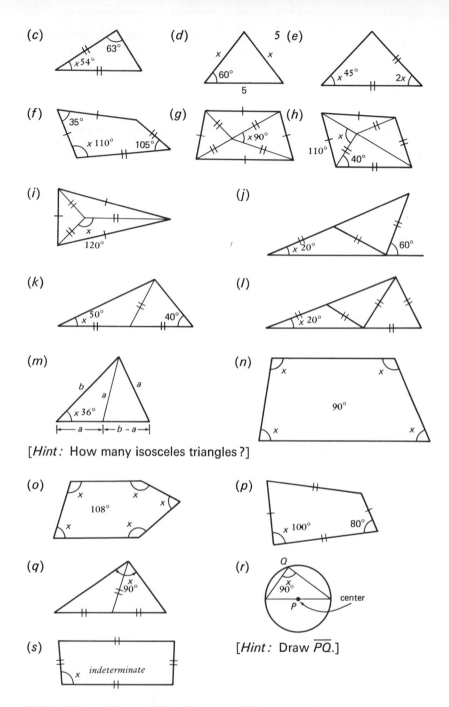

[Hint: How many isosceles triangles?]

[Hint: Draw $\overline{PQ}$.]

[Hint: Never take anything for granted.]

27. If, in the figure below, $AB = CD$ and $AC = DB$, then prove that $BE = CE$.

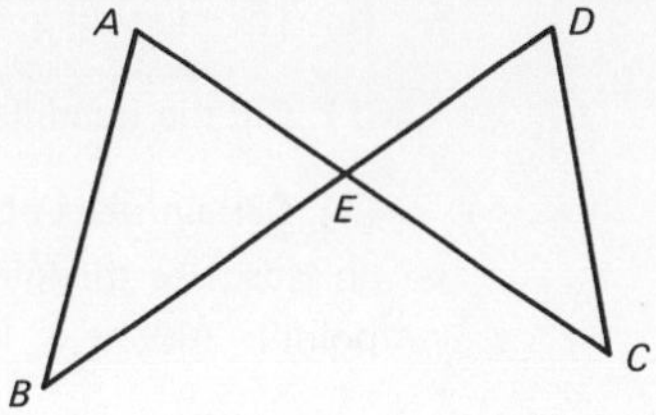

Draw $\overline{BC}$.
$\triangle ABC \cong \triangle DCB$
$\triangle ABE \cong \triangle DCE$

Review Answers

1. (a) Divide a circle into 9 congruent arcs, using your protractor.
(b) Recall what Gauss and Wantzel proved.

2. The cases $n = 7, 9, 14, 27$ are impossible. For example, if you could construct a regular 27-gon, then by connecting every third vertex you would obtain a regular 9-gon, which Gauss and Wantzel proved cannot be constructed using only straightedge and compass.

7. (c) The angle has to be included between the given sides in order that SAS hold.

8. (c) Recall the Triangle Inequality.

9. The case $n = 3$ is impossible (trisecting the angle). The other cases can be done by successive bisections.

10. Copy the given angle as indicated.

12. The circle of radius 2 inches centered at A intersects the circle of radius 1 inch centered at B in two points.

15. The perpendicular bisector of any chord of a circle passes through the center.

16. Choose any two chords and find the intersection of their perpendicular bisectors.

17. (a) Construct a line through P perpendicular to ℓ.

18. Connect A to the reflection of B across ℓ.

19. Intersect the perpendicular bisector of $\overline{AB}$ with ℓ.

20. (b) a sphere (surface of a ball) of radius 2 inches with center A

21. (b) a plane passing through the midpoint of $\overline{AB}$ and perpendicular to $\overline{AB}$

22. $ABCD$ is a parallelogram. What do you know about the diagonals of a parallelogram?

23. By *definition*, a parallelogram is a quadrilateral whose opposite pairs of sides are parallel. That's why $ADBC$ is a parallelogram. Since opposite sides of any parallelogram have equal length, we see that $DA = BC$. Similarly, using the parallelogram $ABCE$, we obtain that $BC = AE$. Can you finish the proof?

26. (b) 70°

(e) The missing angle is also x (why?). Hence $x + x + 2x = 180°$. $x = \underline{}$.

(f) Cut the quadrilateral into two isosceles triangles using a diagonal, $x = 110°$.

(h) Certain pairs of triangles are congruent (SSS). Use the fact that the base angles of an isosceles triangle are equal, and also the fact that the sum of the angles around a point is 360°. $x = 110°$.

(k) $x = 50°$

(m) $x = 36°$

(n) The sum of the angles in any quadrilateral is 360°.

(p) Note that the quadrilateral must be a parallelogram (why?).

(r) Use part (q).

(s) The quadrilateral is some rhombus, but x cannot be determined.

Review Self Test

1. (a) Construct a large *equilateral* triangle.

    (b) Construct the angle bisectors of your triangle.

    (c) Construct the perpendicular bisectors of the sides of the triangle in (a).

    (d) Construct the medians of the triangle in (a).

    (e) What do you observe?   *They are all the same.*

2. Construct the altitude to side $\overline{AC}$ for $\triangle ABC$.

    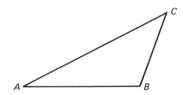

3. Construct the center of the circle:

4. Construct a square with side s: |←———— s ————→|

216   5: Constructions with Straightedge and Compass

5. Construct a parallelogram with all sides of equal length and with a 60° angle.

6. Construct an angle of size 75°.

7. Draw a line segment 9 cm. long with a ruler. Then, using only a straightedge and compass, construct an equilateral triangle of side length 3 cm.

Answers

1. (e) The lines in parts (b), (c), and (d) are the same.

2. Construct a perpendicular from B to $\overline{AC}$.

3. Construct the perpendicular bisectors to two nonparallel chords.

4. There is more than one way; compare your method with that of classmates.

5. Construct two equilateral triangles with a side in common.

6. Construct a 60° angle (by constructing an equilateral triangle). Bisect successively to obtain a 15° angle. Then construct angle of 75° = 60° + 15°.

7. Trisect the segment, then construct the triangle.

Projects

This is an excellent reference.

Encourage a student to give a class report on Archimedes.

1. Read about "Geometrical Constructions" in Chapter 3 of *What is Mathematics?* by R. Courant and H. Robbins (New York: Oxford University Press, 1941).

2. (a) The greatest mathematician of antiquity was Archimedes of Syracuse, who lived about 300 B.C. Here is a description of a method he gave for trisecting an angle, if one is allowed to mark points on the straightedge. Start with a given angle with vertex P:

(i)

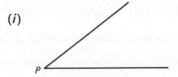

217 Projects

Extend the horizontal side of the angle to the left, and draw a circle (of any radius *r*) with center *P* and *Q* the point where the other side of the angle intersects the circle:

(*ii*)

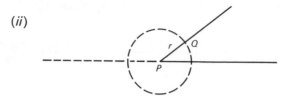

Mark two points *A* and *B* at distance *r* apart on the straightedge. Keeping the mark *B* on the circle, slide the straightedge into a position where *A* lies on the horizontal line and the edge passes through *Q*:

(*iii*)

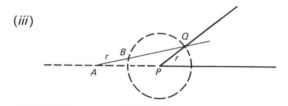

(*b*) Show that the angle at *A* is exactly $\frac{1}{3}$ the size of the original angle. [*Hint:* Draw segment $\overline{BP}$.]   *Use the Exterior Angle Theorem.*

(*c*) Finally, show how to construct two lines through *P* which trisect the original angle.   *Copy the angle at A twice at P.*

(*d*) Why is this *not* a solution of the classical Greek problem of trisecting an angle?   *Marks on the straightedge were used.*

3. Mark two points *A* and *B* on the *bottom edge* of a sheet of paper. Find a method for constructing the midpoint of the line segment $\overline{AB}$ using only a straightedge and compass (all construction lines must be drawn on the paper).

4. Suppose your compass has become rusted and is *fixed* at an opening of 4 cm. Do the following constructions, *using only a straightedge and this rusty compass.*

(*a*) Construct an equilateral triangle of side 4 cm.

(*b*) Bisect a line segment of length 4 cm.

(*c*) Construct an equilateral triangle of side 8 cm.

(d) Construct a line segment of length 10 cm.

(e) Show how to bisect any given line segment, no matter how long it is.

(f) Show how to divide any given line segment into 7 equal segments.

(g) Construct a regular hexagon of side 6 cm.

(h) Show how to bisect any given angle.

(i) Draw a line ℓ and mark a point P on ℓ. Construct a line through P perpendicular to ℓ.

(j) Can you construct a line segment of *any* given length?   no

5. Suppose you are given a line segment of length 1 cm., drawn on a sheet of paper. Explain how you could then construct, using only straightedge and compass, a line segment of length

(a) $\frac{1}{2}$ cm.   (b) $\frac{1}{3}$ cm.   (c) $\frac{1}{7}$ cm.   (d) $\frac{1}{n}$ cm.

(e) $\frac{2}{3}$ cm.   (f) $\frac{2}{5}$ cm.   (g) $\frac{3}{5}$ cm.   (h) $\frac{9}{11}$ cm.

(i) $\frac{m}{n}$ cm.

6. Can you devise a method for constructing an angle of 1°? Explain.
no

7. (a) Draw any segment AB. Now, follow the steps shown in the picture below to bisect $\overline{AB}$.

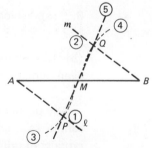

① Draw a line ℓ through A.

② Construct a line m through B parallel to ℓ.

③ Using your compass, draw an arc of a circle with center A, intersecting ℓ at P.

219   Projects

④ Using the *same compass setting*, draw an arc of a circle with center B, intersecting m at Q.

⑤ Draw the line through P and Q.

The point M, where this line intersects $\overline{AB}$, is the midpoint of $\overline{AB}$.

(b) Prove that the above method for bisecting a line segment is valid. That is, prove that M must be the midpoint of $\overline{AB}$. [*Hint:* You want to prove that AM = BM. Can you prove that certain triangles must be congruent?] $\triangle APM \cong \triangle BQM$ by ASA

Which Regular Polygons Can Be Constructed?

In order to understand the work of Gauss on constructing regular polygons with straightedge and compass, we need to know some pertinent facts about prime numbers.

DEFINITION A positive integer, greater than 1, which cannot be represented as a product of smaller positive integers is called a *prime number*, or simply a *prime*.

For example, 2, 3, 5, 7, 11, 13, 17, 19, 23, 29 are primes. The number 15 is *not* a prime number, because 15 can be represented as a product of smaller positive integers: $15 = 3 \cdot 5$. Similarly, $16 = 4 \cdot 4$ is not a prime.

8. (a) Which of the following are primes: 25? 37? 91? 127? 129? 323? 37,127?

(b) The number 2 is a prime and is also an even number. Are there other even primes? Explain. No, all even numbers are divisible by 2.

Any positive integer greater than 1 which is not a prime can be represented as a product of primes. For example, we can represent 72 as a product of primes:

$$72 = 2 \cdot 2 \cdot 2 \cdot 3 \cdot 3$$

9. Represent as a product of primes.

(a) 4   (b) 6   (c) 44   (d) 1400
$2 \cdot 2$ $2 \cdot 3$ $2 \cdot 2 \cdot 11$ $2 \cdot 2 \cdot 2 \cdot 5 \cdot 5 \cdot 7$

A very special kind of prime is crucial to the construction of regular polygons—primes which are "one more than a power of 2." A "power of 2" is a number of the form 2^k for some k. For example, 2, 4, 8, 16, 32, are powers of 2. Then "one more than a power of 2" is a number of the form $2^k + 1$. For example, 3, 5, 9, 17, 33, are each "one more than a power of 2." Notice that some of these numbers are primes, and some are not. For example, 3, 5, and 17 are each "one more than a power of 2" and also primes. Such primes are called *Fermat primes,* after the seventeenth century mathematician Pierre Fermat.

DEFINITION A *Fermat prime* is a number which is both a prime number and of the form $2^k + 1$ for some positive integer k.

10. Which of the following numbers are Fermat primes: 2? 3? 4? 5? 7? 9? 13? 17? 33? 65? 129? 257?
3, 5, 17, 257

We can now state the remarkable result, discovered by Gauss, about constructing regular polygons.

Given an integer $n \geq 3$, a regular n-gon can be constructed using only straightedge and compass if, and only if, n satisfies the following condition:

(*i*) If n is a prime, then it must be a Fermat prime.

(*ii*) If n is not a prime, then when n is represented as a product of primes no odd prime appears more than once, and every odd prime which appears must be a Fermat prime.

(There is no restriction on the number of times the prime 2 appears.)

To illustrate how Gauss's theorem can be used, consider the case $n = 60$. We represent 60 as a product of primes:

$$60 = 2 \cdot 2 \cdot 3 \cdot 5$$

We note that each odd prime which appears is a Fermat prime, and no odd prime appears more than once. Hence, according to Gauss's theorem, a regular 60-gon *can* be constructed using only straightedge and compass. As another example, observe that when 9 is represented as a product of primes, $9 = 3 \cdot 3$, we have an odd prime appearing more than once.

Hence, according to the theorem, a regular 9-gon *cannot* be constructed with straightedge and compass only.

Observe that this settles the ancient Greek question of "constructing a regular 7-gon." Since 7 is a prime, but 7 is *not* a Fermat prime, a regular 7-gon *cannot* be constructed with straightedge and compass. Gauss's discovery brought another surprise. Since 17 *is* a Fermat prime, we see that a regular 17-gon *can* be constructed using only a straightedge and compass. This is something which the ancient Greek geometers never suspected.

Gauss was only 19 years old when he proved this remarkable theorem. The discovery had a profound effect on his life, for he decided at this point that he would become a mathematician by profession. You might be interested in reading further about the life of Gauss in the book *Men of Mathematics,* by E. T. Bell (New York: Simon and Schuster, 1937).

The book of Tietze, referred to in the notes in the Central section, is excellent supplementary reading. Also see chapter 17 of Mathematics, The Man-made Universe, *2nd ed., by S.K. Stein (San Francisco: W. H. Freeman, 1969).*

11. (*a*) Use the condition discovered by Gauss to decide if a regular *n*-gon can be constructed, with only straightedge and compass, for $n = 3, 4, 5, 6, 7, 8, 9, 10, 11, 12, 13, 14, 15, 19, 20, 21, 25, 33, 36, 40, 45, 85, 100, 101, 257, 514.$ 3, 4, 5, 6, 8, 10, 12, 15, 20, 40, 85, 257, 514

(*b*) Find all regular polygons with less than 50 sides which can be constructed using only straightedge and compass.   3, 4, 5, 6, 8, 10, 12, 15, 16, 17, 20, 24, 30, 32, 34, 40, 48

Remarkable Points and Lines of Triangles

12. In the Central section of this chapter we observed, by experimenting with some triangles, that *the altitudes of any triangle intersect in one point.* Prove it! [*Hint:* Given any △*ABC*, through each vertex draw a line parallel to the opposite side, as indicated in the figure below. Prove that the altitudes of △*ABC* are the perpendicular bisectors of the sides of the big triangle.]

Encourage students to do Exercise 13; many review ideas are used in the proof.

13. In the Central section of this chapter we observed experimentally that *the medians of the sides of any triangle intersect in one point.* In this exercise we prove that result and study some interesting properties of that point where the medians intersect.

(*a*) Consider △*ABC* and the medians $\overline{AE}$ and $\overline{BD}$ in the following figure.

P is the point of intersection of $\overline{AE}$ and $\overline{BD}$, F is the midpoint of $\overline{BP}$ and G the midpoint of $\overline{AP}$.

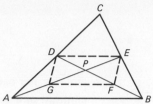

Prove that $DEFG$ is a parallelogram. [*Hint:* What relationship exists between $\overline{DE}$ and $\overline{AB}$? between $\overline{GF}$ and $\overline{AB}$?]

(*b*) Prove that $PE = \frac{1}{3}AE$ and $PD = \frac{1}{3}BD$. In other words, the intersection point of two medians trisects both of them. [*Hint:* The diagonals of a parallelogram bisect each other.]

(*c*) Prove that the median drawn from C to the midpoint of $\overline{AB}$ must pass through P. [*Hint:* Think about applying the result of part (*b*) to the median $\overline{AE}$ and the median joining C to the midpoint of $\overline{AB}$.]

The preceding exercise proves that *the medians of any triangle intersect in a single point.* This point is given a special name, motivated by a certain physical property.

DEFINITION The point where the medians of a triangle intersect is called the *centroid* of the triangle.

Let a student explain Exercise 14 to the entire class, even though others may have done the exercise. More discussion may follow.

14. (*a*) Cut a large triangle out of heavy cardboard and find its centroid. Balance your triangle on a pencil point. What do you observe?

It balances on the centroid.

Put a pencil through a hole drilled at the point where the medians meet. The triangle will rotate more smothly than if the hole were drilled at some other point.

(*b*) Draw the medians of your triangle in part (*a*), if you have not already done so. Let your triangle hang from a vertex, together with a plumb line, as indicated in the figure on the next page.

The plumb line lies on the median.

Do this for each vertex; what do you observe? Can you see why the centroid of a triangle is often called its "center of gravity"?

Miscellaneous Exercises

15. (*a*) Somebody hands you a thin sheet of plain paper with a line segment drawn on it. He asks you to find the perpendicular bisector of the segment. However, you have lost your ruler, compass, and pencil. How would you do it? [*Hint:* You may fold and crease the paper any number of times.]   *Fold the paper so that the endpoints coincide.*

Do not force a student to go through the actual paper folding construction in Exercise 15(b), since this is very tedious. Be content with an understanding of how one could go about doing it if one wanted to.

(*b*) Can you trisect the segment in part (*a*) by folding and creasing the paper?   *yes*

16. The figure below shows a triangle and a median of length *c*. Prove that

$$c < \frac{a+b}{2}.$$

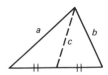

[*Hint:* The triangle is half a certain parallelogram. One diagonal of that parallelogram has length 2*c*.]

Algebra Review

Inequalities

In comparing two numbers on the number line,

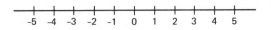

This Algebra Review saves class time in Chapter 11, and makes the student more comfortable with algebra in the remaining chapters.

if a number is to the left of another number, we say, "a is less than b," and write

$$a < b.$$

(Or we may say, "b is greater than a," and write $b > a$.)

1. True or false?

(a) Since $2 < 4$, then $2 + 3 < 4 + 3$.

(b) Since $-2 < 7$, then $-2 - 8 < 7 - 8$.

(c) Since $5 < 13$, then $2(5) < 2(13)$.

(d) Since $-11 < 2$, then $-\frac{11}{3} < \frac{2}{3}$.

(e) Since $4 < 10$, then $4(-3) < 10(-3)$.

(f) Since $-6 < 5$, then $\frac{-6}{-7} < \frac{5}{-7}$.

2. If $x < 10$, which of the following is true? Explain.

(a) $7x < 70$ (b) $x + 19 < 29$ (c) $-6x < -60$

(d) $5 > \dfrac{x}{2}$

3. Observe in Exercises 1 and 2, that the inequality remains true except when we _____ (or _____) by a negative number, in which case the order of the inequality is reversed ("<" becomes ">" and vice versa). Summarizing:

If a student has trouble with inequalities using letters, have him substitute easy numbers in these rules until he understands.

If $a < b$, then

$$a + c < b + c \qquad a - c < b - c.$$

If $a < b$ and $c > 0$, then

$$ac < bc \qquad \frac{a}{c} < \frac{b}{c}.$$

BUT if $a < b$ and $c < 0$, then

$$ac > bc \qquad \frac{a}{c} > \frac{b}{c}.$$

225 Algebra Review

Example 1. Solve for x.

$$-\frac{x}{2} + 7 > 9$$

$$-\frac{x}{2} > 2 \quad \text{(subtract 7)}$$

$$x < -4 \quad \text{(multiply by } -2\text{, reverse the inequality)}$$

4. Solve for x (check your answer).

   (a) $x + 7 < 29$ (b) $x - 14 > 3$ (c) $\frac{x}{4} > 9$

   (d) $5x < 13$ (e) $-2x > 8$ (f) $-5x + 9 < 4$

   (g) $2 - \frac{x}{3} > 1$ (h) $9x - 3 < 8$ (i) $x - 4 > 3x$

5. Fill in with "$<$" or "$>$."

   (a) If $a < b$, then $2a$ ____ $2b$.

   (b) If $a < b$, then $\frac{a}{-3}$ ____ $\frac{b}{-3}$.

   (c) If $a < b$, then $-5a$ ____ $-5b$.

   (d) If $a < b$, then $a - 7$ ____ $b - 7$.

   (e) If $a < b$, then $6a$ ____ $6b$.

   (f) If $a < b$, then $a + 9$ ____ $b + 9$.

6. True or false? Explain.

   (a) If $1 < N$, then $N < N^2$.

   (b) If $0 < N < 1$, then $N < N^2$.

   (c) If $N < 0$, then $N^2 < 1$.

Linear Equations

An equation of the form

$$ax + by + c = 0,$$

where a, b, and c are constants such that a and b are not both 0, is called a *linear* equation, for its graph is a straight line.

7. (a) Find three points (x, y) that satisfy the equation $5x + 3y = 3$.

(b) Plot the points in (a) in the (x, y)-coordinate plane.

(c) Using a straightedge, draw the line through the points in (b). (Two points determine a line; the third point is a check.)

(d) Look at your graph in (c). Name three other points that lie on the graph.

(e) Check that the points in (d) satisfy the equation $5x + 3y = 3$.

8. Do Exercise 7 for the equation $4x + y = 8$ and use the same coordinate axes.

9. Estimate where the lines in Exercises 7 and 8 cross each other.

Simultaneous Linear Equations

We often want to find x and y which *simultaneously* satisfy a given *pair* of linear equations. For example, $x = 3$ and $y = -4$ satisfy *both* the equations in Exercises 7 and 8.

$$\begin{cases} 5x + 3y = 3 \\ 4x + y = 8 \end{cases}$$

We then say we have *solved* the above *simultaneous linear equations*.

10. Graph each pair of equations. Estimate where each pair of lines cross each other. (Use a different set of coordinate axes for each pair.)

(a) $\begin{cases} x - 3y = 7 \\ 2x + 3y = 2 \end{cases}$
(b) $\begin{cases} 5x + 3y + 1 = 0 \\ x = 4 + 3y \end{cases}$

(c) $\begin{cases} y = \dfrac{x}{3} \\ 4x = 15 + 2y \end{cases}$
(d) $\begin{cases} 2x + y = 12 \\ 3x - 2y = 7 \end{cases}$

Comment. You may have found it difficult to estimate where the lines in Exercise 10 intersect each other. An algebraic solution is often faster and always more accurate, as Examples 2 and 3 demonstrate.

Example 2. Addition-subtraction method

Find the point (x, y) of intersection for the lines in Exercise 10(a) using algebra only.

$$\begin{cases} x - 3y = 7 \\ 2x + 3y = 2 \end{cases}$$

$$\begin{cases} 2(x - 3y) = 2(7) \\ 2x + 3y = 2 \end{cases} \quad \text{Multiply by 2.}$$

$$\begin{aligned} \begin{cases} 2x - 6y = 14 \\ 2x + 3y = 2 \end{cases} \\ \overline{-9y = 12} \end{aligned} \quad \text{Subtract lower from upper.}$$

$$\frac{-9y}{-9} = \frac{12}{-9} \quad \text{Divide by } -9.$$

$$y = -\frac{4}{3}$$

$$x - 3\left(-\frac{4}{3}\right) = 7 \quad \text{Solve for } x \text{ by substituting in one of the original equations.}$$

$$x + 4 = 7$$

$$x = 3$$

The point (x, y) of intersection is $x = 3$, $y = -\frac{4}{3}$.

Check:

$$(3) - 3\left(-\frac{4}{3}\right) \stackrel{?}{=} 7 \qquad 2(3) + 3\left(-\frac{4}{3}\right) \stackrel{?}{=} 2$$

$$3 + 4 \stackrel{?}{=} 7 \qquad\qquad 6 - 4 \stackrel{?}{=} 2$$

$$7 = 7 \qquad\qquad\qquad 2 = 2$$

$$\text{yes} \qquad\qquad\qquad\quad \text{yes}$$

Example 3. Substitution method

The previous example could have been solved as follows:

$$\begin{cases} x - 3y &= 7 \\ 2x + 3y &= 2 \end{cases}$$

$$\begin{cases} x &= 3y + 7 \\ 2x + 3y &= 2 \end{cases} \quad \text{Solve for } x.$$

$$2(3y + 7) + 3y = 2 \qquad \text{Substitute the value of } x \text{ from the}$$
$$6y + 14 + 3y = 2 \qquad \text{first equation for } x \text{ in the second}$$
$$9y = -12 \qquad \text{equation.}$$
$$y = -\frac{4}{3}$$

228   5: Constructions with Straightedge and Compass

Substitute $y = -\frac{4}{3}$ in one of the original equations to find x, and check as in Example 2.

11. Solve each pair of equations in Exercise 10(b), (c), and (d) by algebra.

12. Solve for x and y by algebra.

(a) $\begin{cases} 4x - 3y = 13 \\ 2x + y = -1 \end{cases}$ (b) $\begin{cases} 2x - 5y = 29 \\ 4x + y = -19 \end{cases}$

(c) $\begin{cases} x + 2y = 5 \\ 3x + 2y = 17 \end{cases}$ (d) $\begin{cases} 4x + 3y = -9 \\ 5x - 2y = -17 \end{cases}$

13. Solve for x and y.

(a) $\begin{cases} 3(2x-1) - 4(y-3) - 5 = 0 \\ 7(y+2) + 5(2x-1) = 57 \end{cases}$ (b) $\begin{cases} \dfrac{x}{3} - y = 0 \\ \dfrac{x}{2} - \dfrac{y}{2} = 1 \end{cases}$

(c) $\begin{cases} \dfrac{x+3}{2} + \dfrac{y+5}{3} = 7 \\ \dfrac{x+4}{3} - \dfrac{2y-3}{5} = 2 \end{cases}$ (d) $\begin{cases} 2 + \dfrac{x}{7} = \dfrac{y}{4} \\ 3 + \dfrac{y}{5} = \dfrac{x}{3} \end{cases}$

[*Hint:* (b) Clear of fractions, then put in the form $ax + by = c$.]

Answers

1. (a) true (b) true (c) true (d) true (e) false (f) false
2. (a), (b), and (d) are true.
3. multiply, divide
4. (a) $x < 22$ (b) $x > 17$ (c) $x > 36$ (d) $x < \frac{13}{5}$
   (e) $x < -4$ (f) $x > 1$ (g) $x < 3$ (h) $x < \frac{11}{9}$ (i) $x < -2$
5. (a) < (b) > (c) > (d) < (e) < (f) <
6. (a) true (b) false (c) false
7. (a) (0, 1), (3, −4), (−3, 6)   (d) student example
8. (a) (0, 8), (1, 4), (−1, 12)   (d) student example
9. (3, −4)
10. (a) $(3, -\frac{4}{3})$ (b) $(\frac{1}{2}, -\frac{7}{6})$ (c) $(\frac{9}{2}, \frac{3}{2})$ (d) $(\frac{31}{7}, \frac{22}{7})$
12. (a) (1, −3) (b) (−3, −7) (c) $(6, -\frac{1}{2})$ (d) (−3, 1)
13. (a) (2, 4) (b) (3, 1) (c) (5, 4) (d) (21, 20)

Chapter 6 Area and Volume

Area is a number, not a geometric region. This contrasts with the meaning of "area" in daily life as in "troubled area of the world."

Geoboards may be helpful in this chapter. A student can make one by hammering nails into a 1″ pine or ½″ plyboard board at 2-inch intervals, thus dividing the board into 2″ squares. Rubber bands or colored yarn can be used to make different polygons, and the student can then compute the areas.

Volume, introduced before Central Exercise 23, helps build a geometric intuition for space. Bring in (or have the class bring in) physical objects, and find their volumes.

At the end of the chapter discuss the opening puzzle if there are still any questions about it.

Central

Here is an old puzzle which seems to present a paradox. An 8 × 8 square is cut into four pieces, and the pieces are rearranged into a rectangle as indicated in the picture below.

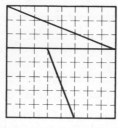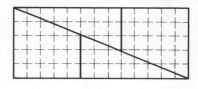

Since the rectangle is formed by simply rearranging the pieces obtained from the square, the area of the rectangle is equal to the area of the square—or is it? Cut this puzzle from a sheet of paper and try it. Compare your work with that of your classmates. *The pieces don't quite fit together, see the comment on p. 241.*

Area

We often measure the size of a flat object in "square inches" or "square feet," and we call the number we obtain the *area* of the object. We emphasize that the area of a geometric figure is a certain *nonnegative*

It needs to be emphasized that the area of a figure is a number associated with that figure. Students may tend to confuse "area" with "region."

number associated with that figure. In this chapter, when we speak of "the area of a polygon" we always mean a certain number associated with the region inside that polygon. For example, "the area of $\triangle ABC$" is a certain number associated with the shaded region in the figure below:

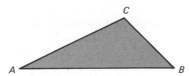

Properties of Area

Properties I, II, and III are axioms. They are so much a part of their intuition that most students are scarcely aware of using them.

I If the sides of a rectangle have lengths b and h respectively, then the area of the rectangle equals bh, the product of b and h.

The area of the rectangle above is bh.

II If two figures are congruent, then they have the same area.

III If a figure is cut into several pieces, then the area of the figure is the sum of the areas of the pieces.

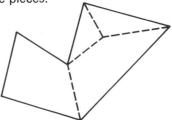

The area of the large polygon above is the sum of the areas of the four smaller polygons obtained by cutting along the dotted lines.

We will see in this chapter how the area of any polygon can be calculated, using only Properties I, II, and III.

You may want to remind students of the meaning of square unit. For example, draw a 3-inch by 4-inch rectangle on the blackboard and divide it into 12 1-inch squares. Each small square represents a "square inch" of area; the area of the rectangle is 12 square inches. Point out how a similar process would show that a 3-foot by 4-foot rectangle has area 12 square feet.

Note that in practical problems some unit is chosen for measuring lengths—and the areas are expressed in square units. For example, if length is measured in *inches,* then area is expressed in *square inches.* If length is measured in *centimeters,* then area is expressed in *square centimeters.* Thus, if a rectangle has sides of lengths 6 inches and 7 inches respectively, then its area is 42 square inches. In most problems in this section, the unit of measurement will not be explicitly mentioned.

Exercise 1(b) provides a good chance to review the notion of square root.

Some algebra is reviewed in Exercise 2.

1. (a) A rectangular floor is 12 feet long and 9 feet wide. What is its area (in square feet)?   *108 sq. ft.*

   (b) A carpenter wishes to build a *square* floor with the same area as the floor in part (a). What should the length of each side of the square be? $\sqrt{108}$ *ft.*

2. A certain rectangle has area 40. Find the length of each side of the rectangle if

   (a) one side has length 5   *8*

   (b) one side is twice as long as another side. $2\sqrt{5}$ *and* $4\sqrt{5}$

3. Give an example of two rectangles that have the same area, but are *not* congruent. (This shows that the *converse* of Property II is false. In other words, it is *not* necessarily true that if two figures have the same area, then they are congruent!)

   The next exercise shows how we can find the area of any parallelogram, using only Properties I, II, and III.

This definition is important. Make sure students understand that the altitude is the length of a line segment perpendicular to both horizontal sides.

4. The parallelogram below has horizontal sides of length *b* (we call *b* the *base* of the parallelogram). The distance between the horizontal sides is *h* (we call *h* the *altitude* of the parallelogram).

   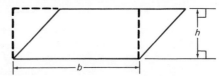

   (a) Show how to cut the parallelogram above into two pieces which can be rearranged to form a *rectangle* with sides of lengths *b* and *h* respectively.

   (b) What is the area of the resulting rectangle?   *bh*

   (c) Then what is the area of the parallelogram?   *bh*

   (d) Discuss with your classmates how Properties I, II, and III helped you find the answer in part (c).

   (e) Complete:

If a parallelogram has base *b* and altitude *h*, then its area is $\underline{\quad bh \quad}$.

---

5. Find the area of each parallelogram.

(a)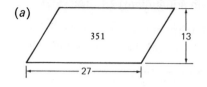

(b)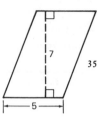

(c)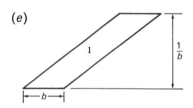

(d)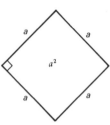

(e)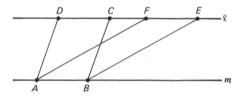

(f)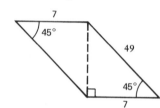

6. In the picture below, ℓ and m are parallel lines, and $AB = CD = EF$.

Which parallelogram has larger area, $ABCD$ or $ABEF$? Compare your answer with those of your classmates. *equal area, since they have the same base and height*

7. Bill has found a flaw in the argument used in Exercise 4 showing that the area of a parallelogram is (base) × (altitude). He says, "What if the parallelogram looks like the one below? Then when we try to cut off a triangle and fit it back to form a rectangle, we run into trouble."

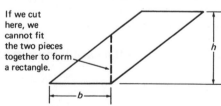

If we cut here, we cannot fit the two pieces together to form a rectangle.

234   6: Area and Volume

Some students may need to experiment by cutting up a parallelogram.

Bill has made a good point here! Discuss with your classmates how to show that the area of a parallelogram like the one above is bh. [*Hint:* Consider the picture below and the explanation following it.]

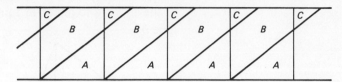

The figure above is obtained by overlapping a row of rectangles with a row of parallelograms. The pieces with the same letters are congruent. Do you see how the pieces A, B, C can be put together to form either a parallelogram or a rectangle having the same base and altitude? *Cut the parallelogram into three pieces: A, B, and C.*

Area of a Triangle

These definitions are important.

Let us now consider the area of a triangle. In the picture below, b is the length of the horizontal side of the triangle (we call b the *base* of this triangle). The distance from the upper vertex to the horizontal side is h (we call h the *altitude* of this triangle).

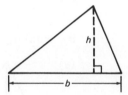

The triangle below also has base b and altitude h:

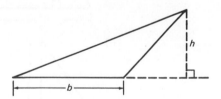

Note that we used the word "altitude" in a different way in Chapter 5 following Central Exercise 55. In that chapter we defined an altitude to be the *line segment* drawn from a vertex perpendicular to the opposite side. Here, we are using the same word to designate the *length* of such a line segment. So the word "altitude" will be used for two different ideas: a certain line segment and also the length of that segment. You will always be able to tell which meaning we have in mind from the context in which it is used, and no confusion should arise.

The next exercise shows how to find the area of a triangle if we know its base and altitude.

8. Given any triangle with base *b* and altitude *h*, we can fit two copies of the triangle together to form a parallelogram having base *b* and altitude *h*:

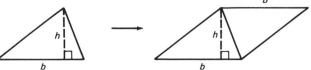

(a) What is the area of the resulting parallelogram? *bh*

(b) Then what must the area of the triangle be? Compare your answer with those of your classmates. $\frac{1}{2} bh$

(c) Complete:

If a triangle has base *b* and altitude *h*, then its area is $\frac{1}{2} bh$ _____.

---

9. Find the area of each triangle.

(a)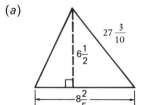

(b)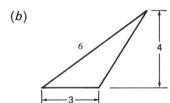

(c)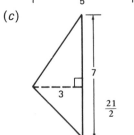

(d)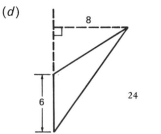

[Ans: 24]

(e)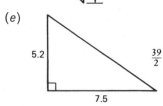

(f)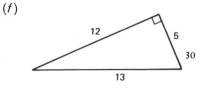

236 6: Area and Volume

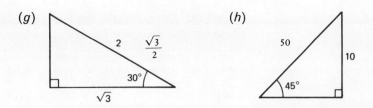

For students having difficulty with Exercise 10, suggest that they imagine D moving along ℓ. The altitude and base do not change; hence the area is unchanged.

10. If ℓ and m are parallel lines, which triangle has larger area, $\triangle ABC$ or $\triangle ABD$? Compare your answer with those of your classmates.   *equal area*

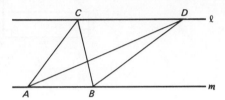

11. Which of the two shaded triangles has larger area? Discuss your answer with your classmates.   *equal area*

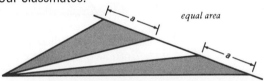

12. Why is the area of $\triangle ABD$ equal to the area of $\triangle CBD$?

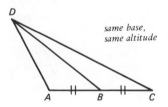

same base, same altitude

13. If $\triangle ABC$ has area 30, what is the area of $\triangle ABE$? Discuss your answer with your classmates.

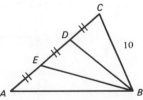

Finding Areas of Polygons

Property **III** of area states that if a polygon is cut into smaller polygons, the area of the polygon is the sum of the areas of the smaller polygons.

237   *Central*

Thus we can compute the area of a complicated polygon by cutting it into simpler polygons for which we know the areas.

On a geoboard unusual polygons can be laid out easily using rubber bands or yarn. Students can make such a board at home and bring it to class.

14. Find the area of quadrilateral *ABCD* in each case.

(a)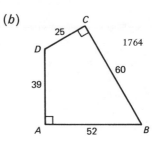

(b)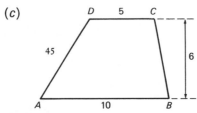

(c)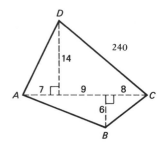

ABCD is a trapezoid. Cut it into two triangles.

(d)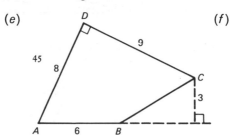

Exercise 14(e): Draw $\overline{AC}$.
(f): Draw lines from the corners to "center point."

(e)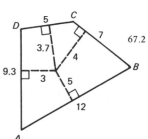

(f)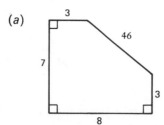

15. Find the area of each polygon below by observing that it is contained in a rectangle or parallelogram whose area is easy to compute.

(a)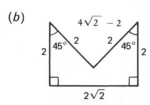

(b)

238   6: Area and Volume

(c) 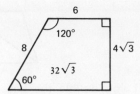   (d)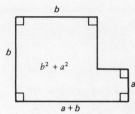

16. (*a*) Using a centimeter ruler, estimate the area of this polygon in square centimeters:

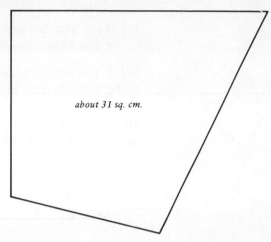

about 31 sq. cm.

(*b*) Compare your answer with those of your classmates.

17. A trapezoid has parallel sides of lengths *a* and *b* at distance *h* apart.

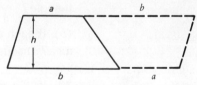

(*a*) Show how to put together two copies of the trapezoid to form a parallelogram of base length $a + b$ and height *h*.

This formula for the area of a trapezoid is important.

(*b*) Deduce from part (*a*) that the trapezoid has area equal to

$$\frac{1}{2} h(a + b).$$

(*c*) Give a different proof of the formula in part (*b*) by drawing a diagonal of the trapezoid and observing that the area of the trapezoid is the sum of the areas of the two triangles (see the figure on the next page).

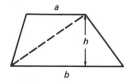

Exercises 18 and 19 provide some preparation for the study of area of similar figures in Chapter 9.

18. (a) If you double the base of a rectangle without changing the altitude, what happens to the area? *doubled*

(b) If you double the altitude of a rectangle without changing the base, what happens to the area? *doubled*

(c) If you simultaneously double each side of a rectangle, what happens to the area? Discuss with your classmates. *multiplied by 4*

(d) If you simultaneously triple each side of a rectangle, what happens to the area? Discuss with your classmates. *multiplied by 9*

19. (a) If each side of a triangle is doubled, without changing the angles, the resulting triangle can be cut into 4 triangles congruent to the original.

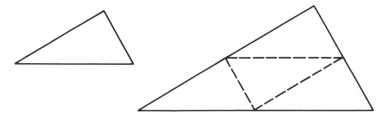

How does the area of the large triangle compare with the area of the small triangle? *4 times larger*

(b) If each side of a triangle is *tripled,* what do you think happens to the area? Discuss with your classmates. *multiplied by 9*

20. In the next chapter we will prove the Pythagorean Theorem, which enables us to find the length of the hypotenuse of any right triangle, given the lengths of its legs. In this exercise a special case of the theorem will be derived. Using properties of area, we will find the length x of the hypotenuse of an isosceles right triangle having legs of given length a.

(a) Show how 4 copies of the triangle fit together to form a square of side x.

(b) What is the area of the square, in terms of x? x^2

(c) The area of the square is 4 times the area of the triangle. Deduce from this that
$$x^2 = 2a^2.$$

(d) In terms of a, $x =$ _____. $a\sqrt{2}$

21. (a) Express in terms of a the length of a diagonal of a square of side a. [*Hint:* Use the preceding exercise.] $a\sqrt{2}$

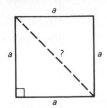

(b) A certain square has side length 5. What is the length of a diagonal of the square? $5\sqrt{2}$

(c) A certain square has a diagonal of length 8. What is the length of a side of this square? $4\sqrt{2}$

(d) What is the area of the square in part (c)? 32

22. Read over the paradox which opened this chapter.

(a) The four small pieces fit together to form an 8×8 square. Hence the sum of the areas of the pieces is _____. Which fundamental properties of area are used here? 64, **I** *and* **III**

(b) The four small pieces also appear to fit together to form a 5×13 rectangle. Hence the sum of the areas of the pieces appears to be _____. 65

(c) What is wrong? (Make a careful drawing using graph paper. Make the small squares about 1 cm. by 1 cm. in your drawing. Discuss with your classmates.) *Property* **III** *seems to be violated.*

Note that a right triangle with legs 5 and 13 is not similar to a right triangle with legs 3 and 8; hence the trapezoidal piece and the triangular piece in the rectangle do not really fit together to form a triangle. Thus there is a gap, in the shape of a parallelogram 1 unit square, along the diagonal of the rectangle.

Properties of Volume

We will now briefly study the concept of *volume* of solid figures. The volume of a solid is a certain *nonnegative number* associated with that solid and having the following fundamental properties:

Compare the properties of volume to the properties of area.

I If a rectangular box has edges of lengths *a, b, c* (as indicated in the picture), then the volume of the box equals *abc*, the product of *a*, *b*, and *c* ("length times width times height").

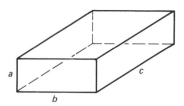

Students should become familiar with three-dimensional objects. Such thinking is difficult for many, even though they live in a three-dimensional world.

The volume of the box above is *abc*.

II If two solids are congruent, then they have the same volume.

III If a solid is cut into pieces, then the volume of the solid is the sum of the volumes of the pieces.

Remind students of the meaning of cubic units by illustrating how a 2-inch by 3-inch by 4-inch box can be divided into 24 1-inch by 1-inch by 1-inch cubes.

Note that in practical problems some unit is chosen for measuring length —and volume is measured in cubic units. For example, if lengths are measured in *inches*, then volume is measured in *cubic inches*. Thus if a box has length 3 inches, width 4 inches, and height 6 inches, then its volume is 72 cubic inches. In most problems in this section, the unit of measurement will not be explicitly mentioned.

23. (*a*) A box is 8 inches wide, 9 inches long, and 3 inches high. What is the volume of the box (in cubic inches) ?   216

(*b*) What is the length of each edge of a *cube* with the same volume as the box in part (*a*) ? (A cube is a rectangular box whose edges are all the same length. It is one of the regular polyhedra, discussed in the Projects section of Chapter 4.) $\sqrt[3]{216} = 6$

24. The base of a rectangular box is a rectangle of area *A*, and the box has height *h*. If *V* is the volume of the box, show that

$$V = hA.$$

In words, the volume of a rectangular box is equal to its height times the area of its base. *Let ℓ and w be the length and width of the base. Then $A = \ell \cdot w$. But $V = h \cdot \ell \cdot w = h \cdot A$.*

Volume of a Prism

In the solid pictured on the next page, the vertical edges are perpendicular to the horizontal triangular base. The figure is called a *right prism* with triangular base.

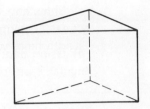

Encourage model building as an aid to spatial visualization.

It is possible to fit together two copies of the above prism to form a new solid with the same height and whose base is a parallelogram. The picture below indicates how this is done. If you have difficulty visualizing this, you should build wood or paper models of prisms and work with them.

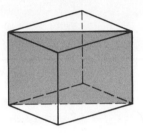

The base of the above solid is a parallelogram. This parallelogram can be cut into pieces which can be rearranged to form a rectangle. If, as we cut up the parallelogram, we make vertical slices through the solid, then we cut the solid into pieces which can be rearranged to form a new solid of the same height, but with a rectangular base. This new solid is a rectangular box. If the original prism has height h and base area A, note that the resulting rectangular box has height h and base area $2A$. Hence the volume of the box is $h(2A) = 2hA$. But the box also has twice the volume of the original prism (why?). Hence the volume of the original prism must have been hA. In other words, if V is the volume of a right prism of height h and with triangular base of area A, then

$$V = hA.$$

25. The solid pictured below is a *right prism* with polygonal base. The base is a polygon, and the vertical edges are perpendicular to the horizontal base.

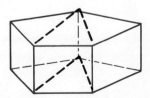

243 *Central*

(a) Show how the solid can be cut up into right prisms of the same height, but each having triangular base.

(b) Using part (a) and the formula for the volume of a prism with triangular base, prove the following statement.

> **The volume V of a right prism with height h and base area A is given by**
>
> $$V = hA.$$

(c) Discuss with your classmates how Properties **I**, **II**, and **III** of volume were used in arriving at the result in part (b). *Properties **II** and **III** were used to preserve volume, Property **I** to compute the volume.*

26. The volume of a certain right prism is 375 cubic units, and its height is 15 units. What is the area of its base (in square units)? *25*

27. The base of a certain right prism is a right triangle with legs of lengths 8 and 12. If the volume of the prism is 336, then what is its height? *7*

Exercise 28: Students should enjoy following the happenings in Polygonia described in the Projects section.

28. In the mythical country of Polygonia they use a trapezoidal coin of the following dimensions:

all measurements are in centimeters!

The coins are manufactured by stamping them out of sheet metal. This metal weighs 2 grams per cubic centimeter, and the sheet metal is $\frac{1}{4}$ cm. thick. How much does each Polygonian coin weigh? *12 gm.*
weight = volume · density

Volume of a Pyramid

The picture below shows a *pyramid.* The horizontal polygon is the *base* of the pyramid. The top vertex, where the triangular faces meet, is the *apex.*

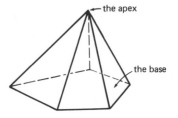

244 6: Area and Volume

If the base is horizontal, then the altitude is vertical.

The *altitude* of a pyramid is the distance from the apex to the base. The pyramid in the picture below has altitude *h*.

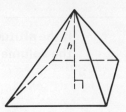

29. The figure below depicts a special kind of pyramid. The horizontal base is a square with dimensions *b* by *b*, and the apex *P* is above one corner of the square, at height *b*.

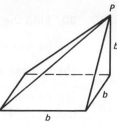

Exercise 29 is valuable for improving intuition of space. It is difficult to visualize the pyramid without making a model.

(a) Construct a model of this pyramid, using a pattern like the one below. *Fold and tape.*

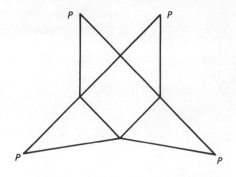

(b) Show how to assemble 3 copies of this pyramid into a cube of edge length *b*. Make a model of this.

(c) What must the volume of this pyramid be (in terms of *b*)? Which of the properties of volume did you use in arriving at your conclusion?
$\frac{1}{3}b^3$; **I, II, and III**

(d) Show that the volume of the above pyramid can be expressed as

$$V = \frac{1}{3} (\text{altitude})(\text{area of the base}).$$

area of base = b^2
altitude = b

245 Central

The formula for the volume of the special pyramid in the preceding exercise actually holds for *any* pyramid.

If the altitude of a pyramid is h and the area of the base is A, then the volume V of the pyramid is

$$V = \frac{1}{3}hA.$$

This formula will be proved in the Projects section of Chapter 9.

30. Find the volume of a pyramid if

(a) its altitude is 5 and its base has area 27.   45

(b) its base is a square of side 3 and its altitude is 10.   130

(c) its base is a right triangle with legs of lengths 8 and 12, and its altitude is 7.   112

31. The picture below shows a pyramid with altitude s. The base is a square whose diagonals have lengths $2s$, and the apex P of the pyramid is directly above the center Q of the square.

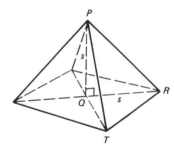

(a) Find PT, PR, and RT in terms of s. [*Hint:* Notice that each is the length of the hypotenuse of an isosceles right triangle with legs of length s. Recall Exercise 20.]   Each equals $s\sqrt{2}$.

(b) Express the volume of the pyramid above in terms of s. $\frac{2}{3}s^3$

32. If two copies of the pyramid studied in the preceding exercise are fitted together along their bases, we obtain a solid called a *regular octa-*

hedron (one of the five regular polyhedra studied in the Projects section of Chapter 4).

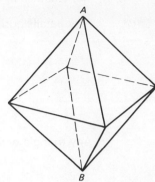

(*a*) The surface of the regular octahedron is made up of 8 triangles. What kind of triangles are they? [*Hint*: Use Exercise 31(*a*).] *equilateral*

If we let d = 2s = the "diagonal" of the octahedron, the volume is $\frac{1}{6}d^3$. It is interesting to contrast this with the formula $\frac{1}{2}d^2$ for the area of a square of diagonal d.

(*b*) If the distance from *A* to *B* is 2*s*, then express the volume of the regular octahedron in terms of *s*. [*Hint*: Use Exercise 31(*b*).] $\frac{4}{3}s^3$

(*c*) Each edge of a certain regular octahedron has length 3 cm. What is the volume of the regular octahedron in cubic cm.? Compare your answer with those of your classmates. $9\sqrt{2}$

Exercise 33 is a little preparation for the study of the volumes of similar solids (Chapter 9).

33. (*a*) If you double the lengths of all the edges of a rectangular box, what happens to the volume? Discuss with your classmates. *multiplied by 8*

(*b*) If you triple the lengths of all the edges of a rectangular box, what happens to the volume? Discuss with your classmates. *multiplied by 27*

Important Ideas in This Chapter

Axioms

Summarize the Central section.

The Fundamental Properties of Area

I The area of a rectangle with base *b* and altitude *h* is *bh*.

II If two figures are congruent, then they have the same area.

III If a figure is cut into pieces, then the area of the figure is the sum of the areas of the pieces.

The Fundamental Properties of Volume

I The volume of a rectangular box with edges *a*, *b*, and *c* is *abc*.

II If two solids are congruent, then they have the same volume.

III If a solid is cut into pieces, then the volume of the solid is the sum of the volumes of the pieces.

Theorems   Any parallelogram with base b and altitude h has area equal to bh.

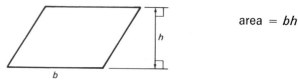

area = bh

Any triangle with base b and altitude h has area equal to $\frac{1}{2}bh$.

area = $\frac{1}{2}bh$

Any trapezoid of height h and parallel sides of lengths a and b respectively has area equal to $\frac{1}{2}h(a + b)$.

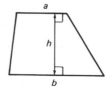

area = $\frac{1}{2}h(a + b)$

The length of the hypotenuse of an isosceles right triangle with legs of length a is $a\sqrt{2}$.

Any prism with height h and base area A has volume equal to hA.

Any pyramid with altitude h and base area A has volume equal to $\frac{1}{3}hA$.

Words area, volume, base, altitude, prism, pyramid, apex, regular octahedron

Summary

In this chapter we studied the fundamental properties of area of figures in a plane and the properties of volume of solid geometric figures. We developed some useful formulas for computing areas and volumes of simple types of geometric figures.

Historical Note

Although Euclid attempted to give a correct axiomatic development of geometry in his treatise the *Elements*, this work contains many flaws. Many definitions in the *Elements* are meaningless, and many of the theorems are not completely proved. A correct treatment of Euclidean geometry did not appear until David Hilbert (1862–1943) published *Foundations of Geometry* in 1899. This book showed clearly what is required for a correct axiomatic approach in geometry and was significant in starting the modern trend toward abstraction in mathematics.

Hilbert was possibly the greatest mathematician of his generation and solved important problems in a variety of fields. At the Second International Congress of Mathematicians at Paris in 1900, he proposed twenty-three unsolved problems whose solution he thought would be of importance for the advancement of mathematics. Many still remain unsolved; the complete solution of the tenth problem was found as recently as 1970. Only the third problem turned out to be easy. In Projects Exercise 5 it is proved that any polygon can be cut into a finite number of smaller polygons which can be rearranged to form a square. Hilbert's third problem asked whether an analogous property held for polyhedra in space. In particular, is it possible to cut a regular tetrahedron into a finite number of polyhedra which can be reassembled to form a cube? One of Hilbert's own students, Max Dehn, soon afterwards proved that it is impossible.

Central Self Quiz

1. Find the area.

(a)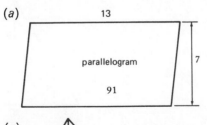

(b) Find the area of a rhombus with diagonals of 16 and 20. [*Hint:* The diagonals of a rhombus are perpendicular.]

(c)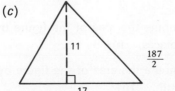

(d)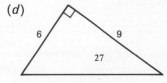

2. Use a centimeter ruler to estimate the area in square centimeters.

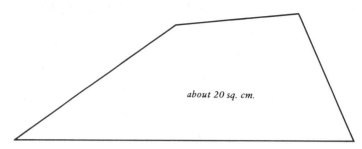

about 20 sq. cm.

3. A certain pyramid is built from stone which weighs 50 pounds per cubic foot. The base of the pyramid is a square with side length 10 feet, and the altitude of the pyramid is 30 feet. How much does the pyramid weigh? $\frac{1}{3} \cdot 10^2 \cdot 30 \cdot 50$

Answers

1. (a) 91 sq. units   (b) 160 sq. units   (c) $93\frac{1}{2}$ sq. units   (d) 27 sq. units

2. 20 sq. cm. (approx.)

3. 50,000 pounds

Review

1. Consider the figure below, made up of four right triangles and a rectangle.

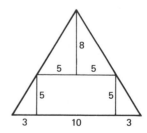

You should not expect any specific responses from students. The problem is really for fun. Note that the big "triangle" is not really a triangle at all; hence multiplying $\frac{1}{2}$ the height with the base length does not yield its area.

(a) Compute the area of the figure by adding together the areas of the smaller pieces.   105

(b) Compute the area of the figure by multiplying $\frac{1}{2}$ the height with the base length.   104

(c) Do you notice anything unusual? If so, how do you explain it?

2. Find the area of each polygon.

(a) 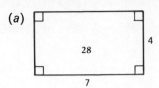   (b)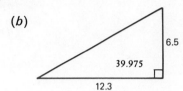

(c) 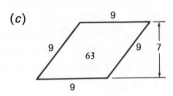   (d)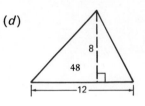

(e) 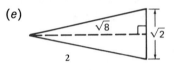   (f)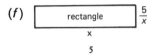

(g) 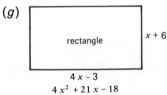   (h)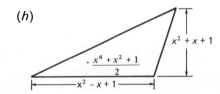

(i) 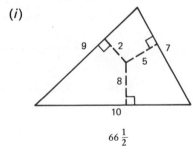   (j)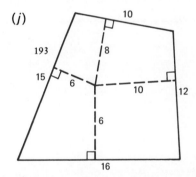

$$area = \frac{18\frac{1}{2} \cdot 17\frac{2}{3}}{9} \text{ sq. yd.}$$
$$cost = area\,(2.25)$$
$$= \$81.708$$

3. A rectangular floor $18\frac{1}{2}$ ft. long by $17\frac{2}{3}$ ft. wide is to be covered with linoleum which costs $2.25 per square yard. What will it cost?

4. Given $\triangle ABC$. Show how to construct, using only straightedge and compass, a point P on segment $\overline{AB}$ such that

(a) area of $\triangle APC$ = area of $\triangle PBC$ $AP = \frac{1}{2}AB$

(b) area of $\triangle APC = \frac{1}{3}$ (area of $\triangle ABC$) $AP = \frac{1}{3}AB$

(c) area of $\triangle APC = \frac{1}{5}$ (area of $\triangle ABC$) $AP = \frac{1}{5}AB$

(d) area of $\triangle APC = \frac{1}{2}$ (area of $\triangle PBC$). $AP = \frac{1}{3}AB$

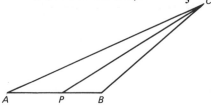

(e) Explain why your constructions are correct.

5. Given $\triangle ABC$, using only straightedge and compass construct a right triangle having the same area.

6. Given $\triangle ABC$, using only straightedge and compass, construct an isosceles triangle with a side of length AB and having the same area as $\triangle ABC$.

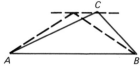

7. The figure below is a "kite" (meaning that the line through C and D is the perpendicular bisector of the segment $\overline{AB}$).

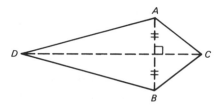

Find the area of the kite if

(a) $AB = 6$ and $CD = 10$.   30

(b) $AB = a$ and $CD = b$. $\frac{ab}{2}$

252   6: Area and Volume

8. Draw a triangle and label its vertices A, B, and C. Now find all those points P such that

(a) area of $\triangle APB$ = area of $\triangle ACB$

(b) area of $\triangle APB$ > area of $\triangle ACB$

(c) area of $\triangle APB$ < area of $\triangle ACB$.

9. In the picture below S represents a ship, and L and M are fixed positions on the shoreline.

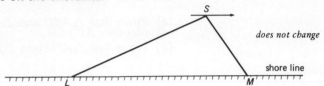

(a) What can you say about the area of $\triangle LSM$ as S moves parallel to the shoreline? Explain.

(b) As S sails parallel to the shoreline, for what position will $LS + SM$ be smallest? Explain.

10. (a) The area of a certain triangle is $a^2 - ab - 2b^2$, and its altitude is $a - 2b$. What is the base (in terms of a and b)? $2(a + b)$

(b) The area of a certain triangle is $x^2 - x - 20$ and its base is $x + 4$. What is its altitude (in terms of x)? $2(x - 5)$

Encourage discussion of Exercises 11, 12, and 13.

11. Draw a parallelogram ABCD. Choose any point P on the diagonal $\overline{BD}$ and draw lines through P parallel to the sides to obtain parallelograms $\mathcal{R}$ and $\mathcal{S}$.

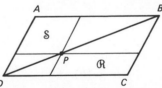

(a) How many pairs of congruent triangles can you find?

(b) Are $\mathcal{R}$ and $\mathcal{S}$ necessarily congruent?

(c) What can you say about the areas of $\mathcal{R}$ and $\mathcal{S}$?

12. In $\triangle ABC$, D is the midpoint of segment $\overline{AC}$ and E is the midpoint of

segment $\overline{BC}$. Prove that the area of quadrilateral DCEF equals that of △AFB.

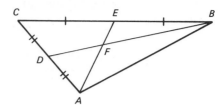

13. Given trapezoid ABCD.

    (a) Prove that △ACD and △BCD have the same area. *same base, equal altitudes since AB ∥ CD*

    (b) Prove that △AED and △BEC have the same area.

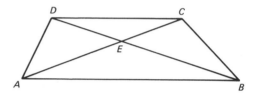

In Exercise 14, (b) and (c), encourage the students to subdivide the polygons in the easiest way possible.

14. By direct measurement, using your centimeter ruler, estimate the area (in square cm.) of each polygon.

    (a)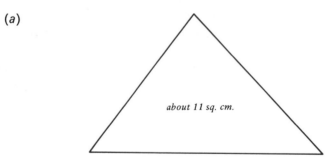

    about 11 sq. cm.

    (b)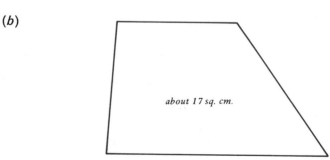

    about 17 sq. cm.

(c)

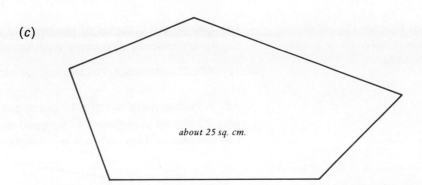

about 25 sq. cm.

Practical computations of area are sometimes done by weighing figures; with uniform thickness the area is proportional to the volume. Another method is to immerse the object in water and note the displacement.

15. Suppose you have a large supply of $\frac{1}{2}$ cm. by $\frac{1}{2}$ cm. squares cut from thin sheet metal of uniform quality. Somebody hands you an irregular figure cut from this same material and asks you to find its area.

(a) How would you do it? (Assume you have access to a laboratory.)

(b) With what accuracy can you find the area?

16. A certain pyramid has a base which is a right triangle with legs of lengths 5 and 8 respectively. The height of the pyramid is 12. What is its volume?

17. Suppose we start with a cube of edge length s and cut off a pyramid by passing a plane through three vertices, as shown below.

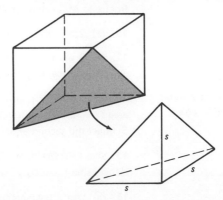

(a) What is the volume of the pyramid (in terms of s)?

Note that Exercise 17 gives another derivation of the result in Central Exercise 32(b).

(b) Show how 8 copies of the pyramid can be fitted together to form a regular octahedron. (You may want to make models to show this.)

(c) What is the volume of this regular octahedron (in terms of s)?

18. A certain metal weighs 10 grams per cubic centimeter. A polygon of area 20 square centimeters is stamped out of a sheet of this metal which is $\frac{3}{4}$ centimeter thick. What is the weight of the resulting piece?

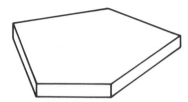

19. Alex Smart asks the class to find the volume of a pyramid with altitude 10 and whose base is a triangle with sides of lengths 55, 61, and 117. What is wrong with Alex's problem?

Review Answers

2. (c) 63 (e) 2 (f) 5 (g) $4x^2 + 21x - 18$ (h) $\dfrac{x^4 + x^2 + 1}{2}$

7. (a) 30 sq. units (b) $\frac{1}{2}ab$ sq. units

8. (a) on the straight line through C parallel to $\overline{AB}$ or on the straight line through C^* parallel to $\overline{AB}$, where C^* is the reflection of C across $\overline{AB}$

(b) outside the strip bordered by the two lines in part (a)

(c) inside the strip in part (b)

9. (b) when $LS = SM$ (Think of this as the problem of finding the shortest path from L to M touching the line through S parallel to $\overline{LM}$.)

10. (a) $2(a + b)$ (b) $2x - 10$

11. (a) 3 pairs (b) no (c) equal areas (The region $\mathcal{R}$ is obtained from $\triangle DCB$ by discarding the two smaller triangles.)

$\triangle ADB$ and $\triangle CDB$: equal base, same altitude. $\triangle CEA$ and $\triangle BEA$: equal base, same altitude.

12. area ($\triangle AFB$) + area ($\triangle AFD$) = area ($\triangle FBE$) + area ($DCEF$), because _____. area ($\triangle AFD$) + area ($DCEF$) = area ($\triangle AFB$) + area ($\triangle FBE$), because _____. Subtract the second equation from the first.

13. area ($\triangle AEB$) + area ($\triangle AED$) = area ($\triangle AEB$) + area ($\triangle BEC$), because _____. Hence, area ($\triangle AED$) = area ($\triangle BEC$).

15. (a) Use a balance scale and find how many of the $\frac{1}{2}$ cm. by $\frac{1}{2}$ cm. squares are needed to balance the figure. If this number is N, then the figure has area $\frac{1}{4}N$ square cm.

256 6: Area and Volume

(since each square has area $\frac{1}{4}$ sq. cm., and the weight will be proportional to the area because of the uniform thickness).

(b) to within $\frac{1}{4}$ sq. cm.

16. 80 cubic units

17. (a) $\frac{1}{6}s^3$ (c) $\frac{4}{3}s^3$

18. 150 grams

19. There is no triangle with sides 55, 61, and 117.

Review Self Test

1. By direct measurement, use a centimeter ruler to estimate the area (compare your answer with those of other students).

(a)

(b)

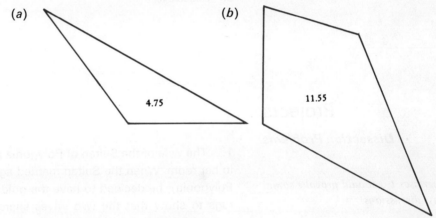

4.75

11.55

2. Find the area of each polygon.

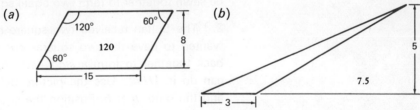

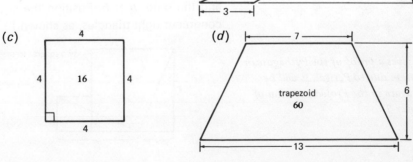

3. The vertical ends of the water trough pictured below are isosceles right triangles with legs 2 feet long. The horizontal edges of the trough are 5 feet long. How many cubic feet of water will the trough hold?

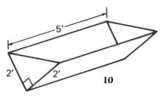

Answers

1. (a), (b) compare answers.
2. (a) 120   (b) $\frac{15}{2}$ (c) 16   (d) 60
3. 10 cubic feet

Projects

Dissection Problems

Exercises 1-3 should provoke some class discussions!

1. The wife of the Sultan of Polygonia had a beautiful square golden rug in her room. When the Sultan married again (polygamy was customary in Polygonia), he decided to have the golden rug cut into two equal square rugs to show that the two wives shared equal favor. As the royal rug cutter, show how the single square rug can be cut into pieces which can be sewn together to form two equal squares.

2. The Sultan received two square rugs of *different* sizes as gifts. He wanted to have the two squares cut into pieces which could be sewn back together to form a *single* square. Show how the royal rug cutter can do it. [*Hint:* Use the picture below. Assume one square is $a \times a$ and the other $b \times b$. Position the squares side by side and cut off two congruent right triangles, as shown.]

This gives a proof of the Pythagorean Theorem due to Perigal. It will be seen again in the Projects section of Chapter 7.

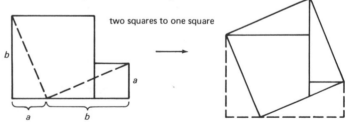

two squares to one square

258 6: Area and Volume

3. After a number of years, the Sultan of Polygonia had acquired a harem of many wives. Each wife had a square rug; however, some wives were favored differently from others, so all the squares were *not the same size.* One day there was a revolution in Polygonia. A democratic legislature was elected and a law abolishing polygamy was enacted. The Sultan was ordered to disband his harem and keep only one wife. The Sultan went to the royal rug cutter. "Guess what," he said. "Yes, I know," sighed the royal rug cutter, "you want me to cut all those little rugs into pieces that can be sewn together to form a single square rug for your one wife." Can the royal rug cutter do it? [*Hint:* Combine two of the squares into one, using the preceding exercise. Next, combine a third square with this, by the same method. Continue.]

The problems of the Sultan and his royal rug cutter are typical "dissection" problems in geometry. Given a polygon, we want to cut it into smaller pieces which can be reassembled to form a polygon of another given shape. For example, the picture below shows how a triangle can be cut into three pieces which can be reassembled to form a rectangle of the same area:

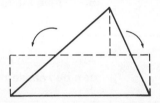

Suppose we are given two parallelograms with the same base *b* and altitude *h*:

Let the students make models; this is not easy to visualize.

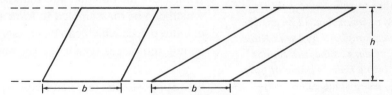

Place the parallelograms base on base and consider the pattern formed by repeating the parallelograms as below:

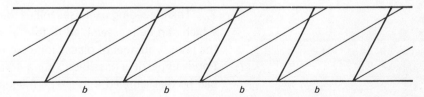

Note that the pattern shows us how to cut one parallelogram into pieces which can be reassembled to form the other.

4. Explain how any rectangle can be cut into pieces which can be reassembled to form a square. [*Hint:* Consider an $a \times b$ rectangle and an $s \times s$ square of the same area, so $s = \sqrt{ab}$. The rectangle can be cut into pieces which can be reassembled to form a parallelogram with base b, altitude a, and two sides of length s (why?).

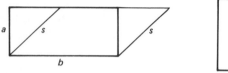

Now explain how the parallelogram can be cut into pieces which can be reassembled to form an $s \times s$ square.]

5. Prove the following *Fundamental Dissection Theorem:*

Any polygon can be cut into a number of smaller polygons which can be reassembled to form a square.

Surprisingly, this theorem does not generalize to solids in space. In 1900 Max Dehn showed, in answer to a question posed by David Hilbert, that a regular tetrahedron cannot be cut into a finite number of polyhedral pieces which can be reassembled to form a cube.

[*Hint:* Cut the polygon into triangles. Each triangle can be cut into pieces which can be reassembled to form a rectangle, and each rectangle can be cut into pieces which can be reassembled into a square. Can these squares be cut into pieces which can be reassembled to form a single square?]

6. Suppose we are given two polygons having the same area. Using the Fundamental Dissection Theorem, explain why it is possible to cut either polygon into pieces which can be reassembled to form the other.

7. The preceding exercises tell us that any polygon can be cut into pieces which can be reassembled to form a square. It is interesting to try to find the fewest number of pieces required in order to do this in particular cases. For example, it is possible to cut a 4×9 rectangle into three pieces which can be reassembled into a 6×6 square as follows:

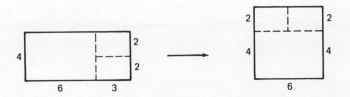

Show how a 4 × 9 rectangle can be cut into *two* pieces, congruent to each other, which can be reassembled to form a square.

Exercise 8: Most students enjoy working on problems of this type.

8. Show how to cut the Greek Cross below into *four congruent* pieces which can be reassembled to form a square.

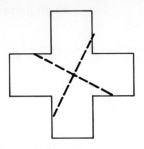

9. Read about dissection problems in Chapter 4 of *The Unexpected Hanging and Other Mathematical Diversions* by Martin Gardner (New York: Simon and Schuster, 1969).

Some Challenging Problems Using Area

The easiest way to do Exercise 10 is to imagine the "nice" case; that is, when P is a midpoint. There are other ways to prove it, all more difficult.

10. If P is any point on $\overline{AB}$ other than A or B, then the area of the quadrilateral $DCEP$ equals the sum of the areas of $\triangle ADP$ and $\triangle PEB$. Prove it! [*Hint:* Let P move along $\overline{AB}$.]

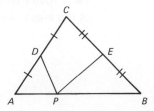

11. $\triangle ABC$ (p. 262) is an equilateral triangle, P is any point inside, and x, y, and z are the lengths of the line segments drawn from P perpendicular to the sides. Prove that $x + y + z$ is always equal to the altitude of $\triangle ABC$.

261 Projects

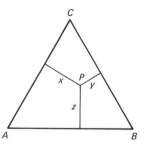

[*Hint:* Draw $\overline{PA}$, $\overline{PB}$, and $\overline{PC}$. Use areas of triangles.]

12. (*a*) Using only straightedge and compass, draw 2 straight lines through point *P* which cut the square below into 3 pieces of equal area.

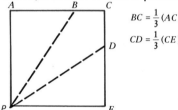

$BC = \frac{1}{3}(AC)$

$CD = \frac{1}{3}(CE)$

(*b*) Discuss your solution with your classmates. Can you prove it is correct?

13. True or false: The medians of any triangle cut it into 6 smaller triangles all having the same area? Prove that your answer is correct. *true*

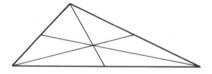

14. Prove that the product of the areas of I and III is equal to the product of the areas of II and IV.

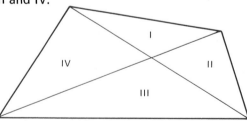

[*Hint:* Triangles I and II have an altitude in common, and triangles III and IV have an altitude in common.]

262   6: *Area and Volume*

Volumes of Some Interesting Solids

15. (*a*) Start with a cube, each of whose edges has length 1, and cut off a corner with a cut which passes through the midpoints of three adjacent edges:

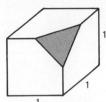

volume of small piece =
$\frac{1}{3}$(area of base)(height) =
$\frac{1}{3}\left(\frac{1}{2} \cdot \frac{1}{2} \cdot \frac{1}{2}\right) \cdot \frac{1}{2}$
= $\frac{1}{48}$ cubic units

What are the volumes of the two pieces obtained this way? [*Hint:* Use a 45°–45°–90° triangle for the base of the cut off pyramid.] $\frac{1}{48}, \frac{47}{48}$

(*b*) If we cut off all eight corners of the cube as above, we obtain a polyhedron known as the *cuboctahedron.*

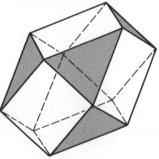

What is the volume of this cuboctahedron? (The original cube had edge length 1.)   volume = $1 - 8\left(\frac{1}{48}\right) = \frac{5}{6}$ cubic units

16. Recall that a *regular tetrahedron,* one of the five regular polyhedra studied in the Projects section of Chapter 4, has four faces, each face an equilateral triangle.

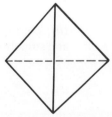

(*a*) If we start with a cube of edge length *a* and join four vertices as indicated below, prove that the solid so formed (light dotted lines) will be a regular tetrahedron.

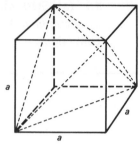

(b) Prove that the volume of the tetrahedron is $\frac{1}{3}$ the volume of the cube.

(c) If a regular tetrahedron has edges of length d, what is its volume, in terms of d? [*Hint:* In the tetrahedron in part (a), $d = a\sqrt{2}$.]

17. Make a model showing how a right prism with triangular base can be cut into three pyramids of equal volume.

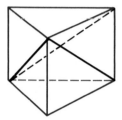

Squaring the Circle with Straightedge and Compass

18. (a) If the line joining A and E is parallel to the line through B and D, what can you say about the areas of $\triangle ECB$ and the quadrilateral $ABCD$?
area $(\triangle ADB)$ = area $(\triangle BED)$
Add these to $\triangle DBC$.

Allow students to demonstrate Exercise 18 using freehand sketches, not actual constructions.

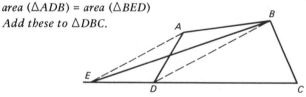

(b) With a *sketch*, show how one could construct, using only compass and straightedge, a triangle having the same area as the quadrilateral below.

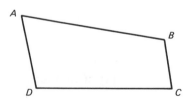

[*Hint:* First draw the diagonal $\overline{BD}$. Next construct the line through A parallel to $\overline{BD}$, and let E be the point where this line intersects the extension of $\overline{CD}$. What does part (*a*) tell us?]

(*c*) With a freehand sketch, show how one can construct, using only straightedge and compass, a 4-gon which has the same area as the 5-gon below.

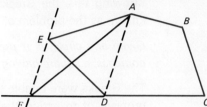

[*Hint:* Construct the line through E parallel to $\overline{AD}$, and let F be the point where it intersects the extension of $\overline{CD}$. Why is the area of quadrilateral $ABCF$ equal to the area of the 5-gon we started with?]   area $(\triangle ADE)$ = area $(\triangle ADF)$, add these to quad. $ABCD$

(*d*) With a freehand sketch, show how one can construct, using only straightedge and compass, a triangle having the same area as the 5-gon in part (*c*). [*Hint:* Proceed in two steps. First construct a 4-gon of the same area, using the method in part (*c*). Then construct a triangle with the same area as the 4-gon.]

(*e*) Given any *n*-gon, explain how you could construct, using only compass and straightedge, an $(n - 1)$-gon with the same area (assume all interior angles are less than 180°). [*Hint:* Use the idea in part (*c*).]

(*f*) Given any *n*-gon (with all interior angles less than 180°), explain how you could construct, using only compass and straightedge, a *triangle* of the same area. [*Hint:* Proceed in successive steps, using part (*e*).]

(*g*) Given any triangle, why is it possible to construct a *rectangle* of the same area, using only straightedge and compass? The picture below gives a hint.

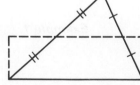

(*h*) In Chapter 10 it will be seen that given any rectangle, it is possible to construct with straightedge and compass a *square* of the same area.

Assuming this for now, explain why given any polygon (with interior angles less than 180°) it is possible to construct with straightedge and compass a *square* of the same area. [*Hint:* Use parts (f) and (g).]

19. The preceding exercise shows that given a polygon, it is possible to construct a square of the same area, using only straightedge and compass. Knowing this, the Greeks asked the following question, which became known as the problem of squaring the circle:

Given any circle, is it possible to construct, using only straightedge and compass, a square having the same area as the circle?

The Greeks were unable to answer this question. Just as with the ancient problem of trisecting the angle, mathematicians tried unsuccessfully for centuries to solve this problem. Then Ferdinand Lindemann, in 1882, proved that no such construction is possible! In other words, given any circle, there is no way to construct a square having the same area, using only straightedge and compass.

Read about this famous problem on page 140 of *What is Mathematics?* by Courant and Robbins (New York: Oxford University Press, 1941), or in Chapter 5 of *Famous Problems of Mathematics* by H. Tietze (New York: Graylock Press, 1965).

Algebra Review

Simplifying Radicals

This Algebra Review prepares for the next one, which deals with the quadratic formula.

We can write $\sqrt{8} \cdot \sqrt{2}$ as 4, and $\dfrac{\sqrt{27}}{\sqrt{3}}$ as 3 because of the following laws:

If *a* and *b* are positive,

$$\sqrt{a}\sqrt{b} = \sqrt{ab} \qquad \text{product law}$$

$$\frac{\sqrt{a}}{\sqrt{b}} = \sqrt{\frac{a}{b}} \qquad \text{quotient law}$$

Example 1.

$$\sqrt{8}\sqrt{2} = \sqrt{16} = 4$$

$$\frac{\sqrt{27}}{\sqrt{3}} = \sqrt{\frac{27}{3}} = \sqrt{9} = 3$$

1. Assuming all letters represent positive numbers, simplify each expression.

   (a) $\sqrt{8} + \sqrt{50}$ (b) $6\sqrt{x} + \sqrt{16x} - 10\sqrt{x}$

   (c) $3\sqrt{45} + 2\sqrt{80}$ (d) $3\sqrt{108ab} + 2\sqrt{75ab}$

   (e) $\sqrt{28} - 2\sqrt{112} + 3\sqrt{63}$ (f) $3\sqrt{2a^3} + \sqrt{8a^3} - \sqrt{32a^3}$

   (g) $(\sqrt{11} - \sqrt{2})(\sqrt{11} + \sqrt{2})$ (h) $(5\sqrt{3} + \sqrt{6})(5\sqrt{2} + \sqrt{2})$

2. Multiply, then simplify.

   (a) $(1 + \sqrt{2})(1 - \sqrt{2})$ (b) $(\sqrt{2} + \sqrt{3})(\sqrt{8} - \sqrt{27})$

   (c) $\sqrt{6}(\sqrt{8} - \sqrt{12})$ (d) $(\sqrt{x} - \sqrt{y})(\sqrt{x} + \sqrt{y})$

Rationalizing the Denominator

We can use the quotient law to save work when we approximate square roots decimally:

Example 2. Approximate $\dfrac{1}{\sqrt{2}}$ to three decimal places, using the square root table on page 549.

$$\frac{1}{\sqrt{2}} \doteq \frac{1}{1.414}$$

$$= 0.707 \text{ (after some tedious division)}$$

There is an easier way, using the quotient law:

$$\frac{1}{\sqrt{2}} \cdot \frac{\sqrt{2}}{\sqrt{2}} = \frac{\sqrt{2}}{2}$$

$$\doteq \frac{1.414}{2}$$

$$= 0.707$$

The "easy way" in Example 2 demonstrates the procedure called *rationalizing the denominator*, which also works for more complicated expressions, as in the following example.

Example 3. Compute to three decimal places:

$$\frac{1}{1-\sqrt{2}} = \frac{1}{1-\sqrt{2}} \cdot \frac{1+\sqrt{2}}{1+\sqrt{2}}$$

$$= \frac{1+\sqrt{2}}{1-2}$$

$$= -1 - \sqrt{2}$$

$$\doteq -2.414$$

3. Use the values given in the table to estimate to three decimal places.

(a) $\dfrac{1}{\sqrt{3}-1}$ (b) $\dfrac{1}{\sqrt{5}+\sqrt{2}}$ (c) $\dfrac{1}{\sqrt{3}}$

$\left[\text{Hint: (a)}\ \dfrac{1}{\sqrt{3}-1} = \dfrac{1}{\sqrt{3}-1} \cdot \dfrac{\sqrt{3}+1}{\sqrt{3}+1} = \dfrac{\sqrt{3}+1}{2}\right]$

4. Write as an expression with the denominator free of square roots (assume $x > 0$).

(a) $\dfrac{1}{\sqrt{x}}$ (b) $\dfrac{1}{\sqrt{x}+1}$ (c) $\dfrac{1}{\sqrt{x}-\sqrt{y}}$ (d) $\sqrt{\dfrac{1}{x}}$

(e) $\sqrt{1\dfrac{1}{2}}$ (f) $\sqrt{1+\dfrac{1}{x}}$ (g) $\sqrt{1-\dfrac{1}{x}}$ (h) $\sqrt{x-\dfrac{1}{x}}$

(i) $\sqrt{2x^2 + \dfrac{x^2}{4}}$

5. (a) Is $\sqrt{x} + \sqrt{y} = \sqrt{x+y}$? Discuss.

(b) Let $x = 16$ and $y = 9$ in (a); was your explanation valid?

6. Approximate to two decimal places.

(a) $\sqrt{125}$ (b) $\sqrt{300}$ (c) $\sqrt{592}$ (d) $\sqrt{549}$

7. Write as an expression with the denominator free of square roots (assume $a > 0$):

$$\sqrt{\frac{b^2}{4a^2} - \frac{c}{a}}$$

Example 4. Since
$$x^2 - 2 = (x - \sqrt{2})(x + \sqrt{2}),$$
the factoring method may be used to solve the equation $x^2 - 2 = 0$:
$$(x - \sqrt{2})(x + \sqrt{2}) = 0$$
Thus
$$x - \sqrt{2} = 0 \quad \text{or} \quad x + \sqrt{2} = 0$$
$$x = \sqrt{2} \qquad\qquad x = -\sqrt{2}.$$
The solutions are $\sqrt{2}$ and $-\sqrt{2}$.

8. Solve for x.

(a) $x^2 - 5 = 0$ (b) $x^2 - 19 = 0$ (c) $x^2 = 7$

(d) $2x^2 - 11 = 0$ (e) $5x^2 = 17$ (f) $x^2 = \frac{3}{11}$

The following exercises concerning "completing the square" will be useful in the next Algebra Review for finding the quadratic formula.

9. Fill in the blanks to make each statement true.

(a) $(x + \underline{})^2 = x^2 + 2x + \underline{}$

(b) $(x - \underline{})^2 = x^2 - 6x + \underline{}$

(c) $(x + \underline{})^2 = x^2 + 10x + \underline{}$

(d) $(x - \underline{})^2 = x^2 - 16x + \underline{}$

10. Fill in the blanks.

(a) $(x + \frac{1}{2})^2 = \underline{} + \underline{} + \underline{}$

(b) $(x - \frac{5}{2})^2 = \underline{} + \underline{} + \underline{}$

(c) $(x - \frac{1}{3})^2 = \underline{} + \underline{} + \underline{}$

11. Fill in the blanks.

(a) $(x - \underline{})^2 = x^2 - x + \underline{}$

(b) $(x + \underline{})^2 = x^2 + 3x + \underline{}$

(c) $(x + \underline{})^2 = x^2 + 5x + \underline{}$

(d) $(x - \underline{})^2 = x^2 - 7x + \underline{}$

12. Discuss with a classmate how you solved Exercises 9 and 11.

13. Fill in the blanks:

$$(x + \underline{})^2 = x^2 + \frac{b}{a}x + \underline{}$$

Answers
1. (a) $7\sqrt{2}$ (b) 0 (c) $17\sqrt{5}$ (d) $28\sqrt{3ab}$ (e) $3\sqrt{7}$
(f) $a\sqrt{2a}$ (g) 9 (h) $30\sqrt{6} + 12\sqrt{3}$

2. (a) -1 (b) $-5 - \sqrt{6}$ (c) $4\sqrt{3} - 6\sqrt{2}$ (d) $x - y$

3. (a) 1.366 (b) 0.274 (c) 0.577

4. (a) $\frac{\sqrt{x}}{x}$ (b) $\frac{\sqrt{x} - 1}{x - 1}$ (c) $\frac{\sqrt{x} + \sqrt{y}}{x - y}$ (d) $\frac{\sqrt{x}}{x}$ (e) $\frac{\sqrt{6}}{2}$

(f) $\frac{\sqrt{x(x+1)}}{x}$ (g) $\frac{\sqrt{x(x-1)}}{x}$ (h) $\frac{\sqrt{x(x^2-1)}}{x}$ (i) $\frac{3x}{2}$

5. (a) no, only when x or y is zero

6. (a) 11.18 (b) 17.32 (c) 24.33 (d) 23.43

7. $\frac{\sqrt{b^2 - 4ac}}{2a}$

8. (a) $x = \pm\sqrt{5}$ (b) $x = \pm\sqrt{19}$ (c) $x = \pm\sqrt{7}$

(d) $x = \frac{\pm\sqrt{22}}{2}$ (e) $x = \frac{\pm\sqrt{85}}{5}$ (f) $x = \frac{\pm\sqrt{33}}{11}$

9. (a) 1, 1 (b) 3, 9 (c) 5, 25 (d) 8, 64

10. (a) $x^2, x, \frac{1}{4}$ (b) $x^2, -5x, \frac{25}{4}$ (c) $x^2, -\frac{2}{3}x, \frac{1}{9}$

11. (a) $\frac{1}{2}, \frac{1}{4}$ (b) $\frac{3}{2}, \frac{9}{4}$ (c) $\frac{5}{2}, \frac{25}{4}$ (d) $\frac{7}{2}, \frac{49}{4}$

12. student discussion

13. $\frac{b}{2a}, \frac{b^2}{4a^2}$

Chapter 7 The Pythagorean Theorem

This chapter focuses on the Pythagorean Theorem, which is the most important result in geometry, for it provides a method of measuring the distance between two points. While the Babylonians knew of it in 2000 B.C., there is no evidence for the legend that Egyptian "rope-surveyors" knew it. The Greeks were the first to prove it.

The proof (Exercise 4) becomes very easy if the students cut the figures out of paper and assemble them like a jigsaw puzzle.

Note that again a chapter concludes with work in 3 dimensions. Students should see physical models in 3-space, beginning with Exercise 13. Preferably, they (not the teacher) should bring in boxes and strings; perhaps each group of four students might provide a model.

Central

A brief class discussion of this problem may be called for. At least, the distance can be measured on the board.

A bird leaves its nest and flies two miles north and then one mile east. If it returns to its nest in a straight line, how far does it fly going back?

1. (a) Make a scale drawing of the bird's trip, using the scale 10 cm. = 1 mile, and measure the length of the return flight as well as you can. *about 22.4 cm.*
(b) Compare your estimate with those of your classmates.

In the preceding exercise, even with the best drawing, you could only estimate the length to one or two decimal places. The next exercise experiments with more right triangles and prepares for the Pythagorean Theorem, which enables us to find any side of a right triangle, if we know the other two sides. The method is purely algebraic; there is no need even to draw the triangle.

2. (a) Make a careful drawing of a right triangle, and measure the lengths of the sides. (Make your triangle big enough to take up a full sheet of paper.) If c is the length of the hypotenuse, and a and b are the lengths of the other two sides, compute $a^2 + b^2$ and c^2. What do you notice? *They are equal.*

(b) Test your observation by measuring some other drawings of right triangles.

(c) How could you compute the length of the hypotenuse of a right triangle if you know the lengths of the other two sides?

The Pythagorean Theorem gives a relationship between the length of the hypotenuse of a right triangle and the lengths of its legs. This theorem, perhaps the most useful in geometry, is named in honor of the Greek mathematician and philosopher Pythagoras who was apparently the first to give a proof. Although Pythagoras (who lived about 500 B.C.) was the first to explain *why* this theorem is true, it was known and used by the Babylonians more than 1000 years earlier.

THE PYTHAGOREAN THEOREM

If a right triangle has legs of lengths *a* and *b* and hypotenuse of length *c*, then

$$c^2 = a^2 + b^2.$$

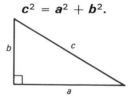

The next exercise will be needed in the proof of the Pythagorean Theorem.

3. In the picture below, the large quadrilateral is a square.

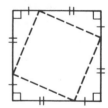

(a) Explain why the four triangles are congruent.   SAS

(b) Explain why the smaller quadrilateral (dotted lines) must be a square. (In other words, explain why all its angles are equal and all its sides have the same length.)

274   7: *The Pythagorean Theorem*

A Proof of the Pythagorean Theorem

Encourage students to make paper models of pictures (i) and (ii) in Exercise 4. They should cut them up and then put them back together, like a jigsaw puzzle.

4. Consider a right triangle with legs of lengths a, b, and hypotenuse c. The picture below shows a square of side $a + b$ cut into

(*i*) four copies of such a triangle and a square of side c,

(*ii*) four copies of such a triangle and squares of sides a and b respectively.

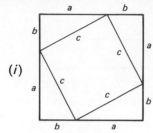
(i)

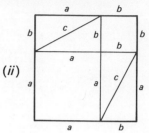
(ii)

(*a*) How do we know that the smaller quadrilateral in (*i*) is actually a square? *Exercise 3*

The smaller square in (i) is obtained by removing four congruent triangles from the large square. The two smallest squares in (ii) are obtained by removing the four triangles in a different way.

(*b*) Use the two preceding pictures to prove the Pythagorean Theorem. Discuss this with your classmates. [*Hint:* The area of the smaller square in (*i*) equals the sum of the areas of the two smallest squares in (*ii*). Why?]

(*c*) In order to help you understand the preceding proof, cut out 4 copies of a right triangle with legs a and b and hypotenuse c, from cardboard or heavy paper. Cut out a square of side c, a square of side a, and a square of side b. Arrange the 4 triangles and square of side c to form a square, as in (*i*). Next, arrange the 4 triangles and the two squares of side a and b respectively to form a square, as in (*ii*). Demonstrate this to a classmate, explaining why this shows that $c^2 = a^2 + b^2$.

Additional proofs of the Pythagorean Theorem given in the Projects section should be enjoyable and instructive.

Euclid's proof involving squares is done in Review; do not teach it here. Exercise 4 is a more gentle approach. Resist the temptation to teach the traditional proof here, as it is not needed.

5. In the picture below, the triangle is a right triangle and the quadrilaterals are squares. What does the Pythagorean Theorem tell us about the relation between the areas of these squares? *I + II = III*

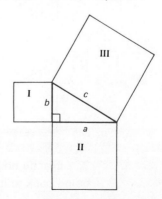

275 *Central*

Using the Pythagorean Theorem

6. Find x, using the Pythagorean Theorem. In cases (a), (b), and (d) check your answer by drawing an accurate scale diagram and measuring x.

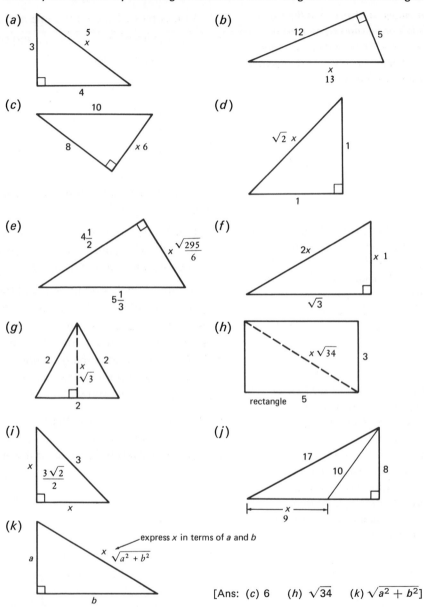

[Ans: (c) 6   (h) $\sqrt{34}$ (k) $\sqrt{a^2 + b^2}$]

7. Exactly how far does the bird at the beginning of this chapter fly in going back to its nest? $\sqrt{5}$ miles

8. Find the area of each polygon.

(a)   (b)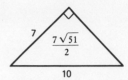

[*Hint:* Exercise 6(g)]

(c)   (d)

[*Hint:* Exercise 6(i)]

[Ans: (a) $\sqrt{3}$ (d) $\frac{9}{2}$]

More About 30°–60° Right Triangles

9. Recall that in any 30°–60° right triangle, the side opposite the 30° angle has length half the hypotenuse.

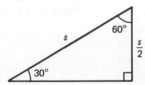

Summarize the relations between the hypotenuse, the short leg, and the long leg of a 30°-60°-90° triangle. Encourage students to memorize them now. They are used repeatedly.

(a) Using the Pythagorean Theorem, show that if the hypotenuse has length s, then the side opposite the 60° angle has length $\dfrac{\sqrt{3}}{2} s$.

(b) Fill in the blanks.

Emphasize these relations:

In any 30°–60° right triangle,

(i) the short leg is $\underline{\dfrac{1}{2}}$ times the hypotenuse,

(ii) the long leg is $\underline{\sqrt{3}}$ times the short leg.

This nice 30°-60°-90° triangle is a memory aid:

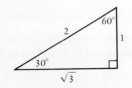   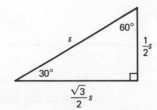

277 Central

10. Find x.

(a)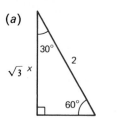

(b)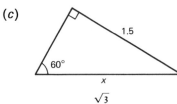

(c)

(d)

[Ans: (b) $3\sqrt{3}$ (d) $4\dfrac{\sqrt{3}}{3}$]

Area of an Equilateral Triangle

Exercise 11(a): You may wish to suggest that the class memorize the formula for the altitude of an equilateral triangle, since it is used so frequently; similarly for the area of an equilateral triangle.

11. (a) The length of each side of a certain equilateral triangle is *s*. What is the length of an altitude (in terms of *s*)?

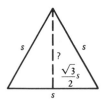

(b) Prove the following.

The area of an equilateral triangle of side *s* is $\dfrac{\sqrt{3}}{4} s^2$.

$$\text{area} = \frac{s \cdot \frac{\sqrt{3}}{2} s}{2} = \frac{\sqrt{3}}{4} s^2$$

---

12. Use the formula derived in the preceding exercise to find the area of each of the following triangles.

(a)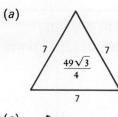

(b)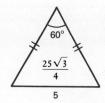

(c)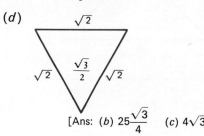

(d)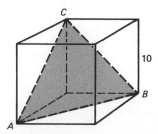

[Ans: (b) $25\dfrac{\sqrt{3}}{4}$ (c) $4\sqrt{3}$]

13. In the picture below, each edge of the cube has length 10.

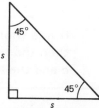

(a) Prove that $\triangle ABC$ is equilateral. *The sides of $\triangle ABC$ are the diagonals of faces of the cube; each has length $10\sqrt{2}$.*

(b) Find the area of $\triangle ABC$. Compare your answer with those of your classmates. $50\sqrt{3}$

45°–45° Right Triangles

Exercise 14: The relation between the sides and the hypotenuse of a 45°-45°-90° triangle should be memorized, since it is used so often.

14. When we cut a square into two pieces along a diagonal, we obtain two isosceles right triangles. In such a right triangle the acute angles are each 45°.

Fill in the blank in the following statement.

The following nice 45°-45°-90° triangle is a good memory aid:

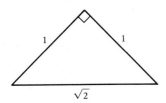

In any 45°–45° right triangle, the hypotenuse is __$\sqrt{2}$__ times either leg.

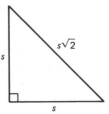

15. Find x.

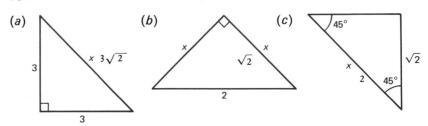

The Converse of the Pythagorean Theorem

Consider a triangle with sides of lengths a, b, and c. The Pythagorean Theorem states:

If the angle opposite the side of length c is a right angle, then $a^2 + b^2 = c^2$.

The *converse* of the last statement reads:

If $a^2 + b^2 = c^2$, then the angle opposite the side of length c is a right angle.

Let us write this converse of the Pythagorean Theorem in its complete form:

CONVERSE OF THE PYTHAGOREAN THEOREM If in a triangle of side lengths a, b, and c we have $a^2 + b^2 = c^2$, then the triangle is a right triangle. The hypotenuse has length c and the legs have lengths a and b.

Is it true? Our previous work on converses should warn us not to jump to conclusions. However, the following exercise shows that it is indeed true.

Although by this time students know that just because a theorem is true the converse is not necessarily true, they are often shocked to find out that a proof of the converse of the Pythagorean Theorem is necessary. Further, the proof seems strange. Perhaps this is because the proof is based on the Pythagorean Theorem itself.

16. Suppose we have $\triangle ABC$ with sides of length a, b, and c such that $a^2 + b^2 = c^2$. The picture below is distorted, because we will prove that $\angle C$ is a right angle.

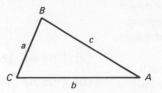

Now consider a *right* triangle $\triangle A'B'C'$ with legs of length a and b (same a and b as above!). Let x be the length of the hypotenuse of this right triangle.

The Pythagorean Theorem applied to $\triangle A'B'C'$ tells us that

$$x^2 = a^2 + b^2.$$

(a) Why is $x^2 = c^2$? *By assumption, $c^2 = a^2 + b^2$.*

(b) What can you say about x and c? *equal, since x and c are positive*

(c) Why is $\triangle ABC \cong \triangle A'B'C'$? *SSS*

(d) What kind of triangle must $\triangle ABC$ be? *right triangle*

(e) Explain to a classmate why the converse of the Pythagorean Theorem is true.

Exercise 17: You might ask, "Is a triangle a right triangle if the sides are a^2, b^2, and $a^2 + b^2$?"

Students may enjoy studying Pythagorean triples (Projects section).

17. Is a triangle a right triangle if it has sides with lengths

(a) 3, 4, 5? *yes* (b) 2, 3, 5? *no* (c) $\sqrt{2}, \sqrt{3}, \sqrt{5}$? *yes*

(d) $\sqrt{2}, \sqrt{2}, 2$? *yes* (e) 16, 34, 30? *yes* (f) $2, \sqrt{3}, 1$? *yes*

(g) $x^2 - 1, 2x, x^2 + 1$, if $x > 1$? *yes*

(h) $x, y, \sqrt{x^2 + y^2}$, if $x > 0$ and $y > 0$? *yes*

18. A triangle has sides of lengths 8, 17, and 15. What is its area? *60*

The Longest Diagonal of a Box

19. Suppose we are given a box 3 feet wide, 4 feet long, and 2 feet high.

You may wish to present Exercise 19 to the entire class, making sure they see the right triangles involved. The classroom itself makes a good model for computing the longest diagonal. Point out to the students that a box has four longest diagonals.

A cardboard box (Exercise 20) also makes a good model, for lines can be drawn to show the triangles and string can be used to show distance.

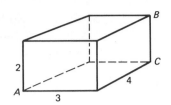

(a) Find AC. [*Hint*: $\overline{AC}$ is the hypotenuse of a certain right triangle.] 5

(b) Find AB, the length of a longest diagonal of the box. [*Hint*: △ACB is a right triangle.] $\sqrt{29}$

(c) Find the length of a longest diagonal of a box which is 2 feet wide, 6 feet long, and 5 feet high. $\sqrt{65}$

20. (a) Bring to class a small cardboard box, and measure the length of the longest diagonal using a piece of string or yarn.

(b) Show on the box the various right triangles you needed when you solved Exercise 19.

(c) Estimate the length of the longest diagonal by measuring the edges of the box and using right triangles. Compare this result with your measurement in part (a).

Exercise 21 provides another good opportunity for doing some 3-dimensional geometry.

21. Explain why the following statement is true.

If *d* is the length of the longest diagonal of a rectangular box of length *a*, width *b*, and height *c*, then

$$d^2 = a^2 + b^2 + c^2.$$

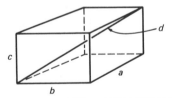

Students should not have too much trouble with square root if they did the Algebra Review in Chapter 6.

22. What is the greatest distance between any two corners of a rectangular box if the dimensions are

(a) 2, 3, and 4?   (b) 4, 9, and 5?   (c) *l*, *w*, and *h*?
$\sqrt{29}$ $\sqrt{122}$ $\sqrt{l^2 + w^2 + h^2}$

23. A telephone pole 50 ft. high stands on one corner of a rectangular

field which is 80 ft. wide and 120 ft. long.

(a) What is the distance from the foot of the pole to the diagonally opposite corner? $40\sqrt{13}$

(b) What is the distance from the top of the pole to the diagonally opposite corner? $10\sqrt{233}$

24. The busy bird leaves its nest, flies 8 miles east, then 4 miles north, and finally 1 mile straight up. It returns in a straight line to its nest. How long is the return journey? *9 miles*

Important Ideas in This Chapter

Theorems

The Pythagorean Theorem   If a right triangle has legs of lengths a and b and hypotenuse of length c, then

$$a^2 + b^2 = c^2.$$

The converse of the Pythagorean Theorem   If the side lengths a, b, and c of a triangle satisfy $a^2 + b^2 = c^2$, then the triangle is a right triangle, with hypotenuse of length c and legs of lengths a and b.

In any 30°–60° right triangle of hypotenuse length s, the side opposite the 30° angle has length $\dfrac{s}{2}$ and the side opposite the 60° angle has length $\dfrac{\sqrt{3}}{2} s$.

memory aid:

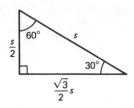

In any isosceles right triangle with legs of length a, the hypotenuse has length $\sqrt{2}\, a$.

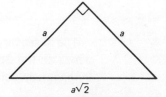

The area of an equilateral triangle of side length s is
$$\frac{\sqrt{3}}{4} s^2.$$

If d is the length of the longest diagonal of a box of length a, width b, and height c, then
$$d^2 = a^2 + b^2 + c^2.$$

Words Pythagoras, The Pythagorean Theorem

Summary We proved the Pythagorean Theorem and its converse, and we used it to compute unknown distances. In particular, we obtained some properties of 30°–60° right triangles. By using appropriate right triangles, we computed the length of the longest diagonal of a box with given edges.

Historical Note Of the many religious and mystical cults that flourished throughout Italy and Greece in 500 B.C., one was particularly interesting because of its use of mathematics. This was the Order of the Pythagoreans, a brotherhood founded by Pythagoras of Samos. The Pythagoreans devoted themselves mainly to philosophical questions, but mathematics played an important role in all their studies. The Golden Ratio (studied in the Projects section of Chapter 10) figured prominently in their geometrical investigations. In the regular five-pointed star, which was the Pythagorean symbol of health, the Golden Ratio appears as the ratio of lengths of various line segments.

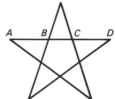

$$\frac{AD}{BD} = \frac{BD}{CD} = \frac{CD}{BC} = \text{the Golden Ratio}$$

The discovery of the regular dodecahedron, whose construction depends on the Golden Ratio, has been attributed to the Pythagoreans. The Pythagorean belief that all nature is based ultimately on the whole numbers was reinforced by their discovery that the harmonic properties of a plucked string depend in a simple way on ratios of whole numbers. Hence the Pythagoreans were shocked when they discovered the existence of irrational numbers, such as $\sqrt{2}$, which cannot be expressed as the ratio of two whole numbers. Legend has it that they tried to keep this a secret because it contradicted their doctrine that all things are expressible in terms of whole numbers. As the story goes, one of their members was executed by drowning for making the discovery public.

Central Self Quiz

Exercise 1 is excellent for a student to check his understanding of the 30°-60°-90° triangle.

Similarly, Exercise 3 helps for the 45°-45°-90° triangle.

1. Fill in the table, referring to the triangle given.

a	b	c
		2
2		
	2	
$\frac{1}{2}$		
	$\frac{1}{2}$	
		$\frac{1}{2}$

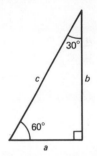

2. Express in words the relation between the sides of a 30°–60°–90° triangle.

3. Fill in the table, referring to the triangle given.

a	b	c
1		
	2	
		2
$\sqrt{3}$		
		1

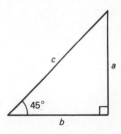

4. Express in words the relation between the sides of a 45°–45°–90° triangle.

5. Find the length of a diagonal in the following figures.

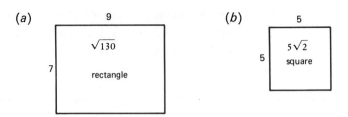

(a) 9, √130, 7, rectangle

(b) 5, 5√2, 5, square

6. Find x and y.

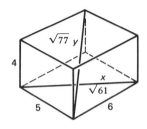

√77 y 4 x √61 5 6

Answers

1.

a	b	c
1	√3	2
2	2√3	4
$\frac{2\sqrt{3}}{3}$	2	$\frac{4\sqrt{3}}{3}$
$\frac{1}{2}$	$\frac{\sqrt{3}}{2}$	1
$\frac{\sqrt{3}}{6}$	$\frac{1}{2}$	$\frac{\sqrt{3}}{3}$
$\frac{1}{4}$	$\frac{\sqrt{3}}{4}$	$\frac{1}{2}$

2. The short leg (the side opposite the 30° angle) is one-half the hypotenuse. The long leg (the side opposite the 60° angle) is √3 times the short leg.

3.

a	b	c
1	1	√2
2	2	2√2
√2	√2	2
√3	√3	√6
$\frac{1}{\sqrt{2}}$	$\frac{1}{\sqrt{2}}$	1

286 7: The Pythagorean Theorem

4. The two legs are equal. The hypotenuse is $\sqrt{2}$ times the length of a leg.

5. (a) $\sqrt{49 + 81} = \sqrt{130}$ (b) $5\sqrt{2}$

6. $x^2 = 5^2 + 6^2$ $\quad y^2 = 4^2 + x^2$
   $ = 61$ $\quad = 16 + 61$
   $x = \sqrt{61}$ $\quad y = \sqrt{77}$

Review

1. Find x.

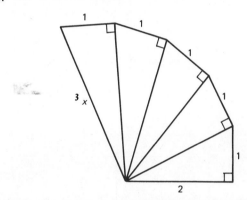

[*Hint:* Compute the length of the hypotenuse of the bottom triangle. Then compute the length of the hypotenuse of the next triangle.]

2. A rope hangs from the top of a flagpole. The rope is 2 ft. longer than the pole. When pulled out taut, the rope reaches a point on the ground 18 ft. from the foot of the pole. How high is the pole?

It is fun to make much of guessing in Exercise 3. The answer is a surprise. You might ask how railroads solve this problem. (There is space between consecutive rails; thus the "clickity-clack" is heard when a train goes by.)

3. A railroad track AB is constructed during the winter using single steel rails 1 mile (5280 ft.) long. In the summer each rail expands 2 ft. and buckles at the midpoint, forming an isosceles triangle ABC with altitude h.

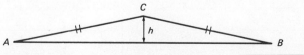

(a) Guess the height h. (b) Calculate h. $h = \sqrt{(2641)^2 - (2640)^2}$

4. (a) A baseball diamond is a square with the distance from home plate to first base 90 ft. How far is it from home plate to second base?

(b) A fielder catches a fly ball on the first base line 30 ft. beyond first base. How far must he throw to get the ball to third base?

Encourage discussion in Exercise 5.

5. Find x.

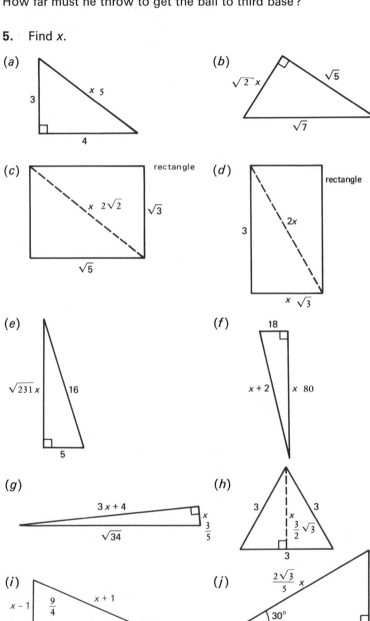

(k)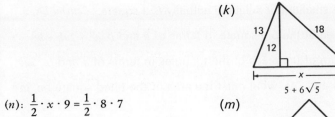
(l)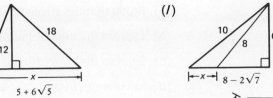

(n): $\frac{1}{2} \cdot x \cdot 9 = \frac{1}{2} \cdot 8 \cdot 7$

(m)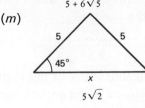

(n)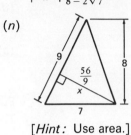

[*Hint:* Use area.]

(p): $\frac{1}{2} \cdot x \cdot 13 = \frac{1}{2} \cdot 5 \cdot 12$

(o)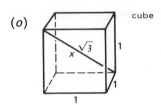

(p)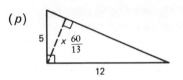

(q): *Draw the hypotenuse of the 60°-30° triangle with legs x and 5.*

(r), (s): *Use the Midpoint Theorem.*

(q)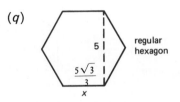

(r)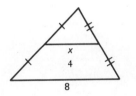

(s)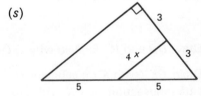

(t)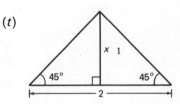

Exercise 6 reviews some algebra.

6. Here is another way to prove the Pythagorean Theorem. In the figure below we have a large square, each side of length $a + b$. The square has been cut up into four right triangles (each having legs of lengths a and b, and hypotenuse of length c) and a smaller square of side length c.

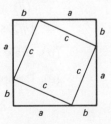

(a) Prove that the smaller quadrilateral actually *is* a square. *Central Ex. 3*

(b) Express the area of the big square in terms of a and b. $a^2 + 2ab + b^2$

(c) Express the sum of the areas of the triangles in terms of a and b. $2ab$

(d) In view of (b) and (c), what must the area of the tilted square be, in terms of a and b? $a^2 + b^2$

(e) From (d), obtain the Pythagorean Theorem! $c^2 = a^2 + b^2$

The most famous proof of the Pythagorean Theorem is that given by Euclid of Alexandria in the ancient textbook now known as the *Elements*. The next exercise demonstrates his proof.

Exercise 7 presents Euclid's proof of the Pythagorean Theorem.

7. In the figure below, $\triangle ABC$ is a right triangle with squares on each side.

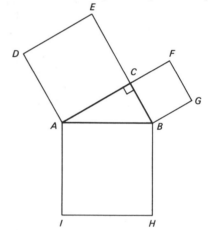

(a) Draw $\overline{DB}$ and $\overline{IC}$. Explain why $\triangle DAB \cong \triangle CAI$. SAS

(b) Draw $\overline{CK}$ (with K on segment $\overline{IH}$) perpendicular to $\overline{IH}$. Let J be the point of intersection of $\overline{CK}$ with $\overline{AB}$. What kind of polygon is $AJKI$? $JBHK$?   rectangles

(c) Compare the areas of
  (i) $\triangle DAB$ and square $ADEC$.   area $(\triangle DAB) = \frac{1}{2}$ area $(ADEC)$
  (ii) $\triangle CAI$ and rectangle $AJKI$.   area $(\triangle CAI) = \frac{1}{2}$ area $(AJKI)$
  (iii) square $ADEC$ and rectangle $AJKI$.   area $(ADEC) =$ area $(AJKI)$

(d) Draw segments $\overline{GA}$ and $\overline{CH}$. Explain why $\triangle GBA \cong \triangle CBH$. SAS

(e) Compare the areas of

(i) △GBA and square CFGB.   area (△GBA) = $\frac{1}{2}$ area (CFGB)

(ii) △CBH and rectangle JBHK.   area (△CBH) = $\frac{1}{2}$ area (JBHK)

(iii) square CFGB and rectangle JBHK.   area (CFGB) = area (JBHK)

(f) Explain why parts (a)–(e) prove the Pythagorean Theorem.
area (ABHI) = area (ADEC) + area (BCFG)

8. In the figure below we have placed an equilateral triangle on each side of a right triangle.

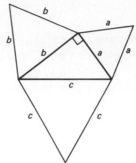

$\frac{c^2\sqrt{3}}{4} = \frac{(a^2 + b^2)\sqrt{3}}{4}$

(by the Pythagorean Theorem)

$= \frac{a^2\sqrt{3}}{4} + \frac{b^2\sqrt{3}}{4}$

Prove that the area of the triangle on the hypotenuse is equal to the sum of the areas of the triangles on the legs.

Exercise 9 is the problem of the shortest path (Chapter 1) again. This time the length of the path is actually computed, using the Pythagorean Theorem.

9. Find the length of the shortest path from A to B which touches the line ℓ.

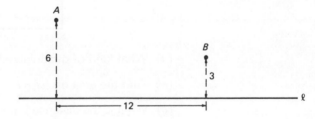

[*Hint:* Consider the reflection B* of B across ℓ. Find a right triangle whose hypotenuse has length AB.*] $AB^* = \sqrt{12^2 + 9^2} = 15$

10. A spider is resting on a very thin pane of glass, 6 cm. below the edge, when he spies a fly on the other side of the pane, 8 cm. below the edge and at a horizontal distance of 12 cm. from the spider. What is the least distance the spider has to creep in order to catch the fly, if the fly does not move? (See the picture on the next page.) $\sqrt{12^2 + 14^2} = 2\sqrt{85}$

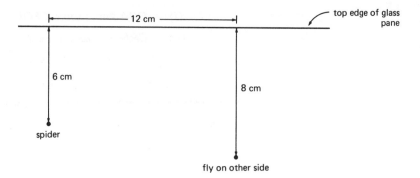

Exercise 11 gives further opportunity for the development of spatial intuition.

11. The picture below represents a solid wooden cube of edge length 4 cm. The points A, B, and C are the midpoints of the edges on which they lie.

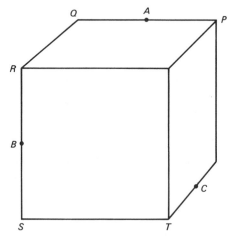

(a) What kind of quadrilateral is *PQST*? Find its area.

(b) Find the area of △*PRT*.

(c) What is the distance from A to B?   B to C?   C to A?   Find the area of △*ABC*.

(d) What is the length of the segment $\overline{PS}$?

(e) A spider sitting at P smells a fly sitting at S. What is the shortest distance the spider will have to travel to reach the fly?

(f) Do part (e) if the spider is at A and the fly is at C.

12. In each of the following cases, state the *converse* of the given statement, and state if the converse is true or false.

(a) If a triangle has a pair of equal angles, then it is isosceles.

(b) If two rectangles are congruent, then they have the same area.

(c) If a quadrilateral is a square, then it has 4 right angles.

(d) If all angles of a triangle are equal, then the triangle is equilateral.

(e) If a rectangle is a square, then its diagonals are perpendicular.

(f) If two triangles are congruent, then their corresponding angles are equal.

(g) If a right triangle has hypotenuse length c and legs of lengths a, b, then $a^2 + b^2 = c^2$.

13. Find x. Explain your answers.

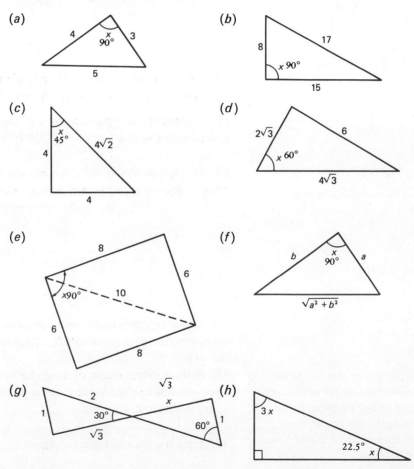

(i) 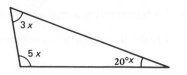   (j)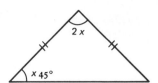

14. If the hypotenuse and one leg of a right triangle are equal in length respectively to the hypotenuse and one leg of another right triangle, then must the triangles be congruent? Explain.

Allow sufficient time for Exercise 15; it is long, but valuable for reviewing the 30°-60°-90° relation.

15. Given the figure below, solve for the remaining letters in each case.

(a) $e = 6$ (b) $d = 4$ (c) $b = 3$ (d) $a = 60$

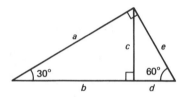

16. (a) Construct a 30°–60° right triangle using only a straightedge and compass.   *Bisect a side of an equilateral triangle.*

(b) Construct an isosceles right triangle using only a straightedge and compass.   *Construct a perpendicular to a line, mark equal lengths.*

17. (a) Given square *ABCD*, use only straightedge and compass to construct a square with twice the area.   *Construct a square with $\overline{AC}$ as side.*

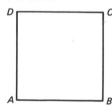

(b) Use only straightedge and compass to construct a right triangle having the same area as square *ABCD*.   *Construct an isosceles right triangle with $\overline{AC}$ as leg.*

In Exercise 18, of course, the merchant would charge for unused remnants. (If he had mastered the dissections in the Projects section of the last chapter, there would be no wasted linoleum.)

18. A floor in the shape of a regular hexagon of sidelength 10 ft. is to be covered with linoleum which costs $2 per square foot. What will it cost to cover the floor?

19. Find the area of each polygon.

Draw auxiliary lines in (c), (e), (f), and (i).

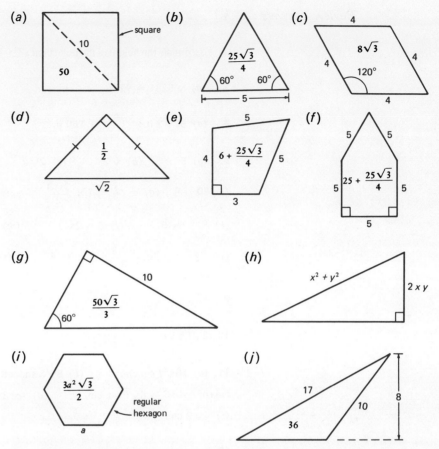

20. What is the length of the longest diagonal of a box of dimensions $\sqrt{3}$, $\sqrt{5}$, and $\sqrt{7}$?

21. What is the length of the longest diagonal of a cube of edge length a?

Exercise 22: This variation of the shortest path problem is impossible to solve without the Pythagorean Theorem (or similar triangles).

22. Find x, if $\angle 1 = \angle 2$ and $AP + PB = 50$. $\sqrt{(50)^2 - (40)^2} = 30$

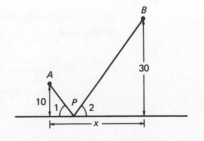

Review Answers

1. $x = 3$

2. If x equals the height of the pole, then $x^2 + (18)^2 = (x + 2)^2$ (why?), so $x = 80$ ft.

3. (b) $\sqrt{5281} \doteq 72$ ft.

4. (a) $90\sqrt{2}$ ft. (b) 150 ft.

5. (a) 5   (b) $\sqrt{2}$ (c) $2\sqrt{2}$ (d) $\sqrt{3}$ (e) $\sqrt{231}$
   (f) 80   (g) $\frac{3}{5}$ (h) $\frac{3\sqrt{3}}{2}$ (i) $\frac{9}{4}$ (j) $\frac{2\sqrt{3}}{5}$
   (k) $5 + 6\sqrt{5}$ (l) $8 - 2\sqrt{7}$ (m) $5\sqrt{2}$ (n) $\frac{56}{9}$ (o) $\sqrt{3}$
   (p) $\frac{60}{13}$ (q) $\frac{5\sqrt{3}}{3}$ (r) 4   (s) 4   (t) 1

9. 15

10. $2\sqrt{85}$ cm.

11. (a) $16\sqrt{2}$ sq. cm.   (b) $8\sqrt{3}$ sq. cm.   (c) $AB = BC = CA = 2\sqrt{6}$ cm.; area of $\triangle ABC = 6\sqrt{3}$ sq. cm.   (d) $4\sqrt{3}$ cm.   (e) $4\sqrt{5}$ cm.
   (f) $4\sqrt{2}$ cm.

12. (a) If a triangle is isosceles, then it has a pair of equal angles. *true*

    (b) If two rectangles have the same area, then they are congruent. *false*

    (c) If a quadrilateral has 4 right angles, then it is a square. *false*

    (d) If a triangle is equilateral, then all its angles are equal. *true*

    (e) If the diagonals of a rectangle are perpendicular, then it is a square. *true* (Prove it!)

    (f) If two triangles have their corresponding angles equal, then they are congruent. *false*

    (g) If a triangle has sides of lengths a, b, and c, such that $a^2 + b^2 = c^2$, then it is a right triangle. *true*

13. (a) 90° (converse of the Pythagorean Theorem)

    (c) The triangle is a right triangle, by the converse of the Pythagorean Theorem. $x = 45°$.

    (h) $x + 3x = 90°$ (i) 20°

14. Using the Pythagorean Theorem, we see that the remaining sides of the triangles must be equal. Hence the triangles are congruent by SSS.

15. (a) $d = 3$, $c = 3\sqrt{3}$, $a = 6\sqrt{3}$, $b = 9$

(b) $e = 8$, $c = 4\sqrt{3}$, $a = 8\sqrt{3}$, $b = 12$

(c) $a = 2\sqrt{3}$, $c = \sqrt{3}$, $d = 1$, $e = 2$

(d) $c = 30$, $b = 30\sqrt{3}$, $e = 20\sqrt{3}$, $d = 10\sqrt{3}$

18. $519.60

19. (a) 50   (b) $\frac{25\sqrt{3}}{4}$ (c) $8\sqrt{3}$ (d) $\frac{1}{2}$

(e) $6 + \frac{25\sqrt{3}}{4}$ (f) $25 + \frac{25\sqrt{3}}{4}$ (g) $\frac{50\sqrt{3}}{3}$ (h) $xy|x^2 - y^2|$

(i) $\frac{3\sqrt{3}}{2} a^2$ (j) 36

20. $\sqrt{15}$

21. $a\sqrt{3}$

22. 30

Review Self Test

1. Complete the table, using the picture.

a	b	c	d	e	f
		2			
	1				

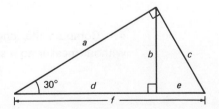

2. Complete the table on the next page, using the picture.

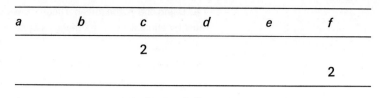

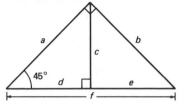

3. Find x (if possible).

(a)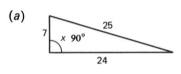

(b)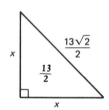

(c)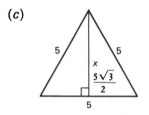

(d)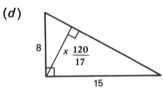

(e)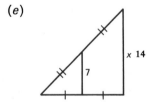

(f)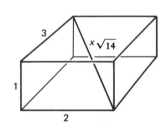

4. Compute the length of the shortest path from A to B by way of ℓ without measuring a scale drawing.

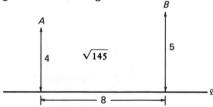

5. Find the area.

(a)
square

(b)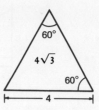

(c)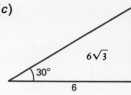

Exercise 5(e) will be difficult for some students.

(d)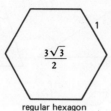
regular hexagon

(e)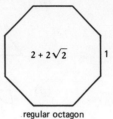
regular octagon

(f)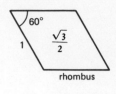
rhombus

Answers

1.

	a	b	c	d	e	f
	$2\sqrt{3}$	$\sqrt{3}$	2	3	1	4
	2	1	$\frac{2}{\sqrt{3}}$	$\sqrt{3}$	$\frac{1}{\sqrt{3}}$	$\frac{4\sqrt{3}}{3}$

2.

	a	b	c	d	e	f
	$2\sqrt{2}$	$2\sqrt{2}$	2	2	2	4
	$\sqrt{2}$	$\sqrt{2}$	1	1	1	2

3. (a) 90°   (b) $\frac{13}{2}$ (c) $\frac{5\sqrt{3}}{2}$ (d) $\frac{120}{17}$ (e) 14

(f) $\sqrt{14}$

4. Find B^*, then use the Pythagorean Theorem.

5. (a) $\frac{1}{2}$ (b) $4\sqrt{3}$ (c) $6\sqrt{3}$ (d) $\frac{6}{4}\sqrt{3}$

(e) $\left(\frac{2}{\sqrt{2}} + 1\right)^2 - 1 = \frac{4}{\sqrt{2}} + 2$ or $2\sqrt{2} + 2$ (f) $\frac{\sqrt{3}}{2}$

Cumulative Review

1. In each picture, find a geometric fact which is violated.

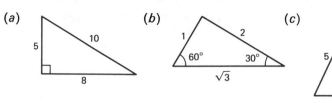

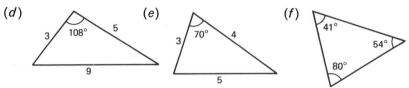

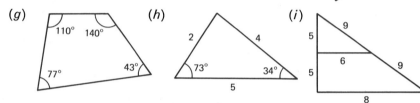

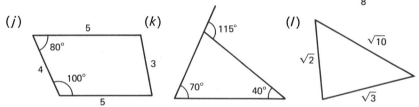

2. A certain equilateral triangle has altitude 6. What is its area?

3. Find the area of $\triangle ABC$.

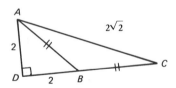

4. A right triangle has hypotenuse of length 14 and a leg of length 8. What is its area?

5. Find AB in each case.

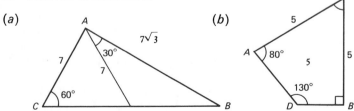

300 7: The Pythagorean Theorem

6. Using only straightedge and compass, construct an angle of size
   (a) 60°   (b) 30°   (c) 15°   (d) 45°   (e) 75°.

7. Find the length of the shortest path from A to B which touches ℓ.

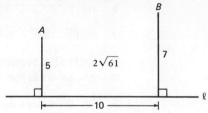

8. Find AB.

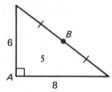

9. You are handed a sheet of paper on which are drawn two line segments of lengths a and b respectively. Explain how you would construct, with straightedge and compass alone, a line segment of length
   (a) $a + b$ (b) $\sqrt{a^2 + b^2}$ (c) $a\sqrt{2}$ (d) $a\sqrt{3}$.

10. Suppose that in $\triangle ABC$ the line segment joining the vertex of $\angle A$ to the midpoint of $\overline{BC}$ has length $\frac{1}{2}BC$. Then prove that $\triangle ABC$ is a right triangle. [*Hint:* Let M be the midpoint of $\overline{BC}$, K the midpoint of $\overline{AC}$, and L the midpoint of $\overline{AB}$. Consider the quadrilateral ALMK.]

Answers

1. (a) The Pythagorean Theorem is violated.   (b) *Pythagorean Theorem*

(c) The other two angles must be 60°, since base angles of isosceles triangle are equal; but the triangle is not equilateral!

(d) Triangle Inequality

(e) Converse of the Pythagorean Theorem

(f) Sum of the angles of any triangle is 180°.   (g) *sum of angles of a quad.*

(h) Notice that the missing angle must be 73°.   (i) *Midpt. Th. and Triangle Ineq.*

(j) The quadrilateral has a pair of opposite sides which are parallel (why?) and have equal length. Hence it is a ──────. But ...?

(k) *Exterior Angle Theorem*   (l) *Triangle Inequality*

2. $12\sqrt{3}$

3. $2\sqrt{2}$ $\quad$ area $(\triangle ABC) = $ area $(\triangle ADC) - $ area $(\triangle ADB)$

4. $8\sqrt{33}$

5. (a) $7\sqrt{3}$ $\quad\quad$ (b) $AB = 5$, because $\triangle ABC$ must be equilateral (why?).

7. $2\sqrt{61}$ $\quad$ $AB^* = \sqrt{(10)^2 + (12)^2}$

8. 5 (The line segment drawn to the midpoint of the hypotenuse from the vertex of the right angle has half the length of the hypotenuse.)

9. (b) Construct a right triangle with legs a and b.

(c) Construct a square of side a and draw a diagonal.

(d) Construct an equilateral triangle of side a, and consider the altitude.

10. Let M be the midpoint of $\overline{BC}$, K the midpoint of $\overline{AC}$, and L the midpoint of $\overline{AB}$.

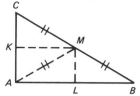

We are given that $AM = MB$. Since $\overline{KM}$ is parallel to $\overline{AL}$ and $\overline{KA}$ is parallel to $\overline{ML}$ (why?), $ALMK$ is a parallelogram. We have $AM = MB = KL$, so the diagonals of parallelogram $ALMK$ have equal length. But if the diagonals of a parallelogram have equal length, then the parallelogram is a rectangle; hence $ALMK$ is a rectangle, and $\angle A$ is a right angle.

Projects

More Proofs of the Pythagorean Theorem

Encourage students to do Garfield's proof.

1. Several years before he became President of the United States, General James A. Garfield discovered a proof of the Pythagorean Theorem based on the following figure. Can you find his proof? [*Hint:* Recall that the area of a trapezoid can be obtained by multiplying its altitude by one-half the sum of the lengths of the two parallel sides. But the area of

area of trapezoid
$= \frac{1}{2}(a+b)(a+b)$
$= \frac{1}{2}(a^2 + 2ab + b^2)$

area of three right triangles =
$\frac{1}{2}ab + \frac{1}{2}ab + \frac{1}{2}c^2$
$= \frac{1}{2}(2ab + c^2)$

Since the areas are the same,
$a^2 + b^2 = c^2$.

the trapezoid below is also the sum of the areas of three triangles.]

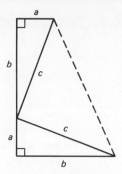

The next exercise gives a proof of the Pythagorean Theorem based on the fact that if you cut up a figure and rearrange the pieces into another figure you do not change the area. This proof was discovered by H. Perigal in 1873, but apparently was known a thousand years earlier by Tâbit ibn Qorra. (A good reference is the book by Howard Eves, *An Introduction to the History of Mathematics* (New York: Holt, Rinehart and Winston, 1964).)

Exercise 2: This proof is of interest in relation to the dissection problems of the Sultan of Polygonia studied in the Projects section of Chapter 6.

2. (a) The polygon below is obtained by placing together a square of side *b* and a square of side *a*. Cut it into 3 pieces which can be rearranged to form a square of side *c*, where *c* is the hypotenuse length of a right triangle with legs of length *a* and *b*. The cuts you have to make are indicated by dotted lines; however, you should explain exactly how they should be made.

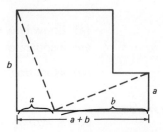

(b) Explain why this shows that the Pythagorean Theorem is true.

This is a beautiful proof devised by Leonardo.

3. The great artist Leonardo da Vinci (1452–1519) found the following interesting way to prove the Pythagorean Theorem. The following figure is obtained by fitting together two copies of a right triangle with legs *a*, *b*, hypotenuse *c*, a square of side *a* and a square of side *b*.

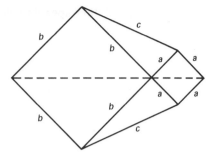

If we cut along the dotted line, turn one piece over, and fit together again, we obtain a new figure.

(a) Show that the new figure consists of two copies of the original triangle and a square of side c fitted together.

(b) Why does this prove the Pythagorean Theorem?

4. The Indian mathematician Bhāskara (1114–1185) gave a proof of the Pythagorean Theorem which consisted simply of the picture below and the statement underneath.

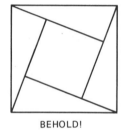

BEHOLD!

(a) Show how 4 copies of a right triangle with legs a, b, and hypotenuse c, and a square of side $(a - b)$, can be arranged to fill up a square of side c.

(b) What do you think Bhāskara had in mind?

Pythagorean Triples

5. A *Pythagorean triple* is a triple of positive *whole* numbers (a, b, c) such that a and b are the lengths of the legs of a right triangle of hypotenuse length c.

(a) Show that $(3, 4, 5)$ and $(5, 12, 13)$ are Pythagorean triples.

(b) Suppose a, b, and c are positive whole numbers such that $a^2 + b^2 = c^2$. Is (a, b, c) a Pythagorean triple? In other words, does there necessarily exist a right triangle with legs of length a and b and hypotenuse length c? Explain. *yes, by the converse of the Pythagorean Theorem*

$$a^2 + b^2 = (m^2 - n^2)^2 + (2mn)^2$$
$$= m^4 - 2m^2n^2 + n^4 + 4m^2n^2$$
$$= (m^2 + n^2)^2$$
$$= c^2$$

(c) Suppose m and n are positive whole numbers, with $m > n$. Let $a = m^2 - n^2$, $b = 2mn$, and $c = m^2 + n^2$. Show that (a, b, c) will always be a Pythagorean triple.

(d) Make a list of 10 different Pythagorean triples obtained by choosing various values for m and n in part (c). *An infinite number of Pythagorean triples may be found using this method.*

(e) Read about Pythagorean triples in *What is Mathematics?* by R. Courant and H. Robbins (New York: Oxford University Press, 1941), pp. 40–42, or in *The Enjoyment of Mathematics* by H. Rademacher and O. Toeplitz (Princeton: Princeton University Press, 1957), pp. 88–95.

Rational and Irrational

6. (a) Give an example of a rectangle, the lengths of whose sides and diagonals are all positive integers. $\boxed{}$ 3
4

(b) Can you find a *square*, the lengths of whose sides and diagonals are all positive integers? *No, if the side is an integer a, the diagonal is $a\sqrt{2}$, which is not an integer.*

7. (a) Can a 1 by $\sqrt{2}$ rectangle be cut into small congruent squares? Explain. [*Hint:* If it were possible, then each small square would have side length $\dfrac{1}{n}$ for some positive integer n (why?). But each square would also have side length $\dfrac{\sqrt{2}}{m}$ for some positive integer m (why?). Thus we would have $\dfrac{\sqrt{2}}{m} = \dfrac{1}{n}$ or $\sqrt{2} = \dfrac{m}{n}$.]

(b) Read about problems of this type in Chapter 7 of *Mathematics, The Man-made Universe*, 2nd edition, by S. K. Stein (San Francisco: W. H. Freeman, 1969).

8. Can a rectangle of altitude h and base b be cut into congruent squares if

(a) $h = 3$, $b = 4$? *twelve 1×1 squares*   (b) $h = 2\frac{3}{4}$, $b = 5\frac{1}{3}$? *2,112 $\frac{1}{12} \times \frac{1}{12}$ squares*

(c) $h = \sqrt{20}$, $b = \sqrt{40}$? *no*   (d) $\dfrac{h}{b}$ is a rational number? *yes*

Explain your answers.

Problems Using the Pythagorean Theorem

9. A ladder is leaning at a slant against a wall. Mike is on the ground and his little brother, Rocko, is at the top of the ladder. When Mike pulls the base of the ladder three feet further away from the wall, Rocko yells,

Considerable algebra is required for Exercise 9.

"Stop! I've gone down one foot on the wall." Mike pulls the base of the ladder another two feet away from the wall. Rocko yells, "Stop! I've gone down another foot on the wall." How long is the ladder? $\dfrac{5\sqrt{58}}{2}$

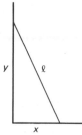

[*Hint:* Let ℓ be the length of the ladder, and x, y the legs of the right triangle formed by the ladder, wall, and floor. This gives $x^2 + y^2 = \ell^2$. From the information given when the ladder is first moved, we obtain $(x+3)^2 + (y-1)^2 = \ell^2$ (why?).]

10. A reed in the center of a circular pool sticks up one foot above the water. When the wind blows, the tip of the reed just touches the edge of the pool. If the radius of the pool is ten feet, how long is the reed? (Assume the reed bends only at its base.) $\dfrac{101}{2}$

11. A certain right triangle has area equal to 42, and the length of the hypotenuse is $\sqrt{193}$. How long are the legs of this triangle?   12, 7

12. Show that the area of a square of diagonal length d is $\tfrac{1}{2}d^2$.

side of square = $\dfrac{d}{\sqrt{2}}$

Therefore area =
$\left(\dfrac{d}{\sqrt{2}}\right)^2 = \dfrac{1}{2}d^2$

13. A parallelogram has sides of lengths a and b and diagonals of lengths c and d. Prove that
$$c^2 + d^2 = 2a^2 + 2b^2.$$

14. Recall our discussion of the regular polyhedra, or Platonic solids, in the Projects section of Chapter 4. In particular, recall that a *regular tetrahedron* is a polyhedron which has 4 faces, each an equilateral triangle.

(a) Build a model of a tetrahedron by folding a pattern like this one:

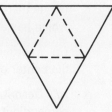

The corners fold up to an apex.

(*b*) The *altitude* of a regular tetrahedron is the distance from a vertex to the opposite face. If each edge of a regular tetrahedron has length 10, then what is the altitude of the tetrahedron? [*Hint:* Apply the Pythagorean Theorem to some appropriate right triangles. In the picture below, note that the line drawn from the top vertex perpendicular to the opposite face passes through the centroid of that face (Projects Exercise 14, Chapter 5).]

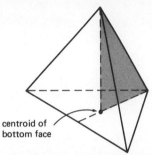

centroid of bottom face

(*c*) Suppose 6 yardsticks are nailed together to form the edges of a regular tetrahedron, and the result is set down on the floor. Then what is the height of the top vertex above the floor? (Give the height in inches, to the nearest inch.)

(*d*) Find the volume of a regular tetrahedron of edge length 10.

15. (*a*) Suppose you are giving a friend, as a birthday present, a regular tetrahedron with edges of length 20 cm. You want to pack the tetrahedron in a box the shape of a cube. What is the smallest size box possible?

(*b*) Suppose you simply want to cover the tetrahedron with paper which costs 10¢ per 100 sq. cm. How much money would you have to spend for paper?   70¢

16. A piece of cheese congruent to a cube of edge length 8 cm. is cut with a knife such that the cut is in a plane that is the perpendicular bisector of a longest diagonal.

(*a*) Describe the shape of the cut and find its area.   *regular hexagon*
$48\sqrt{3}$ *sq. cm.*

(b) Is there any way to cut the cheese into two pieces with a straight cut which has area larger than that in part (a)?

(c) In order to get the largest possible area of a cut, must one cut through the center of the cube? *yes*

(d) Describe the shapes that one gets by cutting at various positions, perpendicular to a longest diagonal. *The only shapes possible are hexagons and equilateral triangles.*

17. Recall that a *regular octahedron* is a regular polyhedron which has 8 faces, each face an equilateral triangle, with 4 faces meeting at each vertex. If each edge of the regular octahedron in the picture below has length 10, what is the distance from A to B? $10\sqrt{2}$

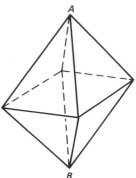

Lattice Polygons

18. If you tile the plane with congruent squares, then the vertices of the squares form a *lattice* as in the picture below. The points are called *lattice points*.

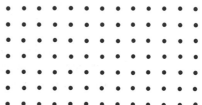

Students will find the use of the geoboard instructive.

A *geoboard* is constructed by driving nails in a board to form a lattice pattern like the one above. By stretching rubber bands around the nails, one can get polygons of various shapes (see top of p. 309). Such polygons are called *lattice polygons*.

In the following exercises you may construct a real geoboard and build lattice polygons on it with rubber bands, or you may find it more convenient to draw a lattice and draw the lattice polygons on paper. For purposes of measuring lengths and areas, assume that each small lattice square has side length 1 unit.

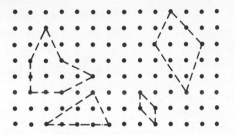

area of hexagon =
  area of square + area of two triangles
  $= 2 \cdot 2 + \frac{1}{2}(2 \cdot 2) + \frac{1}{2}(2 \cdot 2) = 8$
area of triangle $= \frac{1}{2}(4 \cdot 2) = 4$
area of small quad. $= 1 \cdot 1 = 1$
area of large quad. = *area of two triangles* $= \frac{1}{2}(3 \cdot 2) + \frac{1}{2}(3 \cdot 3) = \frac{15}{2}$

(a) Find the area of each polygon in the picture above.

(b) For any lattice polygon, let I be the number of lattice points *inside* the polygon and let B be the number of lattice points which lie on the *border* of the polygon. For example, for the irregular hexagon in the picture above, we have $I = 5$ and $B = 8$. For the parallelogram in the picture above, we have $I = 0$, $B = 4$. A surprising theorem discovered by G. Pick in 1899 tells how one can find the area of any lattice polygon if one knows I and B. (Pick's Theorem is proved in the book by H. S. M. Coxeter, *Introduction to Geometry* (New York: John Wiley & Sons, 1961).)

Pick's Theorem The area of any lattice polygon is $I + \frac{1}{2}B - 1$, where I is the number of lattice points inside and B is the number of lattice points on the border of the polygon.

Check that this is true for each polygon in the picture above.

hexagon: $5 + \frac{1}{2}(8) - 1 = 8$
triangle: $2 + \frac{1}{2}(6) - 1 = 4$
small quadrilateral:
  $0 + \frac{1}{2}(4) - 1 = 1$
large quadrilateral:
  $6 + \frac{1}{2}(5) - 1 = \frac{15}{2}$

(c) Draw some lattice polygons and check that Pick's Theorem holds.

(d) If a lattice parallelogram has no lattice points on its border except at the 4 vertices, and no lattice points inside, then what is its area? 1

(e) Explain why the area of any lattice polygon is a rational number, that is, a number of the form $\frac{m}{n}$, where m and n are integers. $I + \frac{1}{2}B - 1$ *is rational.*

(f) Suppose S is the distance between a pair of lattice points. Explain why S^2 is an integer. [*Hint:* Apply the Pythagorean Theorem to an appropriate right triangle.]

(g) Prove that there does not exist any *equilateral* lattice triangle. [*Hint:* $\sqrt{3}$ is not a rational number.]

(h) Does there exist a regular lattice hexagon? *no*

309 Projects

Heron's Formula for the Area of a Triangle

Recall that Heron devised the shortest path problem used in Chapter 1. The historical facts indicate that Archimedes was the first to discover this formula for the area of a triangle, and Heron simply reported it in one of his books.

19. Heron of Alexandria discovered a beautiful formula for the area of a triangle which involves only the lengths of the 3 sides of the triangle. If the lengths of the sides are a, b, and c respectively, let $s = \frac{1}{2}(a + b + c)$. (Since $a + b + c$ is the perimeter of the triangle, we call s the "semiperimeter" of the triangle.) Heron's Formula states that if A is the area of the triangle, then

$$A = \sqrt{s(s-a)(s-b)(s-c)}.$$

The following sequence proves Heron's Formula.

(a) The figure below shows a triangle with sides of lengths a, b, and c, and altitude h.

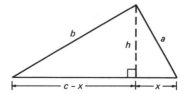

Show that $a^2 - x^2 = h^2 = b^2 - (c-x)^2$. *Pythagorean Theorem*

(b) Use part (a) to obtain

$$x = \frac{a^2 + c^2 - b^2}{2c}.$$

(c) If A is the area of the triangle, show that

$$A^2 = \frac{1}{4}c^2h^2 = \frac{1}{4}c^2(a^2 - x^2) = \frac{1}{4}c^2(a-x)(a+x).$$

(d) Use part (b) to show that

$$a - x = \frac{b^2 - (a-c)^2}{2c} = \frac{(b-a+c)(b+a-c)}{2c} = \frac{4(s-a)(s-c)}{2c}$$

$$a + x = \frac{(a+c)^2 - b^2}{2c} = \frac{(a+c-b)(a+c+b)}{2c} = \frac{4s(s-b)}{2c}.$$

(e) Use parts (c) and (d) to deduce

$$A^2 = s(s-a)(s-b)(s-c)$$

and thus obtain Heron's Formula.

20. (a) Use Heron's Formula to prove that the area of an equilateral triangle of side length a is $\dfrac{\sqrt{3}}{4} a^2$.

(b) Find the area of this quadrilateral:

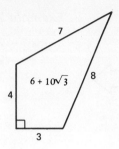

Algebra Review

In deriving the quadratic formula, this Algebra Review uses much of the algebra the student has learned. Few classes can afford to omit it.

The graph of the equation
$$y = ax^2 + bx + c$$
(where a, b, and c are fixed numbers and $a \ne 0$) is a bowl-shaped curve called a *parabola*. In this Algebra Review we want to find where this curve crosses the x-axis, that is, to solve the *quadratic* (second degree) equation
$$ax^2 + bx + c = 0.$$

1. (a) Graph $y = x^2 - 2x - 7$.

(b) From your graph in (a) estimate the values of x such that
$$x^2 - 2x - 7 = 0.$$

Solving Quadratic Equations

Occasionally, a quadratic equation can be solved by factoring.

Example 1. Solve for x: $\quad 4x^2 + 19x - 5 = 0$
$$(4x - 1)(x + 5) = 0$$
Then
$$\begin{aligned} 4x - 1 &= 0 & \text{or} & & x + 5 &= 0 \\ 4x &= 1 & & & x &= -5 \\ x &= \tfrac{1}{4} \end{aligned}$$
and thus $x = \tfrac{1}{4}$ or -5.

Some quadratic equations, such as

$$x^2 - 2x - 7 = 0,$$

cannot be factored. They can be solved by *completing the square*.

Example 2. Solve for x (see Exercise 1):

$$\begin{aligned}
x^2 - 2x - 7 &= 0 \\
x^2 - 2x &= 7 \\
x^2 - 2x + 1 &= 7 + 1 \qquad \text{completing the square} \\
(x - 1)^2 &= 8 \\
x - 1 &= \pm\sqrt{8} \\
x &= 1 \pm \sqrt{8} \\
x &= 1 \pm 2\sqrt{2}
\end{aligned}$$

2. Solve by completing the square.

(a) $x^2 + 6x - 4 = 0$ (b) $x^2 - 12x - 4 = 0$

(c) $x^2 - 8x + 3 = 0$ (d) $x^2 + 11x + 5 = 0$

If the coefficient of x^2 is not 1, you can divide in order to make it so.

Example 3. Solve for x:

$$3x^2 - 2x - 4 = 0$$

Step 1. $x^2 - \frac{2}{3}x - \frac{4}{3} = 0$

Step 2. $x^2 - \frac{2}{3}x = \frac{4}{3}$

Step 3. $x^2 - \frac{2}{3}x + \frac{1}{9} = \frac{4}{3} + \frac{1}{9}$

Step 4. $\left(x - \frac{1}{3}\right)^2 = \frac{13}{9}$

Step 5. $x - \frac{1}{3} = \pm\sqrt{\frac{13}{9}}$

Step 6. $x = \frac{1}{3} \pm \frac{\sqrt{13}}{3}$

Step 7. $x = \frac{1 \pm \sqrt{13}}{3}$

3. Explain each step in Example 3.

4. Solve for x by completing the square.

(a) $x^2 - 2x - 5 = 0$ (b) $2x^2 + 4x - 5 = 0$
(c) $4x^2 - 6x + 1 = 0$ (d) $3x^2 - 2 = x$

The Quadratic Formula

The next exercise derives a formula for solving equations of the type
$$ax^2 + bx + c = 0,$$
where a, b, and c are fixed numbers and $a \neq 0$. The *quadratic formula*,

$$x = \frac{-b \pm \sqrt{b^2 - 4ac}}{2a},$$

will solve a quadratic equation if any solutions exist. It is the general case of completing the square.

5. Explain each of the following steps in solving for x.
$$ax^2 + bx + c = 0 \quad (a \neq 0)$$

Step 1. $x^2 + \dfrac{b}{a}x + \dfrac{c}{a} = 0$

Step 2. $x^2 + \dfrac{b}{a}x = -\dfrac{c}{a}$

Step 3. $x^2 + \dfrac{b}{a}x + \left(\dfrac{b}{2a}\right)^2 = \left(\dfrac{b}{2a}\right)^2 - \dfrac{c}{a}$

Step 4. $\left(x + \dfrac{b}{2a}\right)^2 = \dfrac{b^2}{4a^2} - \dfrac{c}{a}$

Step 5. $\left(x + \dfrac{b}{2a}\right)^2 = \dfrac{b^2}{4a^2} - \dfrac{4ac}{4a^2}$

Step 6. $\left(x + \dfrac{b}{2a}\right)^2 = \dfrac{b^2 - 4ac}{4a^2}$

Step 7. $x + \dfrac{b}{2a} = \pm\sqrt{\dfrac{b^2 - 4ac}{4a^2}}$

Step 8. $x + \dfrac{b}{2a} = \pm\dfrac{\sqrt{b^2 - 4ac}}{2a}$

Step 9. $x = -\dfrac{b}{2a} \pm \dfrac{\sqrt{b^2 - 4ac}}{2a}$

Step 10. $x = \dfrac{-b \pm \sqrt{b^2 - 4ac}}{2a}$ The quadratic formula

6. (a) Write the quadratic formula from memory.

(b) Say the quadratic formula aloud to a classmate.

Example 4. Solve for x: $2x^2 - 3x - 4 = 0$

We observe that $a = 2$, $b = -3$, and $c = -4$; then

$$x = \frac{-b \pm \sqrt{b^2 - 4ac}}{2a}$$

$$= \frac{-(-3) \pm \sqrt{(-3)^2 - 4(2)(-4)}}{2(2)}$$

$$= \frac{3 \pm \sqrt{41}}{4}.$$

Note. Observe that we use parentheses when we substitute for the letters a, b, and c. This prevents many errors when simplifying the expression.

7. Check the solutions to Example 4 by substituting each solution in the original equation.

8. Solve for x using the quadratic formula.

(a) $x^2 + 6x - 4 = 0$ (b) $x^2 - 3x - 3 = 0$

(c) $x^2 - 8x + 3 = 0$ (d) $x^2 - 7x + 11 = 0$

(e) $2x^2 + 3x = 2$ (f) $3x^2 + 4x - 3 = 0$

9. Solve for x by any method.

(a) $2x^2 + 5x - 3 = 0$ (b) $4x^2 + 4x = 3$

(c) $x^2 - 5x = 0$ (d) $x^3 - x^2 - 6x = 0$

10. Check the solutions in Exercise 9 by substituting in the original equations.

11. Solve for x by any method.

(a) $x^2 + 10x + 12 = 0$ (b) $x^2 + 7x = 6$

(c) $x^2 - x = 2$ (d) $3x^2 - 2x = 1$

(e) $6x^2 - 12x = 7$ (f) $5x^2 - 4x - 1 = 0$

Answers   1. (a) graph   (b) 3.8, −1.8

2. (a) $-3 \pm \sqrt{13}$ (b) $6 \pm 2\sqrt{10}$ (c) $4 \pm \sqrt{13}$ (d) $\dfrac{-11 \pm \sqrt{101}}{2}$

3. Step 1. Divide by 3.   Step 2. Add $\frac{4}{3}$.   Step 3. Add $\frac{1}{9}$.
   Step 4. Factor and add.   Step 5. Take square root of both sides.
   Step 6. Add $\frac{1}{3}$.   Step 7. Add.

4. (a) $1 \pm \sqrt{6}$ (b) $\dfrac{-2 \pm \sqrt{14}}{2}$ (c) $\dfrac{3 \pm \sqrt{5}}{4}$ (d) $1, \dfrac{-2}{3}$

5. Step 1. Divide by a. Step 2. Subtract $\dfrac{c}{a}$.   Step 3. Add $\left(\dfrac{b}{2a}\right)^2$.
   Step 4. Factor.   Step 5. Multiply.   Step 6. Add.
   Step 7. Take square root of both sides.   Step 8. Simplify square root.
   Step 9. Subtract $\dfrac{b}{2a}$.   Step 10. Add.

6. student work

7. $2\left(\dfrac{3 + \sqrt{41}}{4}\right)^2 - 3\left(\dfrac{3 + \sqrt{41}}{4}\right) - 4$

$= 2\left(\dfrac{50 + 6\sqrt{41}}{16}\right) - \dfrac{9 + 3\sqrt{41}}{4} - 4$

$= \dfrac{100 + 12\sqrt{41} - 36 - 12\sqrt{41} - 64}{16}$

$= 0$

Similarly, check $\dfrac{3 - \sqrt{41}}{4}$.

8. (a) $-3 \pm \sqrt{13}$ (b) $\dfrac{3 \pm \sqrt{21}}{2}$ (c) $4 \pm \sqrt{13}$ (d) $\dfrac{7 \pm \sqrt{5}}{2}$

   (e) $\frac{1}{2}, -2$ (f) $\dfrac{-2 \pm \sqrt{13}}{3}$

9. (a) $\frac{1}{2}, -3$ (b) $\frac{1}{2}, -\frac{3}{2}$ (c) $0, 5$ (d) $0, 3, -2$

10. student check

11. (a) $-5 \pm \sqrt{13}$ (b) $\dfrac{-7 \pm \sqrt{73}}{2}$ (c) $2, -1$ (d) $1, -\frac{1}{3}$

    (e) $\dfrac{6 \pm \sqrt{78}}{6}$ (f) $1, -\frac{1}{5}$

Chapter 8 Similar Figures

The concept of similarity enormously enriches our geometry and makes significant applications possible for the first time. In this chapter the student studies similar polygons, with particular attention to the properties of similar triangles; the student applies his knowledge of algebra to an extent not possible before. In Chapter 9 he will learn how perimeters and areas of similar figures compare in a plane and will consider similar solids and their volumes. The behavior of the length, area, and volume of similar figures is one of the most practical concepts the student can learn in this course.

Exercise 25 gives another proof of the Pythagorean Theorem. Note that it uses length but not area. A mathematician would prefer this proof because it does not use area, but the students will probably prefer one of the proofs using areas, which are much more visual!

We have chosen a path of least resistance in our treatment of similar triangles in the Central section. We assume the AAA condition for similar triangles and use it to prove what is often called the Basic Proportionality Theorem, namely, the result proved in Exercise 26. This enables us to take up significant theorems while bypassing a number of frustrating stumbling blocks for the student. Most texts first prove the Basic Proportionality Theorem and then proceed to prove the conditions for similar triangles. We give such a self-contained treatment in the Projects section, and a student who is naturally curious about the logical structure of geometry might like to work through these projects.

Central

On a sunny day a tall tree casts a shadow 40 feet long. At the same time, a yardstick casts a shadow 2 feet long when it is held vertically. How tall is the tree?

Two figures can be the same shape without being congruent.

Two figures which have exactly the same "shape" but are possibly of different "size" are called *similar figures*. Similar triangles are the basis for the solution of the preceding problem. The idea of "similarity" is one of the most powerful and useful ideas in geometry, and it has wide application in the sciences, such as physics and biology. Some of these applications are considered in the Projects section of Chapter 9.

We begin by studying similar figures in a plane. The next chapter

Students should be encouraged to develop a strong intuition for similar figures.

Congruent figures are also similar; the ratio of similarity is 1.

considers solid figures in space.

1. In each case compare the given pair of figures. Which pairs would you say have exactly the same "shape"? (Since some of the pictures are distorted, you may wish to carefully redraw them to scale.)

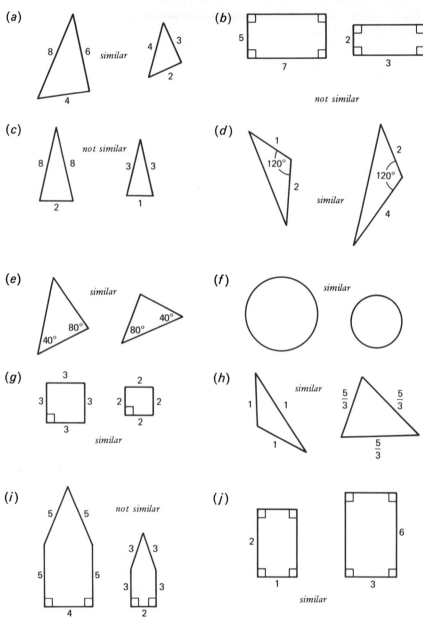

318 8: Similar Figures

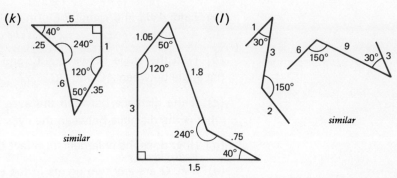

2. Do you think that two figures will necessarily have the same "shape" if they are both

(a) equilateral triangles? yes
(b) squares? yes
(c) rectangles? no
(d) isosceles triangles? no
(e) regular hexagons? yes
(f) regular *n*-gons? yes
(g) congruent figures? yes
(h) quadrilaterals, each having all sides of length 1? no

Two figures are similar if one is a scale drawing, or scale model, of the other. In other words, one figure is an "enlargement" (or "reduction") of the other. But these names do not give a precise definition of similar figures, they only suggest what we mean. We shall give a precise definition of similar polygons later in this chapter, after we have become more familiar with the concept.

Bring a photo and an enlargement of that photo to class. Have the students find the enlargement factor (that is, the ratio of similarity).

3. (a) Where do similar figures occur in photography? prints and negatives
(b) Why are scale drawings useful? *They avoid distortion of figures.*

4. Kate has a photograph of her pet cat, which we have labeled as shown.

In the photograph, $AB = 3''$, $AC = 2''$, $BC = 2''$, and the angle formed by the line segments $\overline{AC}$ and $\overline{BC}$ is 97°. Kate decides to have the picture enlarged to poster size, *six times* its original size. Suppose A' and B' are the tips of the cat's ears and C' the tip of his nose in the poster.

                 18"   12"   12"

(a) Find $A'B'$, $A'C'$, and $B'C'$ (in inches). How do they compare with AB, AC, and BC? *Each is six times the original.*

(b) Find the angle between $\overline{A'C'}$ and $\overline{B'C'}$. How does it compare with the angle between $\overline{AC}$ and $\overline{BC}$? 97°

(c) If the distance between the eyes is $\frac{1}{2}''$ in the original picture, then what is the distance between the eyes in the poster? 3"

(d) How does the enlargement affect distances between points?
multiplied by 6

(e) If P, Q are any two points in the original picture and P', Q' the corresponding points in the poster, then

$$\frac{P'Q'}{PQ} = \underline{6} \quad \text{and} \quad \frac{PQ}{P'Q'} = \underline{\tfrac{1}{6}}.$$

(f) How does the enlargement affect angles? *angles unchanged*

(g) The angle formed by $\overline{AB}$ and the perpendicular bisector of the line segment joining the eyes in the original picture is 90°. What is the size of the corresponding angle in the poster? 90°

You may wish to lead a discussion summarizing Exercises 1-4.

(h) Compare your answers with those of your classmates (especially your answers to parts (f) and (g).)

Ratio and Proportion

In order to precisely define similar figures, we need to understand the concepts of "ratio" and "proportion" for numbers.

DEFINITION   If s and t are any numbers and t is not zero, then the number

$$\frac{s}{t}$$

Generally, it is easier for a student to associate the word "ratio" with the fraction $\frac{p}{q}$, where p and q are integers. However, students need to be reminded that a ratio is a number. It may be an integer, a common or decimal fraction, or an irrational number such as $\sqrt{2}$.

It is necessary that students have a good understanding of ratio and proportion before trying to study similar figures.

is called the *ratio* of s to t.

For example, the ratio of 6 to 2 is $\frac{6}{2}$ or 3. The ratio of 2 to 6 is $\frac{2}{6}$ or $\frac{1}{3}$.

Now consider the sequence of three numbers

$$1, 7, 3$$

and a second sequence

$$2, 14, 6.$$

The ratio of each number in the second sequence to the corresponding number in the first sequence is the same (namely, 2):

$$\frac{2}{1} = \frac{14}{7} = \frac{6}{3}$$

Because of this, we say that the sequences of numbers are "proportional." We define the term "proportional" for more general sequences:

DEFINITION Let

$$a, b, c, d, \ldots \quad \text{and} \quad a', b', c', d', \ldots$$

be two sequences of positive numbers. If

$$\frac{a'}{a} = \frac{b'}{b} = \frac{c'}{c} = \frac{d'}{d} = \ldots,$$

then we say the sequences are *proportional.*

In other words, let $a, b, c, d, \ldots$ and $a', b', c', d', \ldots$ be two sequences of positive numbers. They are proportional if the ratio of each number in the second sequence to the corresponding number in the first sequence is the same.

For example, the two sequences

$$6, 3, 15, 12$$

and

$$2, 1, 5, 4$$

are proportional, since

$$\frac{2}{6} = \frac{1}{3} = \frac{5}{15} = \frac{4}{12}.$$

Check that students realize that the order of the numbers is important.

Notice that the order in which the numbers are written is very important. For example, we have just seen that the sequences 6, 3, 15, 12 and 2, 1, 5, 4 are proportional. But observe that the sequences 6, 3, 15, 12 and 1, 2, 5, 4 (we switched 1 and 2) are *not* proportional, since

$$\frac{1}{6} \neq \frac{2}{3}.$$

Students who are weak in algebra may have difficulty with Exercises 6 and 7.

5. Are the two sequences of numbers proportional?

(a) 1, 2, 3   and   2, 4, 6   *yes*   (b) 14, 35   and   2, 5   *yes*

(c) 2, .5, 3, 2, 1   and   5, 1.25, 7.5, 5, 2.5   *yes*

(d) 3, 7, 2   and   6, 4, 14   *no,* $\frac{6}{3} \neq \frac{4}{7}$

(e) a, b, c and $2a, 2b, 2c$ *yes*

6. (a) Suppose we start with a sequence of positive numbers

$$a, b, c, d, \ldots$$

and multiply each number in the sequence by the same positive number k to obtain the new sequence

$$ka, kb, kc, kd, \ldots$$

Show that the two sequences are proportional. $\frac{ka}{a} = \frac{kb}{b} = \frac{kc}{c} = \cdots = k$

(b) What if we *add* k? Is the sequence 1, 2, 3 proportional to the sequence 2, 3, 4? Discuss your answer with your classmates.   *no,* $\frac{2}{1} \neq \frac{3}{2}$

7. Suppose the sequences $a, b, c, d, \ldots$ and $a', b', c', d', \ldots$ are proportional.

(a) Show that there is a positive number k such that

$$a' = ka, \quad b' = kb, \quad c' = kc, \quad d' = kd, \ldots$$

In other words, show that the second sequence can be obtained by multiplying each number of the first sequence by a positive number k. [*Hint:* Let $k = \frac{a'}{a}$.] $k = \frac{a'}{a} = \frac{b'}{b} = \frac{c'}{c} = \cdots$

(b) Show that the first sequence can be obtained by multiplying each number of the second sequence by a fixed positive number. $\frac{1}{k}$

(c) Since the two sequences are proportional, we know that

$$\frac{a'}{a} = \frac{b'}{b} = \frac{c'}{c} = \frac{d'}{d} = \cdots.$$

Is it also true that

$$\frac{a}{a'} = \frac{b}{b'} = \frac{c}{c'} = \frac{d}{d'} = \cdots ?$$ *yes*

8. (a) Show that the two sequences

$$3, 9, 6, 21 \quad \text{and} \quad 2, 6, 4, 14$$

are proportional. $\frac{2}{3} = \frac{6}{9} = \frac{4}{6} = \frac{14}{21}$

(b) The second sequence is obtained by multiplying each number in the first sequence by _____. $\frac{2}{3}$

(c) The first sequence is obtained by multiplying each number in the second sequence by _____. $\frac{3}{2}$

Exercise 9 is preparation for the definition of similar polygons.

9. The picture below shows two polygons $ABCDE$ and $A'B'C'D'E'$. The polygon on the right has been obtained by "enlarging" the polygon on the left.

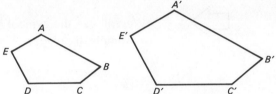

(a) What do you think is true about the five numbers

$$\frac{A'B'}{AB}, \frac{B'C'}{BC}, \frac{C'D'}{CD}, \frac{D'E'}{DE}, \frac{E'A'}{EA} ? \quad \textit{all equal}$$

(b) The two sequences of numbers AB, BC, CD, DE, EA and $A'B', B'C', C'D', D'E', E'A'$ are _____.   *proportional*

(c) What do you think is true about the angles of the two polygons at corresponding vertices?   *equal*

Similar Polygons

We shall now give a precise definition of similar polygons. This definition agrees with our intuitive notion of "enlargement" (or "reduction") in size without changing "shape."

DEFINITION

Let $ABCD \ldots$ and $A'B'C'D' \ldots$ be two polygons. If they satisfy the two conditions

(i) $\quad \dfrac{A'B'}{AB} = \dfrac{B'C'}{BC} = \dfrac{C'D'}{CD} = \ldots$

and

(ii) the corresponding angles of the polygons are equal,

then we say that the polygons are *similar*.

Give a presentation on similar polygons. Make sure the class learns the two conditions.

This definition for similar polygons can be put very concisely using the language of proportions.

> **Two polygons are similar if**
> **(i) corresponding sides are proportional,**
> and
> **(ii) corresponding angles are equal.**

The phrase "corresponding sides are proportional" is a shorter way of saying that the sequence of the lengths of the sides of one polygon is proportional to the sequence of the lengths of the sides of the other polygon.

We will use the symbol "~" in place of the statement "is similar to." For example, if polygon P is similar to polygon P', we will simply write

$$P \sim P'.$$

Encourage much student discussion in Exercises 10 and 11.

10. Without using drawings, decide which of the following pairs of polygons are similar. If you think the polygons are similar, show how they satisfy conditions (i) and (ii). *Caution!* Some of the figures have not been drawn accurately.

(a): $\dfrac{.6}{2} = \dfrac{1.2}{4} = \dfrac{1.5}{5}$

(b): *isosceles trapezoids*

(f): *right triangles*

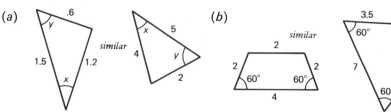

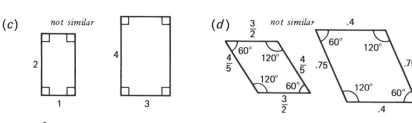

The foundation of this chapter is the two conditions for similarity, which must be referred to repeatedly as students work Exercises 10 and 11.

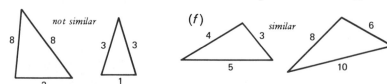

324   8: Similar Figures

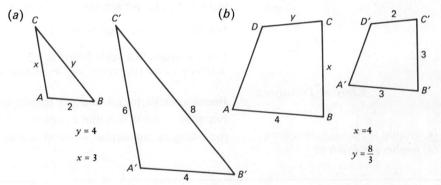

Exercise 11: When the polygons are not similar, the student should give a counterexample.

11. Are the following pairs of polygons similar? Explain, discussing both conditions (*i*) and (*ii*) for similar polygons.

(a) any two squares? *yes* (b) any two rectangles? *no*

(c) any two equilateral triangles? *yes* (d) any two regular hexagons? *yes*

(e) any two regular *n*-gons? *yes* (f) any two congruent triangles? *yes*

(g) any two congruent polygons? *yes*

12. In each case, assume the two figures are similar. Using this fact, find x and y.

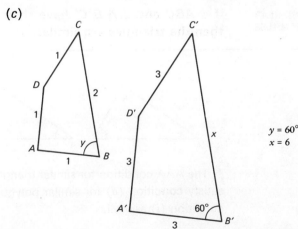

(a) $y = 4$, $x = 3$

(b) $x = 4$, $y = \frac{8}{3}$

(c) $y = 60°$, $x = 6$

Allow time for comparing examples in Exercise 13, which is important preparation for the AAA condition for similar triangles.

13. (a) Make a careful drawing of two polygons which satisfy condition (i) but are *not* similar.   *two rhombuses of side s: one square, one not square*

(b) Make a careful drawing of two polygons which satisfy condition (ii) but are *not* similar.   *two rectangles: one square, one not square*

(c) Compare your work with that of your classmates.

14. (a) If you add one foot to each side of a 3 foot by 3 foot rectangle (a square), is the resulting rectangle similar to the first?   *yes*

(b) If you add one foot to each side of a 5 foot by 7 foot rectangle, is the resulting rectangle similar to the given one? Discuss this with your classmates.   *no, $\frac{5}{6} \neq \frac{7}{8}$*

15. (a) Draw a triangle; draw another triangle of a different size, but having angles the same size as in the original triangle. What do you notice?
They are similar.
(b) Try to find two *triangles* which have their angles respectively equal but which are not similar.   *not possible*

(c) What do you conclude? Discuss this with your classmates.

(d) Compare this with Exercise 13. What is the difference?
Exercise 13 uses polygons with more than three sides.

Similar Triangles

In the case of triangles, condition (ii) implies condition (i).

Recall conditions (i) and (ii) for similar polygons. In the special case of triangles, it turns out that condition (ii) *by itself* is enough to ensure that the triangles are similar. In other words, we have the following.

AAA CONDITION FOR SIMILAR TRIANGLES	If $\triangle ABC$ and $\triangle A'B'C'$ have their corresponding angles equal, then the triangles are similar.

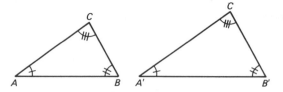

Corresponding angles are equal, hence $\triangle ABC \sim \triangle A'B'C'$.

The AAA condition for similar triangles implies that if a pair of triangles satisfy condition (ii) for similar polygons, then they automatically satisfy condition (i) as well.

The proof of the AAA condition for similar triangles is given in the Projects section of this chapter. The following exercises will help us recognize situations where this condition can be put to use.

16. (a) If two triangles have their corresponding angles equal, then their corresponding sides are _____. *proportional*

(b) If two angles of $\triangle ABC$ are equal to two corresponding angles of $\triangle A'B'C'$, then must $\triangle ABC \sim \triangle A'B'C'$? Discuss your answer with your classmates. *yes*

(c) If one angle of $\triangle ABC$ is equal to one angle of $\triangle A'B'C'$, then must $\triangle ABC \sim \triangle A'B'C'$? Discuss your answer with your classmates. *no*

17. Consider these two triangles:

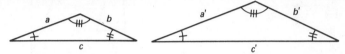

(a) Explain why $\dfrac{a'}{a} = \dfrac{b'}{b} = \dfrac{c'}{c}$. *AAA condition*

(b) Explain why $\dfrac{a}{a'} = \dfrac{b}{b'} = \dfrac{c}{c'}$. *same as (a)*

(c) Is $\dfrac{a}{b} = \dfrac{a'}{b'}$? $\dfrac{a}{c} = \dfrac{a'}{c'}$? $\dfrac{b}{c} = \dfrac{b'}{c'}$? Explain. *yes*

(d) If $a = 2$, $a' = 7$, and $b' = 5$, then find b. $\dfrac{10}{7}$

(e) If $b = 3$, $c = 5$, and $c' = 2$, then find b'. $\dfrac{6}{5}$

Corresponding sides are opposite equal angles.

(a): $\dfrac{x}{7} = \dfrac{y}{9} = \dfrac{8}{4} = 2$

(b): $\dfrac{x}{6} = \dfrac{7}{y} = \dfrac{4}{3}$

18. In each case, find both x and y, or just x in those cases where only x appears. Explain your answers.

(a) $x = 14$, $y = 18$

(b) $x = 8$, $y = \dfrac{21}{4}$

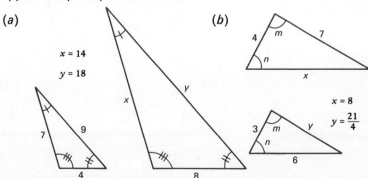

(c): $\frac{x}{4} = \frac{y}{8} = \frac{a}{2a} = \frac{1}{2}$

(d): By the Pythagorean Theorem, $x = 3\sqrt{3}$.

$$\frac{3\sqrt{3}}{2} = \frac{6}{y}$$

(e): By the Pythagorean Theorem, the third side in the left triangle is .5.

$$\frac{x}{1.3} = \frac{y}{1.2} = \frac{1}{.5} = 2$$

(f): $\frac{x}{10} = \frac{8}{y} = \frac{3}{5}$

(g): $\frac{x}{5} = \frac{7}{6}$

(h): $\frac{x}{3} = \frac{2}{1}$

(i): $\frac{x}{7} = \frac{12}{10}$

(j): $\frac{3}{x} = \frac{x}{4}$

(k): $\frac{2}{2+1} = \frac{3}{x}$

(l): $\frac{y}{5} = \frac{3}{2}$

By the Pythagorean Theorem,

$x = \frac{3}{2}\sqrt{21}$

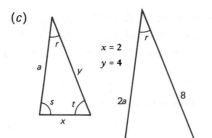

(c) $x = 2$
$y = 4$

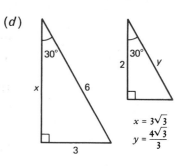

(d) $x = 3\sqrt{3}$
$y = \frac{4\sqrt{3}}{3}$

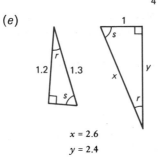

(e) $x = 2.6$
$y = 2.4$

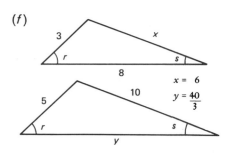

(f) $x = 6$
$y = \frac{40}{3}$

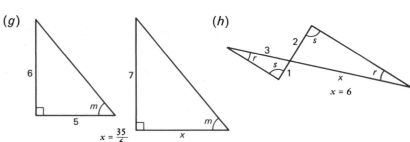

(g) $x = \frac{35}{6}$

(h) $x = 6$

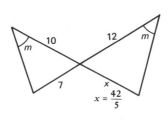

(i) $x = \frac{42}{5}$

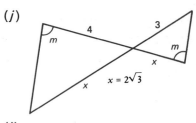

(j) $x = 2\sqrt{3}$

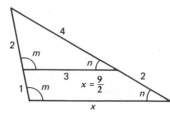

(k) $x = \frac{9}{2}$

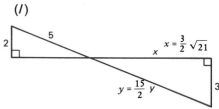

(l) $x = \frac{3}{2}\sqrt{21}$
$y = \frac{15}{2}$

(m): $\dfrac{y}{8} = \dfrac{10}{4}$

(n): *The third side of each triangle is* $\sqrt{x^2 - 4}$.

$\dfrac{2}{\sqrt{x^2-4}} = \dfrac{\sqrt{x^2-4}}{2} = \dfrac{x}{x} = 1$

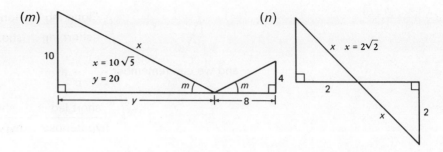

Similar Right Triangles

Similar right triangles frequently occur in applications. The next exercise gives a property of similar right triangles that is useful in trigonometry.

Exercise 19 is an important preparation for trigonometry.

19. Consider these two right triangles with a pair of equal acute angles:

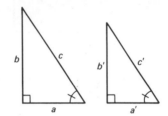

(a) Explain why the triangles must be similar. AAA

(b) Show that $\dfrac{b}{a} = \dfrac{b'}{a'}$, $\dfrac{b}{c} = \dfrac{b'}{c'}$, and $\dfrac{a}{c} = \dfrac{a'}{c'}$. [*Hint:* Exercise 17(c)]

20. Use what you learned in the preceding exercise to solve the opening problem of finding the height of a tree. $\dfrac{x}{3} = \dfrac{40}{2}$

Keep in mind that in two similar right triangles the "short legs" correspond to each other, the "long legs" correspond to each other, and the hypotenuses correspond to each other.

similar right triangles

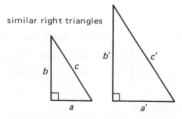

We can remember $\dfrac{b}{a} = \dfrac{b'}{a'}$ as

$$\frac{\text{"long leg"}}{\text{short leg}} = \frac{\text{long leg}}{\text{short leg}},$$

and we can remember $\dfrac{a}{c} = \dfrac{a'}{c'}$ as

$$\frac{\text{"short leg"}}{\text{hypotenuse}} = \frac{\text{short leg}}{\text{hypotenuse}}.$$

21. In each case find $\dfrac{y}{x}$, the ratio of y to x.

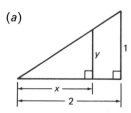

(a) $\dfrac{y}{x} = \dfrac{1}{2}$

(b) $\dfrac{y}{x} = \dfrac{\sqrt{2}}{(1-x)+x} = \sqrt{2}$

(c) $\dfrac{y}{x} = \dfrac{\sqrt{5}}{2}$

As we have seen, if two *right* triangles have a corresponding pair of equal acute angles, then the triangles are similar (because their other angles are then respectively equal). Consider this a hint to the proof in the next exercise.

It may help some students to draw the three triangles side by side in order to see the similarity.

22. The picture below shows a large right triangle which is cut into two smaller right triangles by the altitude drawn to the hypotenuse.

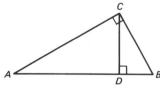

Prove that *the smaller right triangles are similar to each other and also to the large right triangle.* (In other words, prove that $\triangle ADC \sim \triangle CDB$, $\triangle ADC \sim \triangle ACB$, and $\triangle CDB \sim \triangle ACB$.)

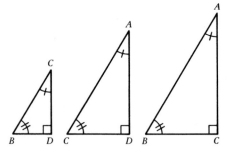

Corresponding Parts of Similar Triangles

When we write $\triangle ABC \sim \triangle XYZ$, we write the letters in the order corresponding to equal angles.

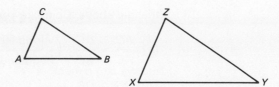

Thus from the statement △ABC ~ △XYZ we know that ∠A = ∠X, ∠B = ∠Y, and ∠C = ∠Z. This helps us keep track of corresponding sides in more complicated situations. For example, in the diagram below we have

△PQR ~ △PSQ.

Notice that corresponding sides appear in the correct order. For example, from

△PQR ~ △PSQ,

we see that

side $\overline{PQ}$ of △PQR corresponds to side $\overline{PS}$ of △PSQ,
side $\overline{QR}$ of △PQR corresponds to side $\overline{SQ}$ of △PSQ, and
side $\overline{PR}$ of △PQR corresponds to side $\overline{PQ}$ of △PSQ.

And corresponding sides are proportional:

$$\frac{PS}{PQ} = \frac{SQ}{QR} = \frac{PQ}{PR}$$

It is of course easier to remember these relationships for right triangles in terms of "short leg," "long leg," and "hypotenuse."

23. Consider this large right triangle cut into two smaller right triangles by the altitude drawn to the hypotenuse.

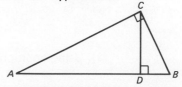

Decide which of the following statements are true, with reference to the

Lead a discussion on corresponding parts of similar figures. The next few exercises will be very confusing if the student does not understand this concept well.

To help keep the correspondence straight, suggest that the students refer to the sides as the long leg, the short leg, and the hypotenuse. Also, they may identify the triangles as small, medium, and large.

331 Central

triangle above. Compare your answers with those of your classmates.

(a) $\dfrac{AD}{AC} = \dfrac{DC}{BC}$ $\left(\dfrac{\text{long leg}}{\text{hypotenuse}} = \dfrac{\text{long leg}}{\text{hypotenuse}}\right)$ true

(b) $\dfrac{BC}{AB} = \dfrac{DB}{BC}$ true   (c) $\dfrac{AC}{BC} = \dfrac{DB}{DC}$ false   (d) $\dfrac{AD}{DC} = \dfrac{DC}{DB}$ true

(e) $(AD)(DB) = (DC)^2$ true   (f) $\dfrac{AC}{CD} = \dfrac{AB}{AC}$ false

Exercise 24 is a long exercise very valuable for student understanding. Do not rush. Some students may have difficulty.

24. In each case, calculate the length indicated, with reference to this diagram.

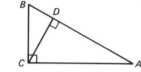

(a) If $BD = 4$ and $BA = 9$, find BC. 6

(b) If $BC = 8$ and $BD = 4$, find DA. 12

(c) If $AD = 24$ and $DB = 18$, find CA. $12\sqrt{7}$

(d) If $CA = 12$ and $DB = 10$, find AB. 18

(e) If $AD = 5$ and $DB = 2$, find CD. $\sqrt{10}$

(f) If $AB = 8$ and $AC = 5$, find CB. $\sqrt{39}$

Proving the Pythagorean Theorem by Using Similar Triangles

Exercise 25 is a good review of the chapter.

Point out to the students the contrast between this proof and that given in the Central section of Chapter 7.

25. In the picture below we have a right triangle with legs of lengths a, b, and hypotenuse c. The dotted line is the altitude drawn to the hypotenuse.

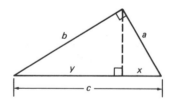

(a) Using appropriate similar triangles, show that

$$\dfrac{x}{a} = \dfrac{a}{c} \quad \text{and} \quad \dfrac{y}{b} = \dfrac{b}{c}.$$

332   8: Similar Figures

(b) Use part (a) to show that

$$a^2 = xc \quad \text{and} \quad b^2 = yc.$$

[*Hint:* Multiply each side of the first equation by *ac*. Multiply each side of the second equation by *bc*.]

(c) Use part (b) to show that

$$a^2 + b^2 = c^2.$$

A Line Dividing the Sides of a Triangle Proportionally

Suppose that a line ℓ intersects two sides of a triangle, as in this picture:

If $\dfrac{a}{b} = \dfrac{c}{d}$, then we say that *the line ℓ divides the two sides proportionally,* or that *the line ℓ divides the two sides into proportional segments.*

26. The line ℓ in the picture below is parallel to the base of the triangle.

(a) Prove that the smaller triangle must be similar to the larger. [*Hint:* Use the AAA condition for similar triangles.] corr. angles formed by a line crossing parallel lines

(b) Prove that $\dfrac{a+b}{a} = \dfrac{c+d}{c}$. [*Hint:* In similar triangles, the corresponding sides are _____.] proportional

(c) Show that $\dfrac{b}{a} = \dfrac{d}{c}$. [*Hint:* Use part (b) and some algebra.]
 Subtract $1 = \dfrac{a}{a} = \dfrac{c}{c}$ from both sides.

(d) Show that $\dfrac{a}{b} = \dfrac{c}{d}$. [*Hint:* Use part (c) and some algebra.] Invert.

(e) Show that $\dfrac{a+b}{b} = \dfrac{c+d}{d}$. [*Hint:* Use part (d) and some algebra.]
 Add $1 = \dfrac{b}{b} = \dfrac{d}{d}$ to both sides.

In Exercise 26(d) we proved the following statement:

> **A line drawn parallel to a side of a triangle cutting the other two sides divides these two sides into proportional segments.**

These memory devices help most students to keep the corresponding parts clearly in mind.

It may be helpful to draw a model and measure the segments.

Here is a way to remember the relationships which hold when a line divides two sides of a triangle into proportional segments. Referring to Exercise 26(b) we see

$$\frac{a+b}{a} = \frac{c+d}{d}$$

and we say to ourselves,

"whole left side over top left segment equals whole right side over top right segment."

Similarly, from part (c)

$$\frac{\text{"bottom"}}{\text{top}} = \frac{\text{bottom"}}{\text{top}},$$

from part (d)

$$\frac{\text{" top"}}{\text{bottom}} = \frac{\text{top"}}{\text{bottom}},$$

and from part (e)

$$\frac{\text{"whole"}}{\text{bottom}} = \frac{\text{whole"}}{\text{bottom}}.$$

If the dotted line in the diagram below is parallel to a side,

then we can solve for x by noting

$$\frac{x}{2} = \frac{6}{4} \qquad \left(\frac{\text{top}}{\text{bottom}} = \frac{\text{top}}{\text{bottom}}\right),$$

hence

$$x = (2)\left(\frac{6}{4}\right) = 3.$$

27. In each diagram the dotted line is parallel to a side. Find x.

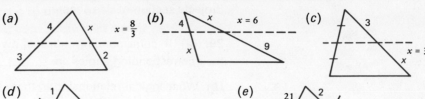

Exercise 27(e): Watch that students do not say that $\frac{2}{5} = \frac{x}{3}$. Point out that in the picture below $s/t = a/(a+b)$; $s/t \neq a/b$!

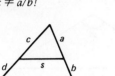

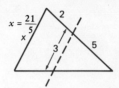

28. If ℓ, m, and n are parallel lines, show that $\dfrac{a}{b} = \dfrac{c}{d}$.

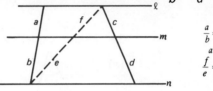

[*Suggestion:* Use the dotted line and Exercise 26(c) or (d) for two different triangles.]

Further Conditions for Similar Triangles

We next consider other conditions that determine if two triangles are similar. Recall conditions (*i*) and (*ii*) for similar polygons. Neither condition *by itself* is enough to ensure that a pair of polygons are similar (Exercise 13). But for *triangles* condition (*ii*) *by itself* ensures similarity (AAA condition for similar triangles). In the special case of triangles, it also turns out that condition (*i*) *by itself* ensures similar triangles. In other words, we have the following.

SSS CONDITION FOR SIMILAR TRIANGLES	If $\triangle ABC$ and $\triangle A'B'C'$ have their corresponding sides proportional, then they are similar.

It follows that if a pair of triangles satisfy condition (*i*) for similar polygons, then they automatically also satisfy condition (*ii*).

The proof of the SSS condition for similar triangles is given in the Projects section of this chapter.

29. (a) If corresponding sides of two triangles are proportional, then their corresponding angles are _____ . *equal*

(b) What logical relationship is there between the statement in part (a) and the statement in Exercise 16(a) ? *converses*

30. Consider these two triangles. Explain why $x = x'$, $y = y'$, and $z = z'$. SSS

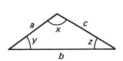 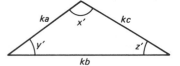

31. Find x. Explain your answers.

(a) 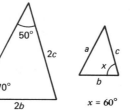 (b)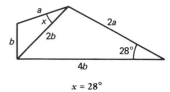

$x = 60°$ $x = 28°$

(c) 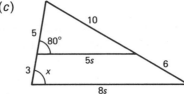 (d)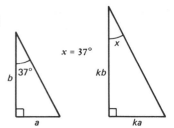

$x = 80°$ [*Hint:* Use the Pythagorean Theorem.]

Ex. 32(a), (b): Since the corresponding sides are proportional, the large triangles are similar. By AAA, the smaller triangles are similar, therefore the corresponding sides are proportional.

32. Find x. Explain your answers.

(a) $x = 6$ (b) $x = kb$

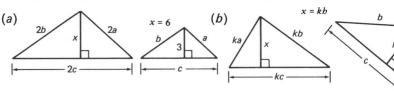

33. Suppose $\triangle ABC \sim \triangle A'B'C'$, with

336   8 : Similar Figures

$$\frac{A'B'}{AB} = \frac{A'C'}{AC} = \frac{B'C'}{BC} = k.$$

Exercise 33 prepares the student for computing areas of similar figures.

In the picture below we have drawn the altitudes from A and A' respectively. Explain each step.

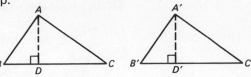

(a) $\triangle ADC \sim \triangle A'D'C'$. AAA

(b) $\dfrac{A'D'}{AD} = \dfrac{A'C'}{AC} = \dfrac{B'C'}{BC} = \dfrac{A'B'}{AB} = k.$ *property of similar triangles*

(c) Complete:

> In similar triangles, the ratio of corresponding altitudes is the same as the ratio of any two corresponding ___sides___.

Having studied the AAA condition and the SSS condition for similar triangles, one suspects that there might be an "SAS condition" for similar triangles.

SAS CONDITION FOR SIMILAR TRIANGLES

> If in $\triangle ABC$ and $\triangle A'B'C'$ we have $\dfrac{AB}{A'B'} = \dfrac{AC}{A'C'}$ and $\measuredangle A = \measuredangle A'$, then $\triangle ABC \sim \triangle A'B'C'$.

Note that in this condition we have a pair of sides of $\triangle ABC$ proportional to a corresponding pair of sides of $\triangle A'B'C'$, and the *included* angles equal. For example, in the picture below we have $\triangle ABC \sim \triangle A'B'C'$, since

$$\frac{AB}{A'B'} = \frac{6}{4} = \frac{AC}{A'C'} \quad \text{and} \quad \measuredangle A = 45° = \measuredangle A'.$$

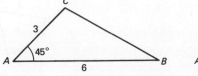 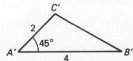

The proof of the SAS condition for similar triangles is given in the Projects section of this chapter. It is used to prove the AAA and SSS conditions for similar triangles.

34. In each case, explain why the two triangles must be similar.

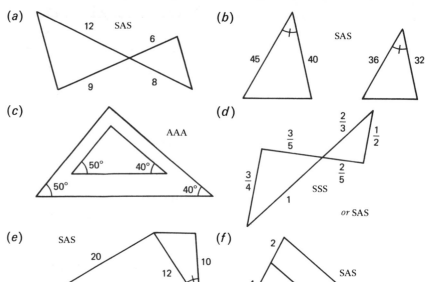

35. Why is the smaller triangle similar to the larger? Discuss with your classmates.

SAS: *common angle and*
$$\frac{3+1}{2} = \frac{4+2}{3}$$

Ratio of Similarity

Consider the similar polygons P and P' in this picture.

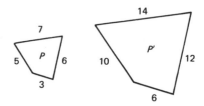

338 8: Similar Figures

The ratio of any sidelength of P' to the corresponding sidelength of P is 2. We say in this case that

"the ratio of similarity of P' to P is 2."

We obtain P' from P by "magnifying 2 times." Thus the ratio of similarity of P' to P measures the "amount of magnification" needed to obtain P' from P.

The ratio of any sidelength of P to the corresponding sidelength of P' is $\frac{1}{2}$. Hence we say

"the ratio of similarity of P to P' is $\frac{1}{2}$."

DEFINITION   If P and P' are similar polygons, then the *ratio of similarity of P' to P* is the ratio of any sidelength of P' to the corresponding sidelength of P.

As a memory aid, point out to students that if s and t are numbers, then the ratio of s to t is s/t. For polygons, the ratio of similarity of P' to P is

$$\frac{\text{side length of } P'}{\text{side length of } P}.$$

36. In each pair of similar polygons, find the *ratio of similarity of Q to P*

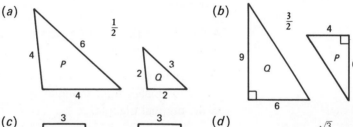

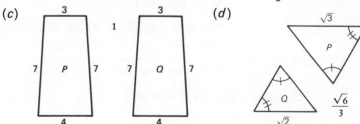

37. If the *ratio of similarity of P' to P* is k, then the *ratio of similarity of P to P'* is _____. Discuss with your classmates. $\frac{1}{k}$

38. If P and Q are similar polygons and the ratio of similarity of P to Q is 1, then P and Q are _____. Discuss with your classmates.
congruent

Important Ideas in This Chapter

Definitions   The *ratio* of s to t, if $t \neq 0$, is $\frac{s}{t}$.

Two sequences of numbers $a, b, c, \ldots$ and $a', b', c', \ldots$ are *proportional* if
$$\frac{a'}{a} = \frac{b'}{b} = \frac{c'}{c} = \ldots.$$

Two polygons are *similar* if their corresponding sides are proportional and their corresponding angles are equal.

If P and P' are similar polygons, then the *ratio of similarity* of P' to P is the ratio of any sidelength of P' to the corresponding sidelength of P.

Theorems

If corresponding angles of $\triangle ABC$ and $\triangle A'B'C'$ are equal, then $\triangle ABC \sim \triangle A'B'C'$.

If corresponding sides of $\triangle ABC$ and $\triangle A'B'C'$ are proportional, then $\triangle ABC \sim \triangle A'B'C'$.

If a pair of sides of $\triangle ABC$ are proportional to a corresponding pair of sides of $\triangle A'B'C'$, and if the included angles are equal, then $\triangle ABC \sim \triangle A'B'C'$.

A line drawn parallel to a side of a triangle divides the other sides into proportional segments.

The altitude drawn to the hypotenuse of a right triangle divides the triangle into two smaller right triangles which are similar to each other and also to the original triangle.

In similar triangles, the ratio of corresponding altitudes is equal to the ratio of any two corresponding sides.

Words

ratio, proportional, similar, ratio of similarity

Symbols

$\sim$ (is similar to)

Summary

This chapter was concerned with similar figures. Particular attention was paid to the AAA, SSS, and SAS conditions for similar triangles, and to similar right triangles. In the next chapter we will learn how the perimeters and areas of similar figures in a plane compare. We will also study similar solids, comparing their volumes and surface areas.

Historical Note

The opening problem of this chapter shows how similar triangles can be used to determine unknown distances. From the simple idea underlying the solution of that problem, the branch of mathematics called trigonometry (literally, "triangle measurement") gradually grew. Trigonometry is essential to fields requiring the measurement of distances to inaccessible objects, as in surveying and astronomy. Therefore it is not surprising that the first trigonometric tables were compiled by an astronomer, Hipparchus of Nicaea, who lived about 150 B.C. A more advanced treatment of trigonometry, based on the writings of Hipparchus, later appeared as part of the *Almagest* by the mathematician and astronomer Claudius Ptolemy (about A.D. 150). This great work, which expounds the Ptolemaic theory of the universe, remained the standard reference in astronomy until A.D. 1600 when Copernicus presented his revolutionary views, placing the sun, rather than the earth, at the center of the solar system.

Central Self Quiz

1. Which pairs of polygons are similar? If they are similar, explain why, and find the ratio of similarity k of Q to P.

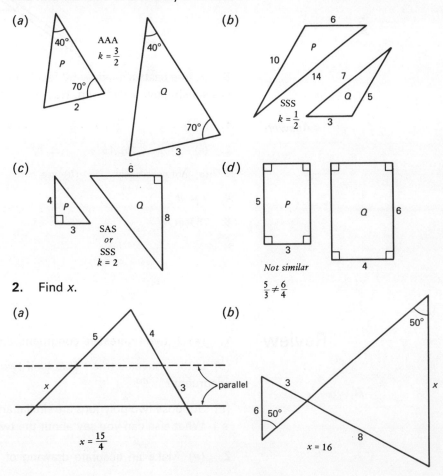

2. Find x.

(c) 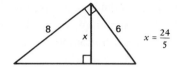 (d)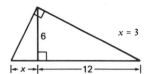

$x = \frac{24}{5}$ $x = 3$

3. (a) Are two parallelograms similar if their corresponding angles are equal?

(b) Are two rectangles similar if their corresponding sides are proportional?

Exercise 4 is a preview of some Projects constructions.

4. The dotted lines are parallel. Find x in terms of a and b.

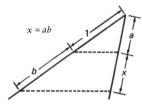

$x = ab$

5. A tree casts a shadow 50 feet long. At the same time a man 6 feet tall casts a shadow 4 feet long. How tall is the tree?

Answers
1. (a) $k = \frac{3}{2}$ (b) $k = \frac{1}{2}$ (c) $k = 2$ (d) not similar, $\frac{6}{5} \neq \frac{4}{3}$
2. (a) $\frac{15}{4}$ (b) 16   (c) $\frac{24}{5}$ (d) 3
3. (a) not necessarily   (b) yes
4. $x = ab$
5. 75 feet

Review
1. (a) If two figures are congruent, then are they similar? *yes*

(b) Is the converse true? That is, if two figures are similar, then are they congruent? *no*

(c) Suppose two polygons are similar and the ratio of corresponding sides is 1. What else can you say about the two polygons? *congruent*

2. (a) Make an accurate drawing of two hexagons which have their

corresponding sides proportional but which are *not* similar.

(*b*) Make an accurate drawing of two hexagons which have their corresponding angles equal but which are *not* similar.

3. (*a*) Make an accurate drawing of a quadrilateral *Q* which is not a rectangle. Can you draw another quadrilateral *Q'* each of whose sides is twice as long as the corresponding side of *Q* and such that *Q'* is not similar to *Q*? If you think so, do it.

(*b*) Make an accurate drawing of a rectangle *R*. Can you draw another rectangle *R'* whose sides are twice as long as the corresponding sides of *R* and such that *R'* is not similar to *R*? Explain.

4. (*a*) If two polygons have corresponding angles equal, then must they be similar?

(*b*) If two triangles have corresponding angles equal, then must they be similar?

(*c*) If two triangles have corresponding sides proportional, then must they be similar?

(*d*) If two polygons have corresponding sides proportional, then must they be similar?

(*e*) What is the condition for two polygons to be similar? What if the polygons happen to be triangles?

5. Which pairs of triangles are similar? In case the triangles are similar, find the ratio of similarity *k* of the triangle on the right to the triangle on the left. Explain why the triangles must be similar, *without* making an accurate drawing. *Note.* The figures are not always accurately drawn!

Conditions (i) and (ii) for similarity should come into the discussion in Exercise 5.

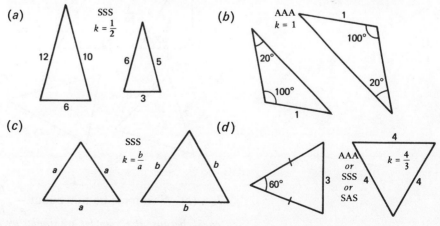

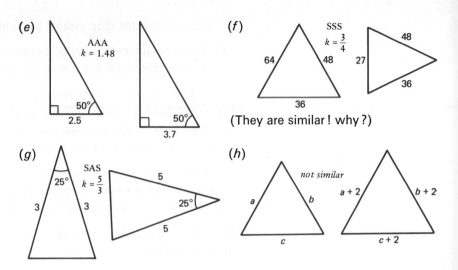

(They are similar! why?)

6. Which pairs of polygons are similar? In case the polygons are similar, find the ratio of similarity of the polygon on the right to the polygon on the left. Explain why the polygons must be similar, *without* making an accurate scale drawing. *Note.* The figures are not always drawn accurately!

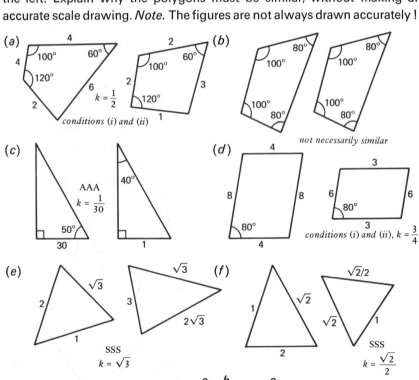

7. (a) What must be true about $\dfrac{a}{a'}$, $\dfrac{b}{b'}$, and $\dfrac{c}{c'}$ in the following figures? equal, because the triangles are similar by AAA

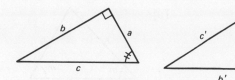

(b) If $a = c'$, $b = 3$, and $b' = 1$, then find the area of each triangle.

8. (a) Explain why if the point P is chosen on line ℓ so that $AP + PB$ is as small as possible, then $\triangle APC \sim \triangle BPD$.
AAA: $\angle 1 = \angle 2$ and right angles

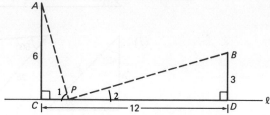

(b) Suppose P has been chosen so that $AP + PB$ is as small as possible. If $x = PD$, then explain why

$$\frac{x}{3} = \frac{12 - x}{6} \quad \begin{array}{l}\textit{corr. sides of}\\ \textit{similar triangles}\end{array}$$

(c) Use part (b) to help you find the length of the shortest path from A to B touching line ℓ. $x = PD = 4$, $PB = 5$, $AP = 10$, $AP + PB = 15$

(d) Find the length of the shortest path from A to B touching line ℓ, *without* using similar triangles. [*Suggestion:* Consider the reflection B^* of B across ℓ. Then $\overline{AB^*}$ is the hypotenuse of a certain right triangle.] $AB^* = \sqrt{9^2 + 12^2} = 15$

9. Without making another drawing, explain what is wrong in each picture.

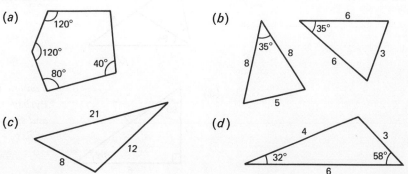

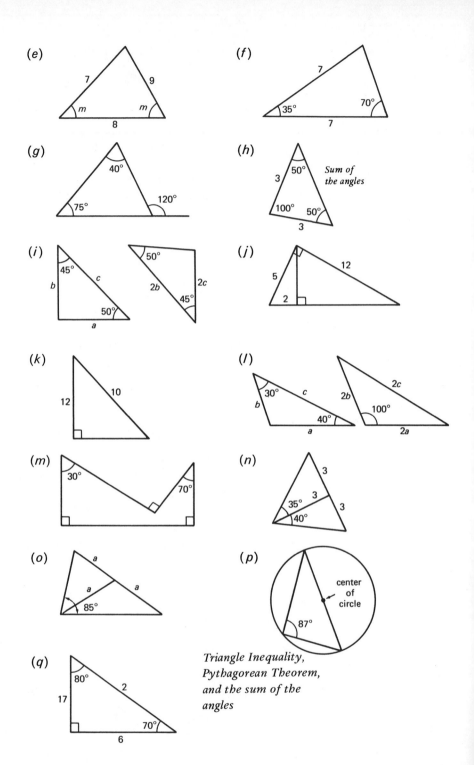

Triangle Inequality, Pythagorean Theorem, and the sum of the angles

10. Find x and y. Explain your answers.

(a)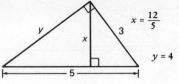

(b)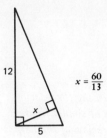

(c)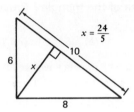

(d)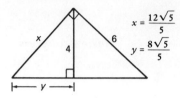

(e)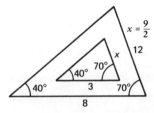

(f)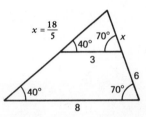

(g)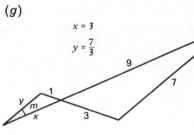

(h)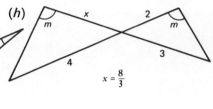

(i)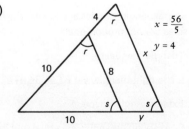

(j)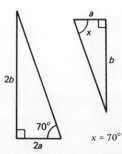

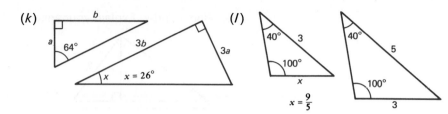

Exercise 11, presenting the relation between the areas of similar figures, is an important preview of the next chapter.

11. If you double the lengths of the legs of a right triangle, what happens to
   (a) the length of the hypotenuse? Prove it.
   (b) the area of the triangle? Prove it.

Review Answers

1. (a) yes   (b) no   (c) They are congruent.

3. (a) Let Q be a rhombus with a 60° angle and all sides of length 1, and let Q' be a rhombus with a 45° angle and all sides of length 2.

   (b) No, corresponding sides proportional and corresponding angles equal imply similarity.

4. (a) no   (b) yes   (c) yes   (d) no

5. (a) similar, corresponding sides proportional, $k = \frac{1}{2}$

   (b) similar, corresponding angles equal, $k = 1$

   (f) similar, corresponding sides proportional $(\frac{27}{36} = \frac{36}{48} = \frac{48}{64})$, $k = \frac{3}{4}$

   (g) similar, corresponding angles equal, $k = \frac{5}{3}$

   (h) not necessarily similar

6. (a) similar, $k = \frac{1}{2}$ (b) not necessarily similar

   (e) similar $\left(\frac{\sqrt{3}}{1} = \frac{2\sqrt{3}}{2} = \frac{3}{\sqrt{3}}\right)$, $k = \sqrt{3}$ (f) similar, $k = \frac{\sqrt{2}}{2}$

7. (b) $\frac{9\sqrt{2}}{8}$ and $\frac{\sqrt{2}}{8}$

9. (a) The sum of the angles in a pentagon is 540°. *The other angle must be* 180°.

   (b) Corresponding angles are equal (why?), so the triangles are similar. But corresponding sides are *not* proportional!

   (c) Triangle Inequality

   (d) It is a right triangle, yet $6^2 \neq 3^2 + 4^2$!   (e) isosceles, but $7 \neq 9$

   (g) Exterior Angle Theorem   (f) isosceles, but $70 + 70 + 35 \neq 180$

   (i) Corresponding angles are equal, but corresponding sides are not proportional.
   (h) *Sum of the angles is* 200°.

(j) $\frac{5}{2} \neq \frac{13}{5}$ similar by AAA

(k) The hypotenuse cannot be *shorter* than a leg!

(l) Since corresponding sides are proportional, the triangles are similar, which implies that the missing angle in the left triangle is 100°. But then the sum of the angles is not 180°.   (m) *There are 550° in the pentagon.*

(n) Isosceles triangles give the missing angles. Look at the sum.

(o): *The figure must be a right triangle.*

(p) Draw the radius to the vertex of the 87° angle. Compare with situation in part (o).

10. (a) $y = \sqrt{25 - 9} = 4$ and $\frac{x}{3} = \frac{y}{5} = \frac{4}{5}$, so $x = \frac{12}{5}$.

(b) $\frac{x}{5} = \frac{12}{13}$, $x = \frac{60}{13}$

(c) The large triangle is a right triangle because $10^2 = 6^2 + 8^2$; $x = \frac{24}{5}$.

(d) $x = \frac{12}{\sqrt{5}}$, $y = \frac{8}{\sqrt{5}}$ (e) $x = \frac{9}{2}$ (f) $x = \frac{18}{5}$

(g) $x = 3$, $y = \frac{7}{3}$ (h) $x = \frac{8}{3}$ (i) $x = \frac{56}{5}$, $y = 4$

(k) $x = 26°$ (l) $x = \frac{9}{5}$

11. (a) doubles   (b) quadruples

Review Self Test

1. Explain why the triangles are similar.

(a)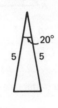

SAS

(b)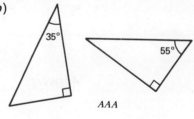

AAA

2. Prove that $\frac{1}{x} = \frac{x}{1-x}$ in the diagram below.

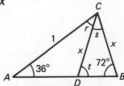

[*Hint:* Show that $\angle t = 72°$. Then $\angle s = 36°$. Then $\triangle ACB$ is isosceles, so $AB = 1$. Why is $AD = x$?]

349 Review

3. A girl looks into a mirror which is lying on the ground and sees the top of a tree. Her eye level is 5′ from the ground and she is standing 27 feet from the tree, which is 24 feet from the mirror. How tall is the tree?

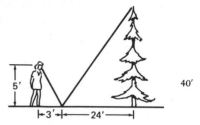

4. Find the area of $\triangle ABC$ if $CD = 3$ and $CA = 6$.

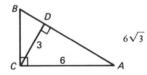

5. In the diagram, $\dfrac{CD}{DA} = \dfrac{3}{2}$.

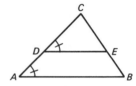

Find (a) $\dfrac{CE}{EB}$ (b) $\dfrac{CA}{CD}$ (c) $\dfrac{EB}{EC}$ (d) $\dfrac{DE}{AB}$.

Answers

1. Corresponding angles are equal in each case.

2. $\angle t = 72°$ because $\triangle CDB$ is isosceles, hence $\angle s = 36°$ and $\angle r + \angle s = 72°$ (sum of angles of $\triangle ACB = 180°$). Thus $\triangle ACB \sim \triangle CDB$ (corresponding angles equal). Note that $AD = DC = x$ (because $\angle r = 36°$, making $\triangle ACD$ isosceles). Then $DB = 1 - x$. By similar triangles, then,

$$\dfrac{AC}{CB} = \dfrac{CB}{BD} \quad \text{or} \quad \dfrac{1}{x} = \dfrac{x}{1-x}.$$

3. 40 feet

4. $6\sqrt{3}$

5. (a) $\dfrac{3}{2}$ (b) $\dfrac{5}{3}$ (d) $\dfrac{3}{5}$ (c) $\dfrac{2}{3}$

350 8: Similar Figures

Projects

Rep-Tiles

Starting with any triangle, we can fit together four copies of that triangle to form a larger similar triangle.

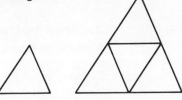

Four copies of any parallelogram fit together to form a similar parallelogram.

Two copies of an isosceles right triangle fit together to form a similar triangle.

A figure is called a *rep-tile* if copies of the figure fit together to form a larger *similar* figure. The "rep-" refers to the fact that the figure "repeats" or "replicates" itself in a larger similar figure. The "-tile" refers to the fact that if copies of the larger figure are fitted together the same way and this is repeated over and over, we tile the plane.

Most students will benefit by working with models in Exercise 1.

1. Show that four copies of the trapezoid below can be fitted together to form a similar trapezoid.

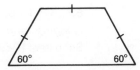

Experiment using freehand sketches. If this still is difficult, make a model with four copies of the trapezoid.

2. A certain rectangle has shorter side of length b and longer side of length *a*. If two copies of this rectangle fit together to form a similar rectangle, prove that

$$\frac{a}{b} = \sqrt{2}.$$

Most students will need to work with models in Exercise 3.

3. Show that four copies of the "Sphinx" pentagon below can be fitted together to form a similar pentagon.

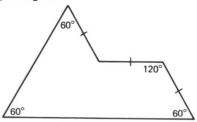

Experiment using freehand sketches. If this is still difficult, make a model with four copies of the above figure.

4. Find a triangle such that *three* copies can be fitted together to form a similar triangle. [*Hint:* It is a well-known triangle!] 30°-60° right triangle

5. For each positive integer *N*, consider a right triangle with legs of lengths 1 and *N*. We can fit together $N^2 + 1$ copies of the triangle to form a similar triangle. The picture below shows the case $N = 3$.

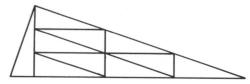

Show the cases for $N = 2$, $N = 1$, and $N = 4$.

6. Read about Rep-Tiles in Martin Gardner's book, *The Unexpected Hanging and Other Mathematical Diversions* (New York: Simon and Schuster, 1969), chapter 19.

More on Pentominoes

A *pentomino* is a figure formed by fitting together five squares all the same size. The twelve different types of pentominoes and some problems concerning them are discussed in the Projects section of Chapter 2. An interesting "triplication problem" with pentominoes is the following.

Select one pentomino. Now, fit nine of the remaining pentominoes together to form a large figure *similar* to the pentomino first selected. (The ratio of similarity will be 3. Why?) For example, the picture below shows how nine pentominoes, different from the Greek Cross pentomino, fit together to form a similar Greek Cross.

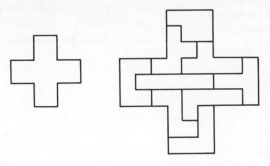

7. Build a set of pentominoes and try the triplication problem with different shaped pentominoes. Read about these problems in Martin Gardner's book, *The Scientific American Book of Mathematical Puzzles and Diversions* (New York: Simon and Schuster, 1959), chapter 13.

Miscellaneous Problems

For a thorough grasp of this problem, a classroom experiment with string and rods of various lengths is needed.

8. Here is an old "trick" question. A 10 ft. pole and a 15 ft. pole are a certain distance apart. Lines are drawn from the top of each to the bottom of the other, as shown in the picture.

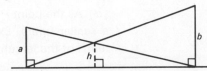

If the lines intersect 6′ above the ground, what is the distance between the poles?

This is a "trick" question because the poles can be *any* distance apart! The two lines always will intersect 6′ above the ground. You may enjoy proving the following more general result. In the following figure,

prove that $h = \dfrac{ab}{a+b}$. [*Hint:* Label the horizontal line segments and use similar triangles.]

Observe that this gives an expression for *h* in terms of *a* and *b* which does *not* depend on the distance between the vertical lines.

Encourage a student to make a pantograph as a project.

Thomas Jefferson made a pantograph in an effort to write more than one letter at a time.

9. A *pantograph* is a drawing instrument which can be used to enlarge or reduce the size of a drawing by copying it in a different scale. In other words, given a figure, the pantograph is used to draw a similar figure. The picture below indicates how the pantograph is constructed. Point *A* is a pivot fastened to the table. A pointer is attached at *P'* and a pen at *P*.

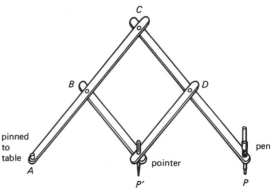

(a) Show that if the instrument is constructed so that in *some* position *A*, *P'*, and *P* are collinear, $\overline{BP'}$ is parallel to $\overline{CP}$, and $\overline{DP'}$ is parallel to $\overline{AC}$, then this will be true when the instrument is in *any* position.

(b) Show that if the instrument is moved so that the pen and pointer are at new positions *Q* and *Q'* respectively, then $\triangle AP'Q' \sim \triangle APQ$, where *P* and *P'* are the original positions as in part (a). [*Hint:* If the new position of *B* is *E* and the new position of *C* is *F*, then

$$\frac{AQ'}{AQ} = \frac{EQ'}{FQ} = \frac{BP'}{CP} = \frac{AP'}{AP}.\text{]}$$

(c) Show that $\dfrac{PQ}{P'Q'}$ is the same for all positions of the pantograph.

(d) As the point *P'* traces out a figure, explain why the point *P* traces out a similar figure. Express how much the pantograph enlarges figures, in terms of the lengths of the arms of the pantograph.

(e) *Extra Credit:* Construct a pantograph which can be used to enlarge drawings 3 times their original size.

(f) How could you use your pantograph to *reduce* the size of drawings? *Interchange the pen and pointer.*

Proofs of Conditions for Similar Triangles

Exercises 10-14 are for the student who does not accept intuitively a few key theorems found in the Central section.

In the following exercises we will prove some of the properties of similar figures which we stated and used in the Central section. Recall that we stated the AAA condition for similar triangles without proving it. We then used this property to prove that *a line drawn parallel to a side of a triangle divides the other two sides into proportional segments* (Central Exercise 26). In Projects Exercise 10 we prove this latter fact using properties of area, *without* using any properties of similar figures. Then we use this in Projects Exercise 11 to prove the AAA condition for similar triangles.

10. In this exercise we will prove:

A line cutting two sides of a triangle and parallel to the third side divides the two sides into proportional segments.

In other words, in the figure below, if the line segment $\overline{DE}$ is parallel to $\overline{BC}$, then $\dfrac{AD}{DB} = \dfrac{AE}{EC}$.

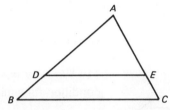

The proof depends on properties of area of triangles.

(a) In the figure below we have drawn the segment $\overline{BE}$ and an altitude of $\triangle ADE$.

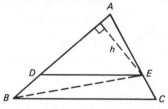

Prove that

$$\frac{\text{area of } \triangle ADE}{\text{area of } \triangle BDE} = \frac{AD}{DB}.$$

[*Hint:* Express the areas of both triangles using the altitude *h*.]

355 Projects

(b) Prove that

$$\frac{\text{area of } \triangle ADE}{\text{area of } \triangle CED} = \frac{AE}{EC}.$$

(c) Prove that

$$\text{area of } \triangle BDE = \text{area of } \triangle CED.$$

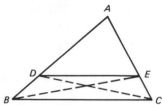

[*Hint:* Do they have the same base and altitude?]

(d) Using parts (a), (b), and (c), show that

$$\frac{AD}{DB} = \frac{AE}{EC}$$

as we wished to prove.

11. We now use the result of the preceding exercise to prove the AAA condition for similar triangles.

 If $\triangle ABC$ and $\triangle A'B'C'$ have corresponding angles equal, then $\triangle ABC \sim \triangle A'B'C'$.

By definition, two polygons are similar if their corresponding angles are equal and their corresponding sides are proportional. Hence the AAA condition for similar triangles can be stated equivalently as follows:

 If $\triangle ABC$ and $\triangle A'B'C'$ have corresponding angles equal, then their corresponding sides are proportional; that is,

$$\frac{AB}{A'B'} = \frac{AC}{A'C'} = \frac{BC}{B'C'}.$$

The last statement will be proved in the following parts of this exercise.

In the following figure we depict the given triangles $\triangle ABC$ and $\triangle A'B'C'$ with $\angle A = \angle A'$, $\angle B = \angle B'$, and $\angle C = \angle C'$. We have chosen points D and E on the sides of $\triangle ABC$ such that $AD = A'B'$ and $AE = A'C'$.

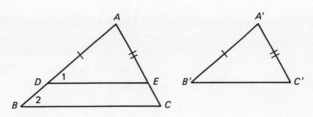

(a) Why is $\triangle ADE \cong \triangle A'B'C'$? SAS

(b) Using part (a), explain why $\angle 1 = \angle 2$. $\angle B = \angle B'$ and corr. angles are equal.

(c) Why is $\overline{DE}$ parallel to $\overline{BC}$? $\angle 1 = \angle 2$

(d) $\dfrac{AD}{DB} = \dfrac{AE}{EC}$ because _____.   *Projects Exercise 10*

(e) Deduce from part (d) that

$$\frac{AB}{AD} = \frac{AC}{AE}.$$

[*Hint:* Deduce that

$$\frac{DB}{AD} = \frac{EC}{AE},$$

then add 1 to each side of the equation and use some algebra. Note that $DB + AD = AB$.]

(f) Using parts (a) and (e), deduce that

$$\frac{AB}{A'B'} = \frac{AC}{A'C'}.$$

(g) In order to complete the proof of the AAA condition for similar triangles, we need to show that

$$\frac{AC}{A'C'} = \frac{BC}{B'C'}.$$

Then we will have proved that $\triangle ABC$ and $\triangle A'B'C'$ have their corresponding sides proportional; that is,

$$\frac{AB}{A'B'} = \frac{AC}{A'C'} = \frac{BC}{B'C'}.$$

Explain how you would go about completing the proof.

(h) In arriving at the result in part (f) we assumed that we could choose

357 Projects

D and E as shown in the figure above. But we might be forced to choose $D = B$ (in case $A'B' = AB$), or we might be forced to choose D on the extension of the side AB past B (in case $A'B' > AB$). Discuss how this would effect our proof. What changes would need to be made in our argument?

12. In this exercise we prove the *converse* of the result in Exercise 10. We shall prove:

If a line divides two sides of a triangle into proportional segments, then it is parallel to the third side.

In other words, in the figure, if $\dfrac{AD}{DB} = \dfrac{AE}{EC}$, then $\overline{DE}$ is parallel to $\overline{BC}$.

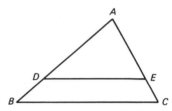

We have drawn the picture below as if $\overline{DE}$ were not parallel to $\overline{BC}$. We have drawn the line through B parallel to $\overline{DE}$, intersecting the line through A and C in C'. We must prove that if $\dfrac{AD}{DB} = \dfrac{AE}{EC}$, then we must actually have $C' = C$.

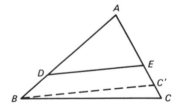

(a) Why is $\dfrac{AD}{DB} = \dfrac{AE}{EC'}$? [*Hint:* Exercise 10] $DE \parallel BC'$

(b) Using part (a) and the given fact that $\dfrac{AD}{DB} = \dfrac{AE}{EC}$, deduce that

$$\dfrac{AE}{EC} = \dfrac{AE}{EC'}.$$

(c) Why must $C' = C$? [*Hint:* What is true about EC and EC'?] $EC = EC'$, so C and C' must be the same point.

13. Let us use the result of the preceding exercise to prove the SAS condition for similar triangles:

If in $\triangle ABC$ and $\triangle A'B'C'$ we have $\dfrac{AB}{A'B'} = \dfrac{AC}{A'C'}$ and $\measuredangle A = \measuredangle A'$, then $\triangle ABC \sim \triangle A'B'C'$.

In the picture below, we have $\triangle ABC$ and $\triangle A'B'C'$ with $\dfrac{AB}{A'B'} = \dfrac{AC}{A'C'}$ and $\measuredangle A = \measuredangle A'$. We have chosen points D and E such that $AD = A'B'$ and $AE = A'C'$.

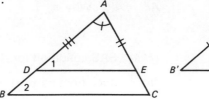

(a) Why is $\triangle ADE \cong \triangle A'B'C'$?

(b) Using the given fact that $\dfrac{AB}{A'B'} = \dfrac{AC}{A'C'}$ and part (a), prove that

$$\dfrac{AB}{AD} = \dfrac{AC}{AE}.$$

(c) Deduce from part (b) that

$$\dfrac{AD}{DB} = \dfrac{AE}{EC}.$$

[*Hint:* Use $AB = AD + DB$, $AC = AE + EC$, and some algebra.]

(d) Deduce from part (c) and the last exercise that $\overline{DE}$ is parallel to $\overline{BC}$; hence $\measuredangle 1 = \measuredangle 2$.

(e) Why is $\measuredangle B' = \measuredangle B$?

(f) Why does it follow that $\triangle ABC \sim \triangle A'B'C'$? [*Hint:* AAA condition for similar triangles]

14. We now can prove the SSS condition for similar triangles:

If $\triangle ABC$ and $\triangle A'B'C'$ have their corresponding sides proportional, then $\triangle ABC \sim \triangle A'B'C'$.

In the figure on the following page we have $\triangle ABC$ and $\triangle A'B'C'$ with

$\dfrac{AB}{A'B'} = \dfrac{AC}{A'C'} = \dfrac{BC}{B'C'}$. We shall prove that $\triangle ABC \sim \triangle A'B'C'$.

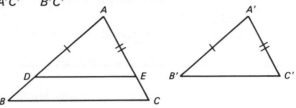

(a) In the above figure, D and E are chosen such that $AD = A'B'$ and $AE = A'C'$. Why is

$$\triangle ADE \sim \triangle ABC ?$$

[*Hint:* SAS condition for similar triangles]

(b) Deduce from part (a) that

$$\dfrac{DE}{BC} = \dfrac{AE}{AC} . \quad \text{condition (i) for similar figures}$$

(c) Why is $\dfrac{AE}{AC} = \dfrac{B'C'}{BC}$? [*Hint:* Replace AE by $A'C'$.]

(d) Deduce from parts (b) and (c) that $DE = B'C'$.

(e) Prove that $\triangle ADE \cong \triangle A'B'C'$. SSS

(f) Why is $\triangle ABC \sim \triangle A'B'C'$? [*Hint:* parts (a) and (e)] AAA

Circular Reasoning

Have a general class discussion here.

Two friends, Bill and Alex, are discussing a proof that the sum of the angles of any triangle is 180°. Their conversation is as follows.

BILL: Prove that the sum of the angles in any triangle is 180°.

ALEX: O.K. Consider a triangle with angles 1, 2, and 3. Let $\angle 4$ be the exterior angle opposite $\angle 1$ and $\angle 2$.

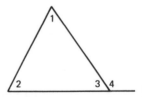

Now $\angle 1 + \angle 2 = \angle 4$, since any exterior angle of a triangle is equal to the sum of the two opposite interior angles. Hence

$\angle 1 + \angle 2 + \angle 3 = \angle 4 + \angle 3 = 180°$. Q.E.D.

BILL: That is a nice proof. But I am a little dissatisfied. I won't really be convinced until you prove the theorem about exterior angles which you used. How do you know that any exterior angle is equal to the sum of the two opposite interior angles?

ALEX: Ah, that is easy to prove. Consider a triangle with exterior angle $\angle 1$ and opposite interior angles $\angle 2$ and $\angle 3$.

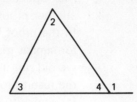

Now, $\angle 1 = 180° - \angle 4$, since $\angle 1 + \angle 4 = 180°$, and $\angle 2 + \angle 3 = 180° - \angle 4$, since $\angle 2 + \angle 3 + \angle 4 = 180°$. Thus $\angle 1 = \angle 2 + \angle 3$, as we wanted to prove.

BILL: Wait a minute. How do you know $\angle 2 + \angle 3 + \angle 4 = 180°$?

ALEX: I just proved a moment ago that the sum of the angles of any triangle is 180°.

BILL: But in your proof that the sum of the angles is 180°, you needed to know that the theorem about exterior angles is true!

ALEX: Of course the theorem about exterior angles is true! I just proved it!

BILL: But in your proof of the theorem about exterior angles, you needed to know that the sum of the angles in any triangle is 180°!

ALEX: Which I proved!

BILL: Now wait a minute . . . !

The argument could go on forever, but Alex is guilty of a logical error called "circular reasoning." He wants to prove a certain statement, call it statement A. In order to prove that statement A is true he needs to know that another statement, statement B, is true. But in showing that statement B is true, he assumes that statement A is true. He says, "Statement A is true, because statement B is true. Statement B is true because statement A is true." He has reasoned in a circle. His reasoning leads him back to where he started from, without proving anything.

15. In proving the fundamental results about similar triangles in Exercises 11–14, we needed to use the result proved in Exercise 10, namely, that *a line drawn parallel to a side of a triangle divides the other two sides into proportional segments*. Alex Smart says, "The proof given in Exercise 10 was very complicated, but the proof given in Central Exercise 26 was simple. Why go through all the trouble of giving the complicated proof in Exercise 10 of this section?"

Alex appears to have a good point here. Can you tell him why we gave another proof in Exercise 10? Discuss this with your classmates. [*Hint:* In the proof in Central Exercise 26, we assumed that the AAA condition for similar triangles was true. In the proof in Exercise 10 of this section, we did *not* use the AAA condition for similar triangles. Keep in mind that the result proved in Exercise 10 was needed to prove the AAA condition for similar triangles in Exercise 12.]

Chapter 9 Perimeter, Area, and Volume of Similar Figures

A firm grasp of the ideas in this chapter will serve the student well no matter what his career may be. Whether one is estimating the cost of a pile of gravel, computing the increase in cost if one doubles the dimensions of a carpet, or investigating the properties of an atomic nucleus, an understanding of distance, area, and volume for two similar figures is essential.

Be sure to give adequate attention to the volumes of similar solids as well as to the area of similar figures. You might bring to class two blocks of wood (or balls) of different size but of the same material, and ask the class to guess their comparative weights.

All students should find Projects Exercises 1-7 most enjoyable. The remaining exercises are chiefly proofs of some theorems used in the Central section. Students who show an interest in the proofs should, of course, be encouraged to go through these exercises.

Central

Very few students will be able to guess the weight of the large statue. We will return to the problem later. If you feel the opening problem has not generated enough interest, you might direct attention to Exercise 26.

Do not give the answer to the statue problem now, even if someone guesses correctly.

Perimeters of Similar Polygons

The perimeter is a number.

A sculptor makes a small marble statue of a man. The statue is 1 foot high and weighs 12 pounds. The sculptor intends to make a marble statue of exactly the same shape but 10 feet high. Guess the weight of the large statue.   *12,000 pounds*

Since the weight of the statue depends on its volume, the solution of the opening problem requires a knowledge of how volume changes when an object is changed in size without changing its shape. We will see that there is a simple way to calculate the ratio of the volumes and other quantities associated with two similar solids, if we know the ratio of similarity.

We begin by considering polygons in a plane, examining how the perimeter of a polygon changes when we change the size without changing the shape.

The perimeter of a polygon is the "distance around" the polygon. The precise definition is given as follows.

365

DEFINITION   The *perimeter* of a polygon is the sum of the lengths of its sides.

1. The picture shows a pair of similar polygons.

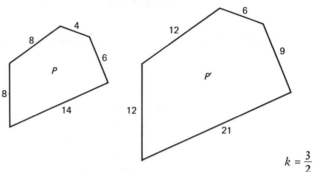

$$k = \frac{3}{2}$$

(a) Find the ratio of similarity of P' to P (sidelength of P' over the corresponding sidelength of P). Does the number depend on which pair of corresponding sides you choose? *The number cannot depend on the corr. sides chosen. If it does, the polygons are not similar.*

(b) Find the ratio of the perimeter of P' to the perimeter of P. How does this number compare with the number obtained in part (a)? $\frac{3}{2}$

2. Suppose $P \sim P'$.

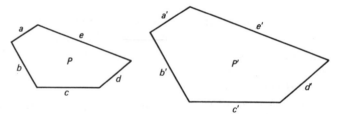

(a) Let $k = \dfrac{a'}{a}$ = the ratio of similarity of P' to P. Why are all the numbers $\dfrac{b'}{b}, \dfrac{c'}{c}, \dfrac{d'}{d},$ and $\dfrac{e'}{e}$ also equal to k? *a condition for similarity*

(b) Using part (a), show that
$$a' + b' + c' + d' + e' = k(a + b + c + d + e).$$

(c) $\dfrac{\text{perimeter }(P')}{\text{perimeter }(P)} = \underline{}$ $\dfrac{a' + b' + c' + d' + e'}{a + b + c + d + e} = k$

(d) Complete the following statement:

366   9: Perimeter, Area, and Volume of Similar Figures

> If two polygons are similar, then the ratio of their perimeters is equal to the ratio of the lengths of any two corresponding *sides*.

3. If P and Q are similar polygons and the ratio of similarity of Q to P is 3, then what is the ratio of the perimeter of Q to the perimeter of P?   3

4. (a) If you "double the sides" of a polygon, what happens to the perimeter?   *doubled*

(b) If you "halve the sides" of a polygon, what happens to the perimeter?
                                                                                                halved

5. (a) What is the ratio of similarity of the small triangle to the larger? $k = \frac{1}{2}$

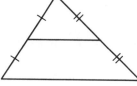

You may wish to summarize Exercises 1-5 before continuing.

(b) What is the ratio of the perimeter of the small triangle to the perimeter of the larger? $\frac{1}{2}$

Areas of Similar Polygons

We now consider how the *area* of a polygon behaves when we enlarge (or reduce) the polygon without changing its shape.

You might think at first that if you "double the sides" then you "double the area," but this is false! For example, consider the two squares below.

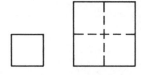

This picture helps students see why area behaves as it does under similarity. It would take twice the string to surround the larger square, but four times as much paint to paint it.

The square on the right has "double the side" of the square on the left (the ratio of similarity is 2). But the square on the right can be cut into 4 squares, each congruent to the one on the left. Hence, the square on the right has *4 times* the area of the square on the left.

6. (a) Draw a square S and a larger square S' such that the ratio of

367   Central

similarity of S' to S is 3. Show how to cut S' into 9 squares, each congruent to S.

$$\frac{\text{area } (S')}{\text{area } (S)} = \underline{} \quad 9$$

(b) Draw an equilateral triangle T and a larger equilateral triangle T' such that the ratio of similarity of T' to T is 2. Show how to cut T' into 4 triangles, each congruent to T.

$$\frac{\text{area } (T')}{\text{area } (T)} = \underline{} \quad 4$$

(c) Draw an equilateral triangle T and a larger equilateral triangle T' such that the ratio of similarity of T' to T is 3. Show how to cut T' into 9 triangles, each congruent to T.

$$\frac{\text{area } (T')}{\text{area } (T)} = \underline{} \quad 9$$

7. In each of the following cases, find the ratio of similarity of the polygon on the right to the polygon on the left, the ratio of the perimeters, and the ratio of the areas.

Exercises 6 and 7, though simple, are important. Students should stop and think about k^2. A brief lecture after they have completed the exercises may be in order.

(a)

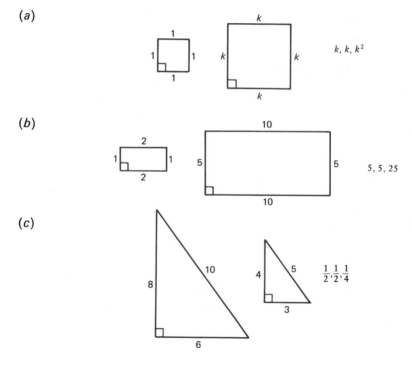

k, k, k^2

(b)

$5, 5, 25$

(c)

$\frac{1}{2}, \frac{1}{2}, \frac{1}{4}$

8. Consider the two similar rectangles R and R' in the picture below:

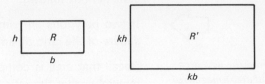

(a) The ratio of the lengths of corresponding sides (that is, the ratio of similarity of R' to R) is _____. k

(b) $\dfrac{\text{area }(R')}{\text{area }(R)} = $ _____ $\dfrac{(kb)(kb)}{bb} = k^2$

The concept of the ratio of similarity is often confusing. It is important to remind students that the ratio of similarity is the ratio of corresponding sides.

(c) If two rectangles are similar, then the ratio of their areas is equal to the square of the _____.   *ratio of similarity*

9. Consider the two similar triangles T and T' in the picture below:

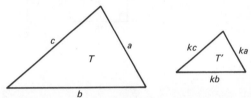

(a) What is the ratio of similarity of T' to T? k

(b) What is the ratio of any altitude of T' to the corresponding altitude of T? [*Hint:* See Chapter 8, Central Exercise 33(c).] k

Let the altitude of T be h.

(c) $\dfrac{\text{area }(T')}{\text{area }(T)} = $ _____ $\dfrac{\frac{1}{2}(kb)(kb)}{\frac{1}{2}bb} = k^2$

(d) If two triangles are similar, then the ratio of their areas is equal to the square of the _____.   *ratio of similarity*

The result of the last exercise holds for polygons in general (the proof is given in the Projects section of this chapter).

If two polygons are similar, then the ratio of their areas is equal to the *square* of the ratio of the lengths of any two corresponding sides.

Keeping in mind that the ratio of similarity is obtained by taking the ratio of the lengths of any two corresponding sides, we can rephrase the above statement as follows:

If two polygons are similar, then the ratio of their areas is equal to the *square* of the ratio of similarity.

Under magnification, the new area of a polygon = (old area)(magnification ratio)².

For students going on in mathematics, this fact is one of the most important learned in this course.

Recall that area is computed in *square* units. Perhaps this will help you remember that the ratio of the areas of similar polygons is the *square* of the ratio of similarity.

10. In each case verify (by computing the areas) that the ratio of the area of P to the area of Q is equal to the square of the ratio of similarity of P to Q.

Exercise 10: Watch out for students who square the area rather than the ratio of similarity!

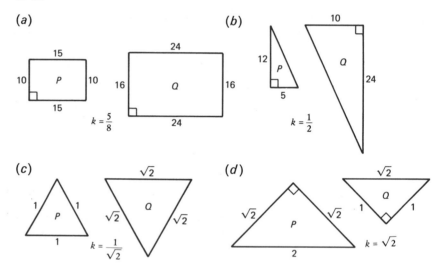

11. What fraction of area is in the top triangle?

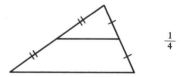

12. Suppose the area of polygon P' is twice the area of polygon P. What is the ratio of similarity of P' to P? $\sqrt{2}$

In Exercise 13, a hint may be in order. This is not an easy problem.

The line parallel to a side must divide the altitude (and each of the sides) such that the altitude of the smaller triangle is $\frac{1}{\sqrt{2}}$ that of the larger triangle.

13. Can you cut $\triangle ABC$ into two pieces of equal area by drawing a line parallel to one of the sides? Exactly how would you find such a line? Discuss your answer with your classmates.

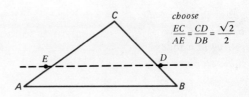

14. The Sultan of Polygonia (whom we have met in the Projects section of Chapter 6) found that the royal playground (which was triangular) was not large enough. He ordered the royal landscaper to design a new playground of the same shape, but with exactly three times as much room in which to play. The royal landscaper thereupon designed the new playground with sides three times as large as the corresponding sides of the original playground. What do you think happened to the royal landscaper? Why? Discuss with your classmates. *The area is nine times as large.*

The fate of the royal landscaper is not recorded in historical annals. See also Projects Exercise 1.

Volumes of Similar Solids

We will not give a formal definition of similarity for solids, but will rely on intuition.

We have been considering only similar figures lying in a plane. But an architect, for example, would be interested in similar solid figures in space, since his scale models are solid figures which are similar to the actual buildings to be constructed. The next exercise will help us understand what "similarity" means for solid figures.

15. Suppose you build a scale model S of a real airplane S' and suppose the linear measurements of the real airplane are 50 times as large as those of your model, that is, the model S would have to be magnified 50 times to obtain S'.

(*a*) What would you guess is the ratio of similarity of S' to S? 50

(*b*) Suppose P and Q are two locations 6″ apart in your model (for example, P and Q might be the wingtips). How far apart are the corresponding locations P' and Q' in the real airplane (in feet)? 25 *feet*

(*c*) Suppose that P and Q are *any* two locations in S and that P' and Q' are the corresponding locations in S'. Then

$$\frac{P'Q'}{PQ} = \underline{\qquad}. \quad 50$$

Let us consider what happens to the volume of a cube when we double the length of each edge. In the picture on p. 372, the cube on the right has "double the edgelength" of the cube on the left (the ratio of similarity or "magnification" is equal to 2).

371 Central

Recall the analogous example used to motivate the behavior of area under similarity. Bring in wooden blocks or sugar cubes to make these models tangible.

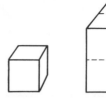

The dotted lines indicate how the cube on the right could be cut into 8 cubes, each congruent to the one on the left. Hence the cube on the right has *8 times* the volume of the cube on the left. Thus, doubling the edgelength has increased the volume by a factor of 8.

For students having real difficulty with these concepts, seeing how the volume of a unit cube changes as the edges are successively multiplied by 2, 3, 4, 5, ... will help.

16. In the picture below is a cube of edge a and a similar cube obtained by "tripling" the edges of the first cube.

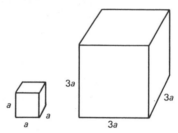

(a) What is the ratio of similarity of the big cube to the small cube? 3

(b) What is the ratio of the volume of the big cube to the volume of the small cube? $3^3 = 27$

17. In the picture below are two similar boxes. The larger box can be obtained by "magnifying" the smaller box $\frac{3}{2}$ times.

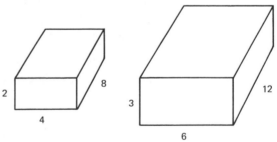

Express the ratio of the volume of the big box to the volume of the small box in terms of the ratio of similarity. $\left(\frac{3}{2}\right)^3 = \frac{27}{8}$

18. The picture on the next page shows two similar boxes B and B'.

372 9: *Perimeter, Area, and Volume of Similar Figures*

The ratio of similarity of B' to B is k.

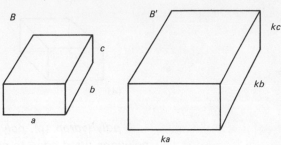

$\dfrac{\text{volume }(B')}{\text{volume }(B)}$ = (ratio of similarity of B' to B)^3 = $\left(\dfrac{ka}{a}\right)^3$ = k^3

Show that

$$\dfrac{\text{volume }(B')}{\text{volume }(B)} = k^3.$$

The result in Exercise 18 holds generally. Let S and S' be similar solids in space. Suppose the ratio of similarity of S' to S is k. Then

$$\dfrac{\text{volume }(S')}{\text{volume }(S)} = k^3.$$

Note. Keep in mind that k gives the "linear" magnification. It is obtained by taking the ratio of any linear measurement in S' to the corresponding linear measurement in S. Recall that volume is computed in *cubic* units. Perhaps this will help you remember that the ratio of the volumes of similar solids is the *cube* of the ratio of similarity.

$\dfrac{\text{volume of large tank}}{\text{volume of small tank}} = 4^3$
$4^3(300) = 19{,}200\ gal.$

19. A certain water tank will hold 300 gallons of water. Suppose a larger water tank is constructed, similar to the smaller tank, with each linear measurement four times as large. Without knowing the shape of the tanks, use the preceding result on volumes of similar solids to find how much water the larger tank will hold. Discuss with your classmates.

Students should enjoy the practical applications of similarity to biology in the Projects section.

20. (a) In the opening problem of the statue, what is the "ratio of similarity" of the larger statue to the smaller? 10

(b) Each linear measurement of the large statue is _____ times as large as the corresponding measurement in the small statue. 10

(c) How heavy is the large statue? Discuss with your classmates.
$12(10)^3 = 12{,}000\ pounds$

Surface Area

The concept of surface area is important in geometry. We first consider surface areas of polyhedra.

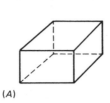 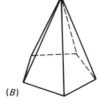 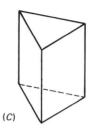

(A) (B) (C)

 A *polyhedron* (pl. *polyhedra*) is a solid whose surface is made up of polygons fitted edge to edge, with each edge of a polygon fitted to the edge of one other polygon. The polygons are called the *faces* of the polyhedron. For example, the box (A) has six faces, each face a rectangle. The pyramid (B) has six faces, five of them triangles and one a pentagon. The prism (C) has five faces, three of which are parallelograms and two triangles. (Other examples, including the five *regular polyhedra*, are found in the Projects section of Chapter 4.)

DEFINITION The *surface area* of a polyhedron is the sum of the areas of its faces.

For example, this cube has six faces, each face a square of area 4.

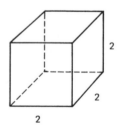

Hence the surface area of this cube is 24.

21. Find the surface area of this square pyramid, all of whose edges have length 5.

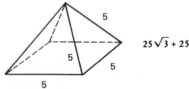

$25\sqrt{3} + 25$

22. If you double the edges of a cube, what happens to the surface area?
quadrupled

374 9: Perimeter, Area, and Volume of Similar Figures

23. Below are pictured similar boxes B and B'. The ratio of similarity of B' to B is k.

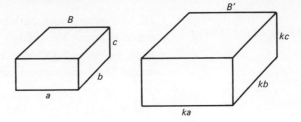

$$\frac{\text{surface area }(B')}{\text{surface area }(B)} =$$
$$\frac{2 \cdot ka \cdot kb + 2 \cdot ka \cdot kc + 2 \cdot kb \cdot kc}{2 \cdot a \cdot b + 2 \cdot a \cdot c + 2 \cdot b \cdot c} = k^2$$

Show that

$$\frac{\text{surface area }(B')}{\text{surface area }(B)} = k^2.$$

24. Suppose that P and P' are similar polyhedra, and the ratio of similarity of P' to P is k. Explain why you would expect that

$$\frac{\text{surface area }(P')}{\text{surface area }(P)} = k^2.$$

Discuss this with your classmates. [*Hint:* Corresponding faces of two similar polyhedra are similar polygons.]

It is possible to talk about the concepts of volume and surface area for solids which are not polyhedra, such as the statue in the opening problem. The results about volume and surface area of similar solids still hold for these more general solids.

If S and S' are similar solids, and the ratio of similarity of S' to S is k, then

$$\frac{\text{surface area }(S')}{\text{surface area }(S)} = k^2$$

$$\frac{\text{volume }(S')}{\text{volume }(S)} = k^3.$$

---

25. Suppose the sculptor in the opening problem of this chapter paints over the surface of the small statue and finds he needs 17¢ worth of paint.

How much would it cost him to paint the large statue? Discuss with your classmates. 17¢ (10)² = $17.00

Exercise 26 is an excellent summary exercise, combining all the ideas of the chapter.

26. An amusement park builds an artificial mountain of dirt covered with artificial snow for skiing. The mountain is cone-shaped with a circular base, contains 1000 tons of dirt, and is covered with 10 tons of artificial snow. The base is surrounded by a fence 200 yards long.

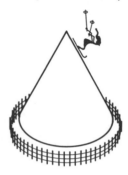

The park enjoys such good business that the operators decide to build a larger skiing mountain of exactly the same shape but twice as high. How much dirt is needed for the new mountain? How much artificial snow? How long a fence is needed to surround the base? Discuss with your classmates.

dirt: $1000(2^3) = 8000$ *tons*
snow: $10(2^2) = 40$ *tons*
fence: $200(2) = 400$ *yards*

Important Ideas in This Chapter

Definitions

The *perimeter* of a polygon is the sum of the lengths of its sides.

Polyhedron, faces of a polyhedron

The *surface area* of a polyhedron is the sum of the areas of its faces.

Theorems

If P and P' are similar polygons, and if s and s' are corresponding linear measurements, then

$$\frac{\text{perimeter of } P'}{\text{perimeter of } P} = \frac{s'}{s} = \text{ratio of similarity of } P' \text{ to } P$$

and

$$\frac{\text{area of } P'}{\text{area of } P} = \left(\frac{s'}{s}\right)^2.$$

If S and S' are similar solids, and if s and s' are corresponding linear measurements, then

$$\frac{\text{surface area of } S'}{\text{surface area of } S} = \left(\frac{s'}{s}\right)^2$$

and

$$\frac{\text{volume of } S'}{\text{volume of } S} = \left(\frac{s'}{s}\right)^3.$$

Summary

You might lead a class discussion about the most important ideas in the chapter and their practical value.

In this chapter we learned how the perimeter and area of a polygon behave when we enlarge or reduce the polygon without changing the shape. We also studied how surface area and volume of a solid change under enlargement or reduction without shape change. This is valuable in practical problems where one needs to estimate the materials required to build structures which are the same shape but different size.

Historical Note

Although there is slight historical evidence to support the claim that Galileo actually murmured "E pur si muove" or even performed the famous experiment from the Leaning Tower, we feel the status of the stories as classics warrants their inclusion.

The applications of similar figures to biology in the Projects section of this chapter are based on ideas from the book *Dialogues Concerning Two New Sciences*, by Galileo Galilei (1564–1642). Galileo lived in an age that broke away from ancient traditions, and he was a major contributor to the scientific revolution that took place. In 1632 he published a *Dialogue on Two Chief World-Systems* in which he showed the superiority of the new Copernican system, in which the earth moves around the sun, over the traditional Ptolemaic system, in which the earth is fixed in space. For this Galileo was charged with heresy and was forced to recant his opinions before the Inquisition in Rome. Galileo did recant, but muttered under his breath afterwards, "E pur si muove" ("And yet it moves"). His greatest contribution to science was his use of the scientific method, with its balance between theory and experiment, which he applied in his study of the laws of motion in nature. In one of his best-known experiments, Galileo dropped unequal weights from the Leaning Tower of Pisa to demonstrate that they fall at the same rate, contrary to Aristotle's assertion that a heavier weight falls faster.

Science came full circle in 1971, from the groundwork laid by Galileo and the profound contributions of Newton to the Apollo 15 lunar expedition. There, Commander David R. Scott performed an amusing experiment as millions on earth watched the televised pictures transmitted from the moon. Scott stood before the television camera with a feather in one hand and a hammer in the other. After recounting Galileo's famous experiment, he let the two objects drop. Feather and hammer hit the lunar surface simultaneously.

Central Self Quiz

1. Suppose P and Q are similar polygons, and the ratio of similarity of Q to P is $\frac{3}{2}$.

(a) If P has perimeter 6, then what is the perimeter of Q?

(b) If P has area 2, then what is the area of Q?

2. If each edge of a box is tripled in length, what happens to

(a) the volume?   (b) the surface area?

3. Find $\dfrac{\text{area}(\triangle CDE)}{\text{area}(\triangle CAB)}$.

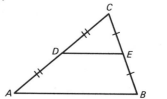

4. The picture shows a regular tetrahedron, that is, a triangular pyramid each of whose four faces is an equilateral triangle. If each edge has length 6 cm., find the surface area.

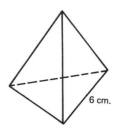

5. Prove that the area of $\triangle CBA$ is four times that of $\triangle CDB$.

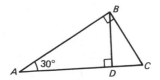

Answers   **1.** (a) 9   (b) $\frac{9}{2}$

2. (a) multiplied by $3^3 = 27$ (b) multiplied by $3^2 = 9$

3. $\frac{1}{4}$

4. $(4)\left(\frac{\sqrt{3}}{4}\right)(36) = 36\sqrt{3}$

5. $\triangle CBA \sim \triangle CDB$ and the ratio of similarity of $\triangle CBA$ to $\triangle CDB$ is $\frac{AC}{BC}$. But $\triangle CBA$ is a 30°–60° right triangle, so $\frac{AC}{BC} = 2$. Since the ratio of similarity is 2, $\triangle CBA$ has 4 times the area of $\triangle CDB$.

Review

1. The picture below shows two "I-beams" often used in construction. The two I-beams are similar, and each linear dimension of the larger is *twice* that of the smaller.

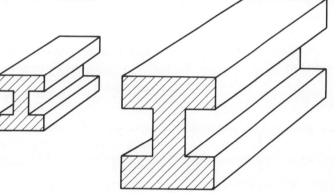

(a) If the cross-sectional area (shaded) of the smaller is 2 square feet, what is the cross-sectional area of the larger?

(b) If the volume of the smaller is 6 cubic feet, what is the volume of the larger?

Before doing Exercise 1(c), let the students use their intuition. Say, "You could lift the smaller one. Could you lift the larger?"

(c) If the smaller weighs 50 pounds, how much does the larger weigh?

2. Suppose polygon P is similar to polygon Q, and the ratio of similarity of Q to P is 3.

(a) If P has an edge of length 2, how long is the corresponding edge of Q?

(b) If P has perimeter 12, what is the perimeter of Q?

(c) If P has area 8, what is the area of Q?

3. Suppose polygon P is similar to polygon Q and the area of P is three times the area of Q. What is the ratio of similarity of Q to P?

4. Suppose △ABC ~ △A'B'C' in the picture below. If △ABC has area 6, what is the area of △A'B'C'?

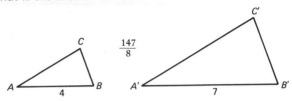

$\frac{147}{8}$

5. What is the ratio of the area of the small triangle to the area of the large triangle?

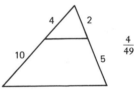

$\frac{4}{49}$

Exercise 6 is difficult. Many students will need help.

6. A triangular piece of material is to be cut into three pieces of equal area with two straight cuts parallel to one of the sides. How should the cuts be made?

choose $\frac{AB}{AD} = \frac{\sqrt{3}}{3}$

and $\frac{AC}{AD} = \frac{\sqrt{6}}{3}$

7. If you double each edge of a cube, what happens to
(a) the surface area?   (b) the volume?

8. The picture shows two similar triangular pyramids P and Q.

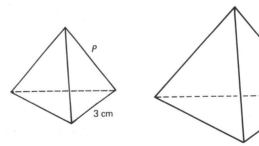

(a) If the surface area of P is 36 square cm., then what is the surface area of Q? $36\left(\frac{4}{3}\right)^2 = 64$

(b) If the volume of P is 27 cubic cm., then what is the volume of Q?
$(27)\left(\frac{4}{3}\right)^3 = 64$

9. The right triangular prism below has 3 square faces, and 2 faces which are equilateral triangles.

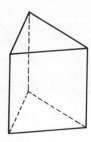

Assuming each edge has length 4 cm., find

(a) the surface area   (b) the volume.

10. Consider a prism similar to that in the preceding exercise but with each edge 2 cm. long. Use the results of Exercise 9 and properties of similar solids to compute the surface area and volume of this prism.

11. (a) To double the area of a square, multiply each edge by _____.

(b) To double the volume of a cube, multiply each edge by _____.

12. It costs 50¢ to paint the surface of a certain box. How much would it cost to paint the surface of a similar box whose edges are twice as long?

This application of similarity is under study at the College of Engineering of the University of Illinois.

13. An engineer has proposed that railroad tracks should be twice as far apart as they now are, and that boxcars be twice as long, twice as wide, and twice as high as they now are.

(a) How would the cost of building these larger boxcars compare with the cost of building the present boxcars? (Assume that cost is determined by surface area.)

(b) How would the cargo space of the larger boxcars compare with that of the present boxcars? (Assume that this is determined by volume.)

Review Answers

1. (a) 8 square feet   (b) 48 cubic feet   (c) 400 pounds

2. (a) 6   (b) 36   (c) 72

3. $\dfrac{1}{\sqrt{3}} = \dfrac{\sqrt{3}}{3}$

4. $\dfrac{147}{8}$

5. $\dfrac{4}{49}$

6. $\dfrac{AB}{AD} = \dfrac{\sqrt{3}}{3}$, $\dfrac{AC}{AD} = \dfrac{\sqrt{6}}{3}$

7. (a) multiplied by 4 (b) multiplied by 8

8. (a) 64 sq. cm. (b) 64 cubic cm.

9. (a) $48 + 8\sqrt{3}$ sq. cm. (b) $16\sqrt{3}$ cubic cm.

10. Surface area is $\tfrac{1}{4}(48 + 8\sqrt{3}) = 12 + 2\sqrt{3}$ sq. cm., and volume is $\tfrac{1}{8}(16\sqrt{3}) = 2\sqrt{3}$ cubic cm.

11. (a) $\sqrt{2}$ (b) $\sqrt[3]{2}$

12. $2

13. (a) four times as much (b) eight times as much

Review Self Test

It is interesting to test the principle underlying Exercise 1(a) for correspondence with reality by using pairs of students. Unfortunately, it is rather difficult to find a mathematically similar pair of students! The closest one gets to similarity is in the case of identical twins, when we have congruence (or mirror images) and the problem is trivial.

1. Two boys, A and B, have the same "build" (shape), but A is 5 feet tall and B is 6 feet tall.

   (a) If A weighs 125 pounds, guess how much B weighs.

   (b) If A and B are dressed the same and the material for A's clothes cost $10, guess how much the material for B's clothes cost.

2. Suppose P and Q are similar polygons, and the ratio of the length of an edge of P to the corresponding edge of Q is $\tfrac{1}{3}$.

   (a) What is the ratio of the perimeter of P to the perimeter of Q?

   (b) What is the ratio of the area of P to the area of Q?

3. In the diagram on the next page, suppose that $\triangle PQR$ has twice the area of $\triangle ABC$. Find $\dfrac{PQ}{AB}$.

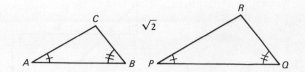

4. Find the ratio of the area of $\triangle ADC$ to the area of $\triangle CDB$.

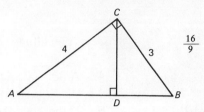

Answers
1. (a) $(125)(\frac{6}{5})^3 = 216$ pounds (b) $14.40
2. (a) $\frac{1}{3}$ (b) $\frac{1}{9}$
3. $\sqrt{2}$
4. $\frac{16}{9}$

Cumulative Review

1. Bill and Frank work in neighboring buildings and they want to connect a telegraph between their offices. Bill's second floor office is 20 feet above the ground and Frank's is 28 feet above the ground. The horizontal distance between their offices is 30 feet. How long a wire will they need?
$\sqrt{(30)^2 + (8)^2} = 2\sqrt{241}$

2. Find x.

(a): *converse of the Pythagorean Theorem*
(b): *The short leg of the small right triangle is* 1. $(x+1)^2 + 2^2 = (2x)^2$

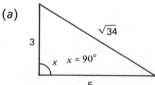

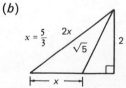

383 Cumulative Review

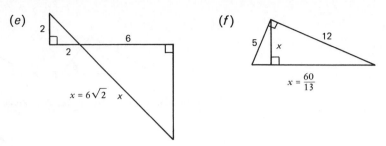

3. The altitude of an equilateral triangle is 3. What is its area?

4. A 30°–60° right triangle has area $6\sqrt{3}$. Find the length of each side.

area $(\triangle ABC)$ = *area* $(\triangle ADE)$ +
area $(BCED)$ = 4(*area* $(\triangle ADE)$)
Ratio of similarity of $\triangle ABC$ to
$\triangle ADE$ is 2.

5. If the area of trapezoid $BCED$ is three times the area of $\triangle ADE$, and $BC = 8$, find DE.

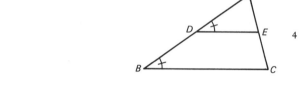

6. Find PB, PC, PD, and PE.

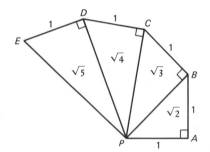

7. Is P congruent to Q? If so, why?

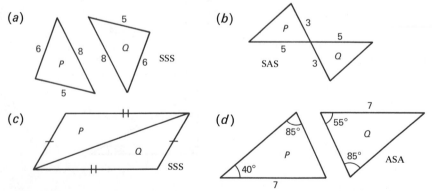

384 9: Perimeter, Area, and Volume of Similar Figures

(e) *not necessarily congruent*

(f) ASA

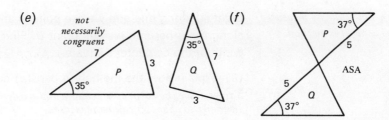

8. If two similar polygons have the same area, then they are in fact _____. *congruent*

9. Is $P \sim Q$? If so, find the ratio of similarity of P to Q.

(a) SAS $k = \frac{2}{3}$

(b) *not similar* $\frac{2}{4} \neq \frac{4}{6}$

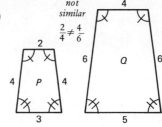

(c) AAA $k = \frac{1}{2}$

(d) AAA $k = \frac{3}{4}$

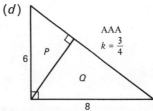

(e) SSS $k = 1$

(f) AAA $k = \frac{\sqrt{3}}{4}$

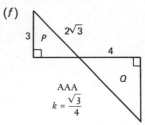

10. (a) A carpenter's square is placed on a vertical yardstick as shown below:

385 Cumulative Review

Exercise 10(b): Let students measure some distances using this method.

Sighting along one arm we see point A and measure $AQ = 2$ ft. Sighting along the other arm we see point B. Find the distance QB. (Neglect the width of the carpenter's square.) $\triangle AQP \sim \triangle PQB$, $\frac{QB}{3} = \frac{3}{2}$

(b) Explain how the method in part (a) can be used to find the width of a river. *Set the carpenter's square and yardstick so that QB is the width of the river and so that AQ can be measured.*

11. A ladder 20 feet long reaches a windowsill 16 feet above the ground. How far from the house is the foot of the ladder?

12. Show that $(a + b)(a - b) = (c + d)(c - d)$.

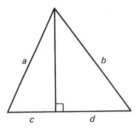

[*Hint:* Express the length of the common side of the two right triangles in two ways, using the Pythagorean Theorem.]

13. The lengths of all sides of a polygon are tripled, but the angles remain unchanged. What happens to the area of the polygon?

14. A statue of a man is made from a certain kind of metal. The statue is 6 inches tall and weighs 8 pounds. A larger statue of exactly the same shape is made using 27 pounds of the same kind of metal. How tall is it?

15. Find the area of each polygon.

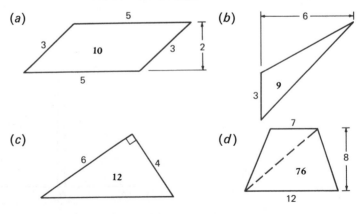

386 9: Perimeter, Area, and Volume of Similar Figures

16. What is the length of the hypotenuse of an isosceles right triangle of area 288 square inches?

17. The area of a rectangle is $3a^2 - 7ab + 2b^2$ and its base has length $a - 2b$. Find its altitude.

18. How many sides does a regular polygon have if each angle has size

(a) 60°?   (b) 90°?   (c) 120°?   (d) 108°?

19. Given: $AD = AE$, $DC = EB$
Prove: $\angle B = \angle C$

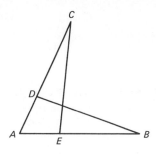

20. Draw a line segment of length 6 inches on a sheet of paper. Now, using only straightedge and compass, construct

(a) a line segment of length 3 inches

(b) a line segment of length 2 inches

(c) an equilateral triangle, with side of length 4 inches

(d) a square, with side of length 3 inches

(e) a rectangle of area 32 square inches.

21. Using only straightedge and compass, construct a similar triangle whose area is $\frac{1}{9}$ times the area of the triangle below.

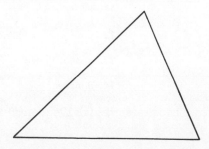

Cumulative Review

22. Which figures will tile the plane?

(a) equilateral triangle (b) rectangle

(c) regular pentagon *no* (d) regular hexagon

(e) any triangle (f) any quadrilateral

23. In the figure below, *ABCD* is a square. Prove that *PQRS* is also a square.

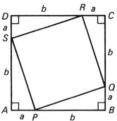

Answers

1. $2\sqrt{241}$ feet

2. (a) 90°   (b) $\frac{5}{3}$ (c) $\frac{8\sqrt{3}}{3}$ (d) $\sqrt{2}$ (e) $6\sqrt{2}$ (f) $\frac{60}{13}$

3. $3\sqrt{3}$

4. $6, 2\sqrt{3}, 4\sqrt{3}$

5. 4

6. $\sqrt{2}, \sqrt{3}, \sqrt{4}, \sqrt{5}$

7. (a) SSS   (b) SAS   (c) SSS   (d) ASA   (e) not necessarily congruent   (f) ASA

8. congruent

9. (a) $\frac{2}{3}$ (b) not similar   (c) $\frac{1}{2}$ (d) $\frac{3}{4}$ (e) 1   (f) $\frac{\sqrt{3}}{4}$

10. (a) $4\frac{1}{2}$ ft.

11. 12 ft.

12. $\sqrt{a^2 - c^2}$ = altitude = $\sqrt{b^2 - d^2}$

13. multiplied by 9

14. 9 inches

15. (a) 10   (b) 9   (c) 12   (d) 76

16. $24\sqrt{2}$ inches

17. $3a - b$

18. (a) 3   (b) 4   (c) 6   (d) 5

19. $\triangle AEC \cong \triangle ADB$ by SAS

21. Trisect a side and draw parallel line.

22. all but (c)

23. Prove that all sides are equal, all angles equal 90°.

Projects

Similar Figures in Fables

1. According to legend, a certain town in ancient Greece was stricken with a deadly plague. The inhabitants consulted the oracle at Delphi and were told that the god Apollo was angry with them and hence had sent the plague. The altar to Apollo in this town was a cube of solid gold, but Apollo was angry, said the oracle, because he wanted his altar to be twice as big. So the people made a new altar with edges twice the length of the old one. The plague worsened. The people then realized that Apollo meant the *volume* of the new altar should be exactly twice that of the old. In order to placate Apollo, how should the edge length of the new altar compare with the edge length of the old? $\sqrt[3]{2} \doteq 1.26$ *times as large on each edge*

2. During his travels, Gulliver, the hero of Jonathan Swift's classic *Gulliver's Travels*, relates how he is imprisoned by the tiny inhabitants of the country of Lilliput after being shipwrecked on their shores. The Lilliputians, being only a twelfth the stature of Gulliver, recognize that he needs a great deal more food for nourishment than they. Gulliver recounts the Lilliputians' solution:

> The Emperor (Golbasto Momaren Evlame Gurdilo Shefin Mully Ully Gue) stipulates to allow me a quantity of meat and drink sufficient for the support of 1728 Lilliputians. Some time after, asking a friend at the court how they came to fix on that determinate number, he told me that his Majesty's mathematicians, having taken the height of my body by the help of a quadrant, and finding it to exceed theirs in the proportion of twelve to one, they concluded from the similarity of their bodies that mine must contain at least 1728 of theirs, and consequently would require as much food as was necessary

to support that number of Lilliputians. By which the reader may conceive an idea of the ingenuity of that people, as well as the prudent and exact economy of so great a prince.

(a) How did his Majesty's mathematicians arrive at the number 1728?

(b) How much more cloth would be required to clothe Gulliver than to clothe a typical Lilliputian in similar garb? (The answer is *not* 1728 times as much!) 144 *times as much cloth*

Similar Figures in Biology

3. The picture below shows two bones *B* and *B'* of the same shape.

On Growth and Form, *by D'Arcy Thompson (Cambridge: Cambridge University Press, 1942), is a beautiful work which deals with geometry in biology; chapter 2 considers applications of similarity.*

The shaded disks denote cross sections of the two bones, cut at corresponding positions on each. It is known that the strength of a bone, in its ability to withstand compression, depends on the *area of the cross section*. For example, if the cross section of *B'* has twice the area of the corresponding cross section of *B*, then *B'* will be twice as strong (in withstanding compression).

(a) Suppose *B'* is *k* times as long as *B* (the ratio of similarity of *B'* to *B* is *k*). How does the cross-sectional area of *B'* compare with that of *B*? (Keep in mind that corresponding cross sections of *B* and *B'* are similar *plane* figures with ratio of similarity *k*.) k^2 *times as large (This applies to any area measure.)*

(b) Explain why doubling all linear measurements of a bone (without changing shape) increases its strength 4 times. *It has 4 times the cross-sectional area.*

(c) If *B'* is *k* times as long as the similar bone *B*, then *B'* is _____ times as strong. k^2

4. Consider two animals *A* and *A'* of the same shape, but different size.

 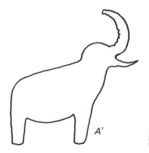

390   9: Perimeter, Area, and Volume of Similar Figures

(a) If each leg of A' is twice the length of the corresponding leg of A, then explain why A' weighs eight times as much as A.

(b) If the ratio of similarity of A' to A is k, then A' weighs _____ times as much as A. k^3

5. In the preceding exercise, if we suppose each linear measurement of A' is twice the corresponding linear measurement of A, then A' weighs 8 times as much as A. But since each bone of A' has only 4 times the strength of the corresponding bone of A (Exercise 3(b)), the skeleton of A' has to support 8 times the weight with only 4 times the strength of A. *Hence the ability of A' to support its own weight is only half that of A!*

Literary-minded students might want to do some research on Gulliver's Travels *to see if the Brobdingnagians could possibly have supported their own weight.*

(a) Suppose the ratio of similarity of A' to A is k. Explain why the ability of A' to support its own weight is $\dfrac{1}{k}$ that of A.

(b) In view of part (a), explain why the bones of large animals are much thicker, compared to their length, than the bones of smaller animals.

6. Why might you expect the ratio "thickness to height" to be larger for a taller tree?

7. Consider two animals A and A'. Suppose they are the same shape, and A' is twice as tall as A.

(a) Explain why A' can store 8 times as much water as A.

(b) Explain why 4 times as much water evaporates from the surface of A' as from the surface of A, in the same length of time.

(c) From (a) and (b), we see that A' can store 8 times as much water as A, yet loses it only 4 times as fast as A. Therefore A' loses water half as fast as A *relative to his own capacity.* Explain, then, why birds and insects are predominant in damp regions. Explain why camels are used for journeys in the desert.

Area of Similar Polygons

In Exercises 8 and 9 we prove that if polygon P is similar to polygon P', and the ratio of similarity of P' to P is k, then

$$\frac{\text{area }(P')}{\text{area }(P)} = k^2.$$

8. Suppose the two polygons $P = ABCDE$ and $P' = A'B'C'D'E'$ are

similar, with

$$\frac{A'B'}{AB} = \frac{B'C'}{BC} = \ldots = \frac{E'A'}{EA} = k = \text{the ratio of similarity of } P' \text{ to } P.$$

Exercises 8 and 9 are for students who like challenging proofs. Students should not be forced to do them unless they are sincerely interested.

In the picture below, we have cut the polygons into corresponding triangles.

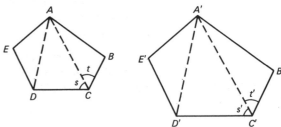

Explain each step.

(a) $\triangle ABC \sim \triangle A'B'C'$ and the ratio of similarity is k. [*Hint:* Since $P \sim P'$, we have $\angle B = \angle B'$ and $\frac{A'B'}{AB} = \frac{B'C'}{BC}$. Why are the triangles similar? Why is the ratio of similarity $= k$?] SAS, $\frac{A'B'}{AB} = k$

(b) $\dfrac{\text{area } (\triangle A'B'C')}{\text{area } (\triangle ABC)} = k^2$. [*Hint:* Central Exercise 9]

(c) area $(\triangle A'B'C') = (k^2)(\text{area } (\triangle ABC))$

(d) $\triangle DAC \sim \triangle D'A'C'$ and the ratio of similarity is k. [*Hint:* $\angle s = \angle C - \angle t = \angle C' - \angle t' = \angle s'$ (why?) and $\dfrac{A'C'}{AC} = \dfrac{C'D'}{CD}$ (why?). Why are the triangles similar? Why is the ratio of similarity $= k$?]

(e) area $(\triangle D'A'C') = (k^2)(\text{area } (\triangle DAC))$

(f) $\triangle AED \sim \triangle A'E'D'$ and the ratio of similarity is k.

(g) area $(\triangle A'E'D') = (k^2)(\text{area } (\triangle AED))$

(h) area $(P') = (k^2)(\text{area } (P))$ [*Hint:* Use (c), (e), (g) and the fact that area $(P') = $ area $(\triangle A'B'C') + $ area $(\triangle D'A'C') + $ area $(\triangle A'E'D')$.]

(i) $\dfrac{\text{area } (P')}{\text{area } (P)} = k^2$

(j) If the ratio of corresponding sides of two similar polygons is k, then the ratio of their areas is _____ . k^2

9. Our proof in Exercise 8 demonstrated the case of two similar pentagons. Discuss how this method of proof can be used for similar polygons with any number of sides.

10. (*a*) Using only straightedge and compass, construct a triangle similar to the one below and having four times the area.

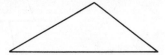

(*b*) Discuss how you could construct, using only straightedge and compass, a polygon similar to the one below and having four times the area.

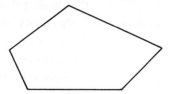

(*c*) Discuss how you could construct, with only straightedge and compass, a polygon similar to the one in part (*b*) and having twice the area. [*Hint:* Given any segment, an isosceles right triangle having that segment as one leg has hypotenuse of length $\sqrt{2}$ times the length of the given segment.]

Cavalieri's Principle

To illustrate the generality of Cavalieri's Principle, it is even more effective to do this with a stack of cards of different shapes and sizes.

Imagine a deck of cards on a table, stacked in the shape of a box. Now suppose the cards are pushed into a stack of a different shape.

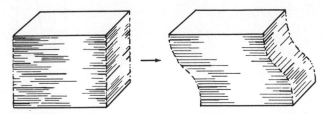

The new stack has the same volume as the original. If we think of the two stacks as two different solids, we see that if both solids are cut simultaneously by a horizontal plane, the two cross sections have the same area. This idea motivates a more general property called *Cavalieri's Principle:*

If each horizontal plane cutting one solid intersects another solid in a cross section of the same area, then the two solids have the same volume.

In the picture below, imagine a horizontal plane moving up and down, cutting the two solids as it moves. Suppose the area of the cross section on the left-hand side is always equal to the area of the cross section on the right-hand side.

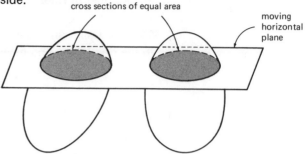

Cavalieri's Principle tells us that the two solids must have the same volume.

Cavalieri's Principle is an axiom *in our presentation of geometry.*

Cavalieri's Principle is very useful for computing volumes of solids. It will be used in Exercise 15 to show that any two pyramids of the same base area and altitude have the same volume. This in turn will enable us to prove that the volume of any pyramid of base area A and altitude h is given by

$$V = \tfrac{1}{3} hA.$$

Cavalieri's Principle appears again in the Projects section of the next chapter, where the volume of a ball is computed.

Volume of a Pyramid

Imagine a triangular pyramid cut with a plane parallel to the base plane:

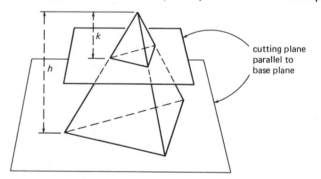

The intersection of the cutting plane with the pyramid is a triangle. The next exercise proves that this triangle is similar to the base triangle. Moreover, if the top vertex is at distance k from the cutting plane and distance h from the base plane, then

the ratio of similarity of the small triangle to the larger is $\dfrac{k}{h}$.

11. In the diagram, $PQ = h =$ the altitude of the pyramid. A plane parallel to the base plane cuts the pyramid in a triangle, $\triangle A'B'C'$. The distance from P to the cutting plane is $PQ' = k$.

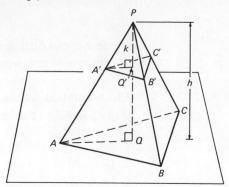

We want to prove that $\triangle ABC \sim \triangle A'B'C'$ and that the ratio of similarity of $\triangle A'B'C'$ to $\triangle ABC$ is $\dfrac{k}{h}$. Explain each of the following steps.

(a) $\triangle PQ'A' \sim \triangle PQA$ [*Hint:* similar right triangles]

(b) $\dfrac{PA'}{PA} = \dfrac{k}{h}$ [*Hint:* Obtain $\dfrac{PA'}{PA} = \dfrac{PQ'}{PQ}$ from part (a).]

(c) $\dfrac{PB'}{PB} = \dfrac{k}{h}$

(d) $\triangle PA'B' \sim \triangle PAB$ and the ratio of similarity of $\triangle PA'B'$ to $\triangle PAB$ is $\dfrac{k}{h}$. [*Hint:* The SAS condition for similar triangles is proved in the Projects section of Chapter 8. Another way to see that the triangles are similar is to note that $\overline{A'B'}$ is parallel to $\overline{AB}$.]

(e) $\dfrac{A'B'}{AB} = \dfrac{k}{h}$

(f) $\dfrac{B'C'}{BC} = \dfrac{C'A'}{CA} = \dfrac{k}{h}$ [*Hint:* Proceed as in parts (a)–(e).]

(g) $\triangle ABC \sim \triangle A'B'C'$ and the ratio of similarity of $\triangle A'B'C'$ to $\triangle ABC$ is $\dfrac{k}{h}$, as we wanted to prove. [*Hint:* parts (e) and (f)]

12. A triangular pyramid is cut with a plane parallel to its base. Let S be the area of the base triangle and S' the area of the triangle where the cutting plane intersects the pyramid. Explain why

$$\frac{S'}{S} = \left(\frac{k}{h}\right)^2,$$

where k is the distance from the top vertex to the cutting plane, and h is the altitude of the pyramid.

13. A pyramid whose base is a polygon of area equal to A is cut by a plane parallel to the base in a polygon of area A'.

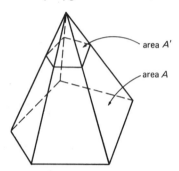

If k is the distance from the top vertex to the cutting plane, and h is the altitude of the pyramid, prove that

$$\frac{A'}{A} = \left(\frac{k}{h}\right)^2.$$

[*Hint:* Cut the base into triangles and use these to cut the pyramid into triangular pyramids. Then apply the preceding exercise.]

14. Consider two pyramids *having bases of the same area lying in the same plane and having the same altitude.*

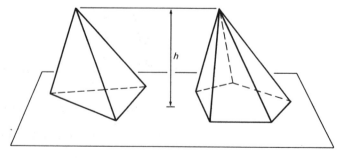

Prove that if the two pyramids are cut by a plane parallel to the base plane, then the two cross sections have the same area. [*Hint:* Exercise 13]

15. Prove that any two pyramids with the same base area and same altitude have the same volume. [*Hint:* You may assume the bases are in the same plane. Apply the preceding exercise and Cavalieri's Principle.]

Encourage students to build a model for Exercise 16. Give them time; the exercise is worth it.

16. The diagram below shows a right triangular prism with base $\triangle ABC$. The prism is shown cut into 3 triangular pyramids.

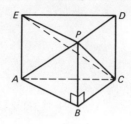

(*a*) Consider the (upside down) pyramid with base $\triangle EPD$ and opposite vertex *C*. Explain why this pyramid has the same volume as the pyramid with base $\triangle ABC$ and top vertex *P*. [*Hint:* Apply the preceding exercise.]

(*b*) Consider the pyramid with base $\triangle EPA$ and opposite vertex *C*. Explain why this pyramid also has the same volume as the pyramid with base $\triangle ABC$ and top vertex *P*. [*Hint:* The pyramid with base $\triangle ABC$ and opposite vertex *P* can also be viewed as a pyramid with base $\triangle BPA$ and opposite vertex *C*. Then the two pyramids in question have the same altitude *BC*. Do their bases have the same area?]

(*c*) Conclude from parts (*a*) and (*b*) that the pyramid with base $\triangle ABC$ and top vertex *P* has volume one-third that of the original prism.

(*d*) Consider the prism with base $\triangle ABC$ and top vertex *P*. If the altitude *PB* of this prism is equal to *h*, and the area of the base $\triangle ABC$ is equal to *A*, then explain why the volume *V* is given by

$$V = \frac{1}{3}hA.$$

17. Prove the following, using Exercises 15 and 16(*d*).

Any pyramid with altitude *h* and base area *A* has volume equal to

$$V = \frac{1}{3}hA.$$

Chapter 10 Circles

The opening problem of the football field-goal kicker offers many opportunities for learning by discovery. After the class has worked on it individually or in groups, you might have volunteers draw on the board a point from which the angle is 30° or more. Before going on in the chapter, the class should see that the possible positions for kicking the football lie inside a circle. Do not tell them; let them find out for themselves.

The remainder of the chapter covers the traditional material. The pupil in trouble should be encouraged to use a centimeter ruler and see that the theorems seem true. This will help him read the theorems. Careful reading is essential to learning algebra and geometry; often a student does poorly in mathematics because his reading ability is low. If the class is divided into small groups, such a handicapped student can be aided by his classmates and also get more attention from the teacher.

Central

The field goal kicker for a football team finds (because of his peculiar kicking style) that he kicks most accurately if the goalposts subtend an angle of at least 30°. That is, if the goalposts are at *A* and *B* on the goal line, then he kicks best if his position *P* is such that ∡1 is at least 30°.

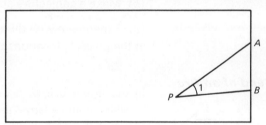

What field positions are available to him so that ∡1 is at least 30°?

Notation for Angles

Since much of our work in this chapter will deal with angles, it is convenient to introduce a concise notation for them. We use the notation ∡*APB* to denote angles such as this:

399

The letter denoting the vertex of the angle is placed between letters denoting points on the sides of the angle. For example, in the triangle below we may write ∡1 = ∡CAB, ∡2 = ∡ABC, ∡3 = ∡BCA.

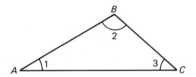

The measure of an inscribed angle is the cornerstone of the chapter. Allow the students time to experiment in the early exercises.

1. (a) Mark a pair of points A and B on a sheet of paper (or cardboard) and, using a 30° angle cut out of heavy paper, find a number of different points P such that ∡APB = 30°. (You might stick in pins firmly at A and B and let the angle move between the pins, with the sides of the angle against the pins.)

 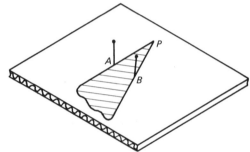

(b) Make a conjecture about those locations of P such that ∡APB = 30°. *They lie on a circle.*
(c) Experiment with different size angles. What do you find? [*Hint:* Look at the possible locations of P.] *They lie on different size circles.*

Bend of an Arc

We use "bend" rather than "degree measure of an arc" because the term seems more descriptive. We can speak of the amount of bend, which can be demonstrated by bending a yardstick. The amount of bend (measured in degrees) does change; the length does not. Guard against confusion of "bend of an arc" and "arc length."

In the figure below, a circle with center O is divided into two arcs (a smaller arc and a larger arc) by two points A and B on the circle. Then ∡1 is called the *central angle* determined by the smaller arc, and ∡2 is the central angle determined by the larger arc.

Since the central angle of an arc measures how much we "turn" when we travel along the arc from one endpoint to the other, we will call the size of the central angle the *bend* of the arc:

DEFINITION   The *bend* of an arc of a circle is the size of its central angle.

For example, the bend of the smaller arc from A to B in the figure below is 78°, and the bend of the larger arc is 282°.

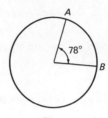

⇨ We will often use the symbol $\widehat{AB}$ to denote an arc of a circle from A to B. It will be clear from the context if we mean the smaller arc or the larger arc. For example, in the above picture, "the bend of $\widehat{AB}$ = 78°."

2. In the circle below, what is the bend of

(a) the smaller arc from A to B? 55°   (b) the smaller arc $\widehat{CD}$? 150°

(c) the larger arc from B to D? 190°   (d) the larger arc $\widehat{AD}$? 245°

(e) the smaller arc from A to D? 115°   (f) the smaller arc $\widehat{BD}$? 170°

(g) the larger arc from A to A? 360°

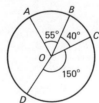

3. A circle is divided into *n* congruent arcs. What is the bend of each arc if

(a) $n = 4$? 90°   (b) $n = 27$? $13\frac{1}{3}°$ (c) $n = 100$? 3.6°

(d) *n* is any positive integer? $\left(\frac{360}{n}\right)°$

4. The figure on the next page shows two *concentric* circles (concentric means "same center").

Exercise 4 should help students see that bend represents angle measure, not length of arc.

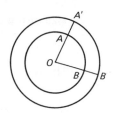

Which arc has the larger bend, the arc of the smaller circle from A to B or the arc of the larger circle from A' to B'? Discuss with your classmates.

The arcs have the same bend because the central angles are the same.

We are relying on intuition in Exercise 5. Length of arc has not been defined and will be discussed later in the chapter.

5. (a) What is the bend of an arc of a circle if the ratio of the length of the arc to the length of the circle is

(i) $\frac{1}{4}$? 90° (ii) $\frac{1}{2}$? 180° (iii) $\frac{1}{3}$? 120° (iv) $\frac{3}{5}$? 216° (v) 1? 360°

(b) Draw pictures to illustrate (i)–(v).

(c) Does the size of the circle make any difference? *no*

Inscribed Angles

Recall that a line segment joining two points on a circle is called a *chord* of the circle, and that a *diameter* of the circle is any chord containing the center of the circle.

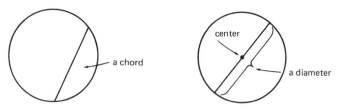

DEFINITION

The definition of inscribed angle must be very clear. The rest of the chapter depends on it.

An *inscribed angle* of a circle is an angle whose vertex lies on the circle and whose sides determine chords of the circle. For example, in the figure below, ∡APB is an inscribed angle.

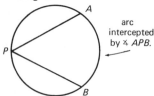

The arc of the circle from A to B which does not contain the vertex P is called *the arc intercepted by the inscribed angle*.

6. (a) Carefully draw a large circle and mark two points A and B on

your circle such that the bend of the smaller arc from A to B is 90°. Choose at least four different points P on the larger arc from A to B and measure in each case the inscribed angle ∡APB using your protractor.
∡APB = 45°

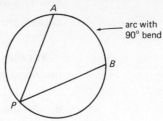

(b) Compare and discuss your results with your classmates.

(c) What do you think must be true about the size of ∡APB as you let P move along the larger arc from A to B? *always 45°*

(d) Do the experiment in part (a) using an arc with 60° bend, with 80° bend, with 130° bend. *Angles are always 30°, 40°, 65°.*

(e) Make a guess about the relationship between the size of an inscribed angle and the bend of the arc intercepted by that angle. Discuss your conjecture with your classmates.
size of inscribed angle = $\frac{1}{2}$ (bend of intercepted arc)

The Inscribed Angle Theorem

In the next few exercises we will prove that the relationship always holds that you might have guessed in Exercise 6(e). Exercise 7 considers the case where one side of the inscribed angle passes through the center of the circle.

7. In the figure below is an inscribed angle ∡APB with one side passing through the center O of the circle.

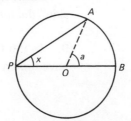

(a) Explain why ∡PAO = x. [Hint: The base angles of ___*an isosceles triangle are equal*___.]

(b) Why is a = 2x? [Hint: Apply the Exterior Angle Theorem, Central section of Chapter 2, to △POA.] *a = x + ∡PAO = 2x*

403 Central

(c) From part (b) obtain $\angle APB = (\frac{1}{2}) \angle AOB$.

(d) The size of $\angle AOB$ is the _____ of the smaller arc $\widehat{AB}$. *bend*

(e) The inscribed angle $\angle APB$ equals _____ the bend of its intercepted arc $\widehat{AB}$. $\frac{1}{2}$

We now consider the case where the center of the circle is "inside" the inscribed angle.

8. In the figure below we start with an inscribed angle $\angle APB$, and then draw the diameter $\overline{PQ}$ and the radii $\overline{OA}$ and $\overline{OB}$.

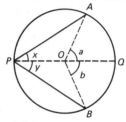

(a) $x = (\frac{1}{2})a$ (why?) (b) $y = (\frac{1}{2})b$ (why?) *Exercise 7(a), (b)*

(c) Using parts (a) and (b), prove that

$$\angle APB = \left(\frac{1}{2}\right) \angle AOB.$$

[*Hint:* Add the equations.] $\angle APB = x+y = \frac{1}{2}a + \frac{1}{2}b = \frac{1}{2}(a+b)$

(d) Part (c) gives a relationship between an inscribed angle and the bend of its intercepted arc: The inscribed angle $\angle APB$ equals _____ the bend of its intercepted arc $\widehat{AB}$. $\frac{1}{2}$

Finally, we consider the situation where the center of the circle is "outside" the inscribed angle.

9. In the figure below we are given an inscribed angle $\angle APB$, and we have drawn the diameter $\overline{PQ}$ and the radii $\overline{OA}$ and $\overline{OB}$.

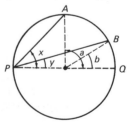

(a) $x = (\frac{1}{2})a$ (why? Don't forget Exercise 7!)   (b) $y = (\frac{1}{2})b$ (why?)

(c) Using parts (a) and (b), prove that

$$\angle APB = \left(\frac{1}{2}\right) \angle AOB.$$

[Hint: $x - y = \frac{1}{2}a - \frac{1}{2}b = \frac{1}{2}(a - b)$.]

Summarize the proof of the Inscribed Angle Theorem. The method is used several times in the chapter.

(d) Part (c) showed that $\angle APB$ equals _____ the bend of its intercepted arc $\widehat{AB}$. $\frac{1}{2}$

10. (a) Exercises 7, 8, 9 have shown that the following holds for any inscribed angle in any circle:

THE INSCRIBED ANGLE THEOREM	**Any inscribed angle of a circle equals half the bend of its intercepted arc.**

(b) In the figure below, O is the center of the circle. How do $\angle APB$ and $\angle AOB$ compare in size? Why?
$\angle APB = \frac{1}{2}(\angle AOB)$

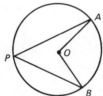

(c) In the figure below, A and B are points fixed on a circle. How does the size of $\angle APB$ change as P is allowed to move along an arc of the circle from A to B (with A and B fixed)?

Stays the same, as it must always be $\frac{1}{2}$ the bend of $\widehat{AB}$.

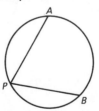

Explain, using the Inscribed Angle Theorem.

In the Projects section of this chapter you will find a general theorem relating angles and the bends of their intercepted arcs. The Inscribed

Angle Theorem is a special case of that more general theorem.

Encourage discussion by students as they work Exercises 11 and 12.

11. Find x. (The letter O indicates the center of the circle where it appears in this exercise.)

(a) 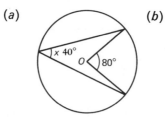 (b) 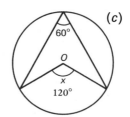 (c)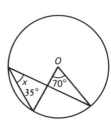

(d) 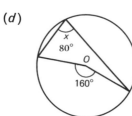 (e) 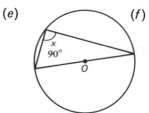 (f)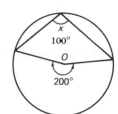

(i): x is half the sum of the bends of the arcs intercepted by the inscribed angles 50° and 30°.

(g) 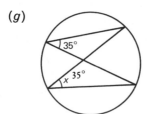 (h) 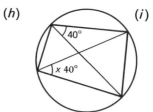 (i)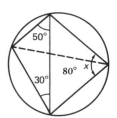

(j): The central angle is an angle of an equilateral triangle.
(k): The 110° angle is a central angle of the circle determined by the outermost vertices of the quadrilateral.
(l): The small central angle is 40°, $x = \frac{1}{2}(360° - 40°)$.
(m): $2x + x = \frac{1}{2}(360°)$

(j) 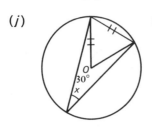 (k) 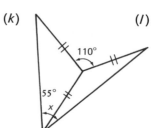 (l)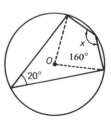

(m) 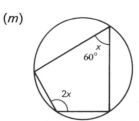 (n) 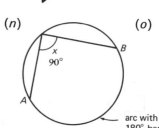 (o)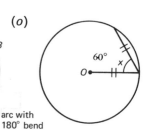

406   10: Circles

(p)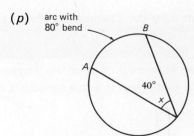
arc with 80° bend

12. Consider the field goal kicker in the opening problem of this chapter.

(a) Show that he can make ∡1 equal to 30° by standing anywhere on an arc of an appropriate circle passing through A and B.

(b) How would you draw that circle? Discuss with your classmates.

(c) What are the positions that make ∡1 *greater than* 30°? Discuss with your classmates.

The Theorem of Thales

The Theorem of Thales is easily proved and very important. It is probably the most important and commonly applied special case of the Inscribed Angle Theorem.

Thales (Thā'lēz), a Greek mathematician who lived about 600 B.C., is given credit for being the first man to prove some of the basic geometrical properties we have been studying. For example, it is said that he first proved that *the base angles of any isosceles triangle are equal.*

It is also said that he proved the following, often called the Theorem of Thales:

Any angle inscribed in a semicircle is a right angle.

DEFINITION The endpoints of any diameter of a circle divide it into two congruent arcs. Each arc is called a *semicircle*.

a semicircle

13. What is the bend of a semicircle?   180°

14. The figure below shows an angle inscribed in a semicircle.

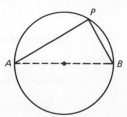

407 Central

Explain why the Theorem of Thales is a special case of the Inscribed Angle Theorem. $\angle APB = \frac{1}{2}(180°) = 90°$

If the semicircle is true, when the right angle of the square is placed at any point on the circle, the sides of the square should meet the endpoints of the semicircle.

15. (a) How might a carpenter's square be used to test whether a casting such as that below is a true semicircle?

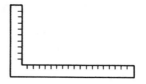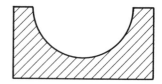

(b) Discuss with your classmates the relationship of part (a) with the Theorem of Thales. *The validity of the test in part (a) depends on the Theorem of Thales.*

Intersecting Chords of a Circle

We now experiment with chords and then derive a remarkable property of chords of a circle.

16. (a) Draw a large circle (with radius about 3 inches) and mark a point P inside and different from the center. Draw a chord through P. Very carefully measure the two segments (into which P divides the chord) and compute the product of their lengths. Do this for three more chords through P. Your measurements must be extremely accurate in this exercise.

(b) Do you observe anything special about the products? Discuss your observations with your classmates. *The product of the lengths of the parts into which P divides the chords stays the same.*

(c) From your experiments, what would you guess should be true about the products *ab* and *cd* in the following figure? *ab = cd*

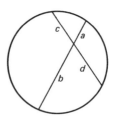

Discuss your guess with your classmates.

(d) Would your guess in part (c) be correct if P happened to be the center of the circle? *yes, (radius)² = (radius)²*

Now let us see why this property of chords of a circle which you might

408   10: Circles

have guessed in the last exercise is true.

17. In the figure below we have an arbitrary point P inside a circle, and two chords, $\overline{AB}$ and $\overline{CD}$, drawn through P.

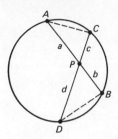

(a) Why is $\triangle APC \sim \triangle DPB$? [*Hint:* What is true about $\angle CAB$ and $\angle CDB$?] $\angle CAB = \angle CDB$ because they intercept the same arc.
$\triangle APC \sim \triangle DPB$ by AAA

(b) Why is $\dfrac{a}{d} = \dfrac{c}{b}$? (If two triangles are similar, then their corresponding sides are _____.)   *proportional*

(c) Show that $ab = cd$.

Summarize Exercises 16 and 17.

(d) State in words what part (c) says about chords of a circle passing through the same point:

If two chords intersect, then the product of the segments of one chord is equal to *the product of the segments of the other chord*.

---

18. Find x.

(a): $x \cdot 8 = 4 \cdot 6$
(b): $x^2 = 4 \cdot 8$
(c): The third angle in the triangle with angle x is 40°.

(a)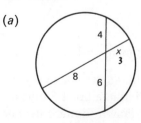

(b)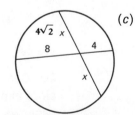

(c)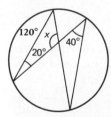

(d): $(x + 2)4 = (2x)x$
(e): $\dfrac{x}{4} = \dfrac{2}{3}$, since the triangles are similar.
(f): $(x + 4)2 = (x + 1)3$

(d)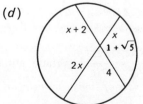

(e)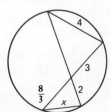

(f)

19. Two runners, *A* and *B*, are running around a circular track. Their coach stands fixed at *P* inside the circle. The runners *A* and *B* run so that the chord joining them always contains *P*. How does the product $(PA)(PB)$ behave as *A* and *B* run around the track? *It remains the same.*

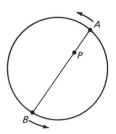

Extended Chords of a Circle

We have learned that if $\overline{AB}$ and $\overline{CD}$ are chords intersecting at a point *P* inside a circle, then $(PA)(PB) = (PC)(PD)$.

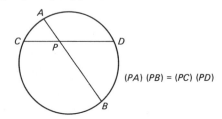

We now examine what happens if $\overline{AB}$ and $\overline{CD}$ are chords whose "extensions" intersect in a point *P* outside the circle, as in the picture below.

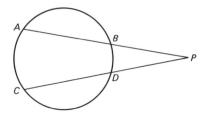

We use the term "extended chord" because a secant is a line.

We call the segments $\overline{PA}$ and $\overline{PC}$ *extended chords* of the given circle.

20. Consider the figure above, where we have two extended chords of a circle intersecting in a point *P* outside the circle. In the figure on the next page, we have drawn (dotted) the chords $\overline{AD}$ and $\overline{BC}$.

(a) Why is $\triangle ADP \sim \triangle CBP$? (Recall the AAA condition for similar triangles!) *The inscribed angles with arc $\stackrel{\frown}{BD}$ are equal.*

(b) Why is $\dfrac{PA}{PC} = \dfrac{PD}{PB}$? (Which are the corresponding sides?)
Corresponding sides of similar triangles are proportional.

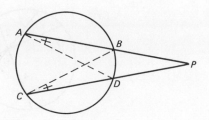

(c) Show that $(PA)(PB) = (PC)(PD)$.

(d) If we call $\overline{PB}$ the *external segment* of the extended chord $\overline{PA}$ and $\overline{PD}$ the external segment of the extended chord $\overline{PC}$, the result of part (c) can be stated as follows:

If two extended chords intersect outside a circle, then the product of the external segment and the whole segment of one extended chord is equal to $\dfrac{\text{the product of the external segment and the}}{\text{whole segment of the other}}$.

When we say "product of the external segment and the whole segment," we mean the product of the *lengths* of these segments.

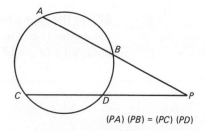

$(PA)(PB) = (PC)(PD)$

Compare this property with the property of chords intersecting *inside* a circle. Remember that when we have extended chords intersecting *outside*, we use "external segment times whole segment."

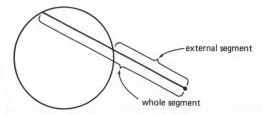

21. Find *x* and *y* in the figures on the next page.

411 Central

(a): $3 \cdot y = 4 \cdot 2$
$x(y + 1) = 4 \cdot 2$
(b): $(x + 4)4 = 10 \cdot 3$

(a)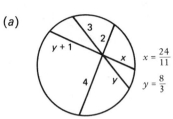

(b)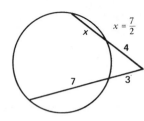

(c): $(x + 2)x = 8(1)$
(d): $(x + 3)3 = (2x + 2)2$

(c)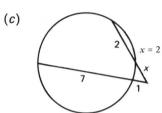

[Ans: $x = 2$]

(d)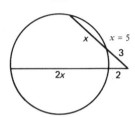

(e): $(x + 3)x = (x + 3)3$
(f): $(y + 7)y = 10(6)$
Since the triangles are similar,
$\dfrac{x}{8} = \dfrac{6}{y}$.

(e)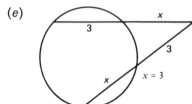

(f)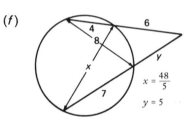

22. In the figure below, the circle has radius r and the distance from P to the center O of the circle is d. Prove that $(PA)(PB) = d^2 - r^2$.
$(PA)(PB) = (d - r)(d + r)$
$= d^2 - r^2$

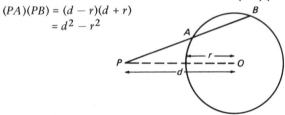

23. The coach in Exercise 19 walks to the *outside* of the circular track and stands fixed at a point P. His runners, A and B, run toward each other so that their extended chord always passes through P.

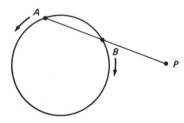

(a) How does the product $(PA)(PB)$ behave as A and B run this way on the track? *It stays the same.*

(b) The runners A and B meet suddenly at a point Q. What do you think is true about $(PA)(PB)$ and $(PQ)^2$ in the figure below?
$(PA)(PB) = (PQ)^2$

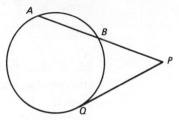

Tangent Lines to Circles

If you have a circle and a straight line lying in a plane, the straight line will intersect the circle in either two points, one point, or not at all. The three possible cases are illustrated below.

Discuss tangent and secant.

In the first case, where we have two points of intersection, the line is called a *secant line* or simply a *secant.* In the second case the line is called a *tangent line* or simply a *tangent,* and the point where it meets the circle is often called the *point of contact.* In the third case there is no special name for the line.

24. (a) Draw a circle and draw a tangent line to the circle as accurately as you can. Draw the radius of the circle from the center to the point of contact of the tangent line. What do you observe about the angle between the radius and the tangent? Compare your result with those of your classmates. $90°$

(b) Draw a circle and choose any point P on it. Draw the radius from the center to P and then draw a line through P perpendicular to the radius. What kind of line do you get? Discuss your result with your classmates.
tangent line

(c) The following important theorem about tangent lines of a circle will be proved in the Projects section of this chapter:

> Any tangent line of a circle is _perpendicular_ to the radius at the point of contact.
>
> Conversely, a line that intersects a circle at a point and is perpendicular to the radius at that point is a _tangent_ line of the circle.

You may wish to summarize Exercises 24-26. In Exercise 26 a common error is that of merely guessing the radius rather than actually constructing the circle.

25. Draw a circle and mark a point P on the circle. Now, using only a straightedge and compass, construct a tangent line to the circle at the point P. *Draw the line from P to the center of the circle. Construct a line perpendicular to this line at P.*

26. (a) Draw a line ℓ and mark a point O not on ℓ. Now, using only a straightedge and compass, construct a circle with center O such that ℓ is a tangent line to the circle. *Construct a line through O perpendicular to ℓ.*

(b) Compare your method with those of your classmates.

27. In the figure below, ℓ is tangent to the circle with center O, OA = 13, and OB = 5.

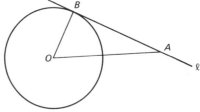

(a) What kind of triangle is $\triangle OBA$? *right* (b) Find AB. *12*

28. In the figure below, $\overline{PA}$ and $\overline{PB}$ are *tangent segments* to the circle, meaning that they are portions of tangent lines to the circle, whose points of contact are A and B respectively.

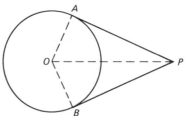

(a) What kind of triangle is $\triangle OAP$? $\triangle OBP$? *right triangles*

(b) Using the Pythagorean Theorem, express PA in terms of OP and OA.
$PA = \sqrt{(OP)^2 - (OA)^2}$

Express *PB* in terms of *OP* and *OB*. $PB = \sqrt{(OP)^2 - (OB)^2}$

(c) Using part (b), prove that *PA* = *PB*. *PA* = *PB* since *OA* = *OB*

(d) The two tangent segments drawn to a circle from an external point are _____ . equal

29. Find *x*. Explain your answers. (The letter *O* denotes the center of the circle where it appears in this exercise. Also, you may assume that lines which look like tangents *are* tangents.)

(*a*): *Tangent lines are perpendicular to the radii, 90° + 90° + 40° + x = 360°.*

(a)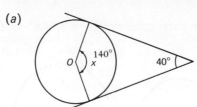

(b)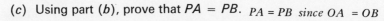

Tell students to split up and label the line segments in (d) and (e).
(d): $8 - x = 9 - 3$
(e): *Let the two segments of x be a and x − a.* $(5 - a) + (8 - (x - a)) = 7$

(c)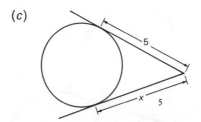

(d)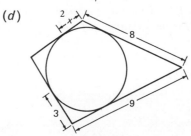

(*f*): *The triangle is isosceles*, $2x + 50° = 180°$.

(e)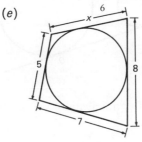

(f)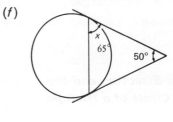

(*g*): *The triangle is equilateral.*
(*h*): $90° + 90° + 120° + x = 360°$

(g)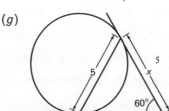

(h)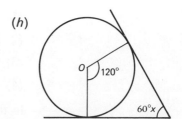

30. Prove that *PA* = *PC* in the figure on the next page.

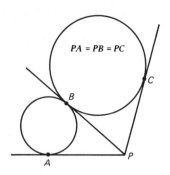

You may wish to give hints or summarize the more difficult parts of Exercise 31.
(a): Let P be the point of intersection of $\overline{AB}$ and $\overline{CD}$: PA = PC and PD = PB, thus AB = PA + PB = PC + PD = CD.
(b): Let P be the point of intersection of the lines through $\overline{AB}$ and $\overline{CD}$: PB = PD and PA = PC, thus AB = PB − PA = PD − PC = CD.
(c): Use (b).

31. In each case, explain why AB = CD.

(a) 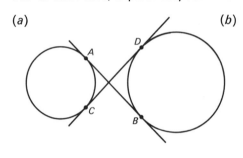 (b)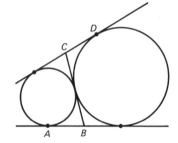

(c)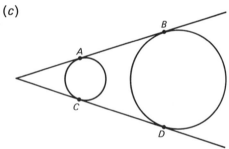

The Angle Bisectors and the Inscribed Circle of a Triangle

In Chapter 5, Central section, we observed the following facts about any triangle:

(1) The perpendicular bisectors of the sides intersect in one point.

(2) The altitudes intersect in one point.

(3) The medians intersect in one point.

(4) The angle bisectors intersect in one point.

In the Central section of Chapter 5 we proved (1), and in the Projects section we proved (2) and (3). However, we never did prove (4); we will now do so, and the next exercise prepares us for the proof.

32. In the figure below, $\overline{PA}$ and $\overline{PB}$ are tangent segments to the circle, and O is the center of the circle.

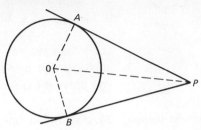

(a) Prove that $\triangle OAP \cong \triangle OBP$. [*Hint:* SSS for congruent triangles]

(b) Why is $\angle APO = \angle BPO$?   *corresponding parts of congruent triangles*

(c) The bisector of $\angle APB$ passes through the _____ of the given circle.   *center*

(d) If two tangents to a circle are drawn from an external point, then their angle bisector passes through the _____ of the circle.   *center*

33. Given any triangle, $\triangle ABC$, let O be the point of intersection of the bisectors of $\angle A$ and $\angle B$. Let D, E, and F be the endpoints of line segments drawn from O perpendicular to sides of the triangle, as shown below.

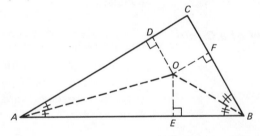

Exercise 33: Difficulty in understanding this proof is to be expected.

(a) Prove that $OE = OF$. [*Hint:* Why is $\triangle OBE \cong \triangle OBF$?]   ASA

(b) Prove that $OE = OD$. $\triangle OAD \cong \triangle OAE$ by ASA

(c) Why does $OF = OE = OD$? $OF = OE$ and $OE = OD$

(d) Let $r = OE$. Explain why the circle with center O and radius r passes through D, E, and F.   *All are distance r from O.*

(e) Why must the sides of $\triangle ABC$ each be tangent to the circle in part (d)?
[*Hint:* Exercise 24(c)]   *Any line that intersects a circle at a point and is perpendicular to the radius at that point is a tangent line to the circle.*

(f) Why must the bisector of $\angle C$ pass through O? Discuss with your classmates. [*Hint:* Exercise 32(d)]   *If two tangents are drawn from an external point, then their angle bisector passes through the center of the circle.*

(g) The circle you found in part (d) is called the *inscribed circle* of △ABC. The angle bisectors of any triangle intersect in one point, and this point is the _____ of the inscribed circle of the triangle. *center*

(h) When we say "the" inscribed circle of a triangle, we seem to be implying that a triangle has only *one* inscribed circle. Is this true? Can you prove it? Discuss with your classmates. *yes*

A common mistake is to guess at the radius rather than construct it. As in Exercise 33, OF, OE, or OD must be constructed.

34. (a) Using only straightedge and compass, construct the inscribed circle of this triangle:

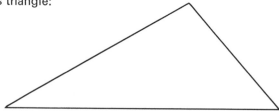

Remark. Be sure your method of finding the radius is precise.

(b) Compare your method with those of your classmates. Check their methods to make sure each step makes correct use of straightedge and compass.

Perimeter of a Circle

Encourage students to do Projects Exercise 8 on the Isoperimetric Theorem.

Do you recall the definition of the perimeter of a polygon? (If you have forgotten, see the Central section of Chapter 9.)

The definition of perimeter for closed curves which are *not* polygons, such as a circle or something complicated as pictured below, is more difficult.

The intuitive idea, however, is the same. The perimeter is the "distance around" the curve—the distance a tiny bug would have to crawl in going once around the curve. Another way to think about it is to imagine the curve as a loop of string. Then the perimeter is the length of this string, which can be measured by cutting it and straightening it out.

35. When a wheel rolls (without skidding) through one revolution, the distance it travels is equal to its perimeter.

(a) Use this idea to measure the perimeter of a circle by cutting a circular disk out of cardboard.

Go slowly in Exercise 35. Let the students discover. This exercise is not too trivial to consider for most students. A great deal depends on the teacher's attitude.

(b) Use this method to find the approximate perimeter of a circle of diameter

(i) 4 cm.   (ii) 5 cm.   (iii) 6 cm.   (iv) 10 cm.
about 12.6 cm.   15.7 cm.   18.8 cm.   31.4 cm.

(c) In each of the cases (i)–(iv) in part (b), compute the ratio

$$\frac{\text{perimeter}}{\text{diameter}}.$$

What do you find?   *ratio ≐ 3.14*

(d) Make a guess about the ratio $\frac{\text{perimeter}}{\text{diameter}}$ for any circle.   *about 3.14*

Exercise 36: Stress that C_1 and C_2 are the lengths of the circles.

36. Consider a circle of perimeter C_1 and diameter D_1, and a second circle of perimeter C_2 and diameter D_2 (here we use the term "diameter" to mean the *length* of a diameter).

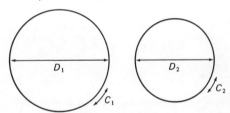

Assume the very plausible (and indeed true!) statement that any two circles are similar figures. Recalling the situation with polygons, explain why it is plausible that

$$\frac{C_1}{D_1} = \frac{C_2}{D_2}.$$

The perimeters of similar polygons are proportional to the other linear measurements.

You may wish to summarize Exercises 35-37, or you may prefer to lead a discussion as the class works through them together.

In the preceding exercise it was pointed out that for all circles the ratio $\frac{C}{D}$ of perimeter to diameter is always the same number. This number is denoted by the symbol π, whose Greek name is *pi* (pī). Thus, if C is the perimeter and D the diameter of any circle, then

$$\frac{C}{D} = \pi.$$

Students will find the Projects on π both instructive and enjoyable.

This number π is not rational (that is, there are *no* integers m and n such that $\pi = \dfrac{m}{n}$), and π is larger than 3, but smaller than $\tfrac{22}{7}$. (King Solomon and wise men of his day used the number 3 as an approximation for π.)

The perimeter of a circle is commonly referred to as the *circumference* of the circle. Using this terminology, we have the definition:

DEFINITION   The ratio of the circumference of any circle to its diameter is the number π.

37. Prove the following:

If a circle has radius r and circumference C, then

$$C = 2\pi r.$$

Since $D = 2r$, $C = \pi D = 2\pi r$.

38. The figure below shows a regular hexagon *inscribed* in a circle of radius r.

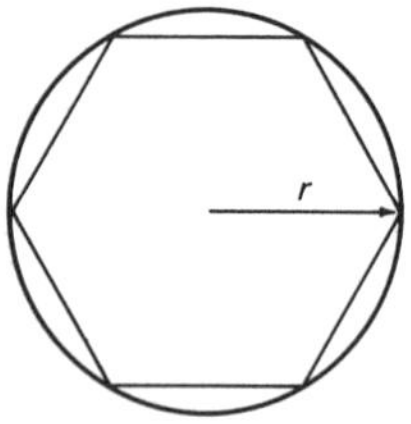

The regular hexagon is constructed by marking arcs around the circle equal to the radius r. Thus each side of the hexagon is r.

(a) Prove that the perimeter of the hexagon is $6r$.

(b) Taking for granted the fact that the perimeter of the circle is greater than the perimeter of the hexagon, prove that $\pi > 3$. $C = 2\pi r > 6r$

39. You want to circle the world, nonstop at constant speed along the equator, in 80 days. Will you have to travel faster than 10 miles per hour? (Assume that the diameter of the earth is 8000 miles.) Exactly how fast should you travel?   yes, $\dfrac{8000\pi \text{ miles}}{1920 \text{ hours}} \doteq 4.17\pi \text{ mph} \doteq 13.1 \text{ mph}$

40. A loop of string is wrapped tightly around the equator of the earth. Suppose that one yard is added to the length of the string and the lengthened string is formed into a larger circle going around the earth above the equator.

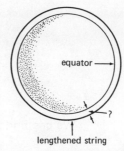

Exercise 40 may be a surprise to the intuition.

(a) Make a guess as to how high above the surface the new circle will be.

(b) Compute how high the new circle will be above the surface (diameter of earth = 8000 miles). *about 6"*

Length of an Arc of a Circle

Give a presentation on length of an arc.

It is easy to compute the length of an arc of a circle, given the size of the central angle and the radius. For example, the arc $\widehat{AB}$ below has central angle 60°, and the radius is 4.

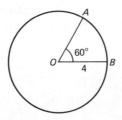

The fraction of the entire circumference taken up by $\widehat{AB}$ is the same as the fraction of the whole angle around O taken up by the central angle. Hence we have

$$\frac{\text{length of } \widehat{AB}}{\text{circumference of circle}} = \frac{60°}{360°}$$

or

$$\frac{\text{length of } \widehat{AB}}{8\pi} = \frac{60°}{360°} = \frac{1}{6},$$

so

$$\text{length of } \widehat{AB} = \left(\frac{1}{6}\right)(8\pi) = \frac{4}{3}\pi.$$

In general, consider an arc with central angle θ (Greek letter *theta*) and radius r.

421 Central

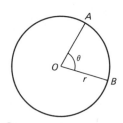

Stress the given formula:
$$\frac{\text{length of arc}}{2\pi r} = \frac{\text{bend of arc}}{360°}.$$

We can find the length of $\overset{\frown}{AB}$ using the formula:

$$\frac{\text{length of arc}}{2\pi r} = \frac{\theta \text{ (degrees)}}{360°}$$

41. Find the length of the arc of a circle if
(a) the central angle is 160° and the radius is 1 inch
(b) the central angle is 225° and the radius is 3 feet $\frac{15}{4}\pi$ *feet*
(c) the central angle is 35° and the radius is 6 cm.
$\frac{7}{6}\pi$ *cm.*   [Ans: (a) $\frac{8\pi}{9}$ inches]

42. A piece of wire 3π inches long is to be bent into an arc of a circle with radius 10 inches. What will be the size of the central angle of the arc?

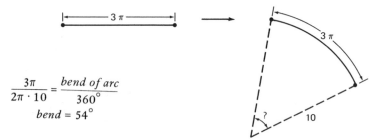

$$\frac{3\pi}{2\pi \cdot 10} = \frac{\text{bend of arc}}{360°}$$
$$\text{bend} = 54°$$

43. If the circumference of the circle below is 10π, what is the length of the smaller arc $\overset{\frown}{AB}$?
$\frac{20\pi}{9}$

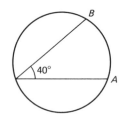

Area of a Circle

The picture on the following page shows a circle of radius *r* with a *circumscribed* regular 8-gon.

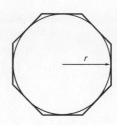

The computation of the perimeter of the octagon is in Projects Exercise 20.

Imagine drawing a circle on a piece of paper and cutting off corners of a circumscribed polygon successively with cuts tangent to the circle, yielding better and better approximations to the circle.

(The polygon is said to be "circumscribed" about the circle because each side of the polygon is tangent to the circle.) The perimeter of the 8-gon is a fairly good approximation to the perimeter of the circle. In fact, the perimeter of the 8-gon is about $(6.62)r$, while the perimeter of the circle is about $(6.28)r$. If we wanted to obtain a better approximation to the perimeter of the circle, we could use the perimeter of a circumscribed regular n-gon with a larger value of n. As we choose larger and larger values for n, the perimeters of the circumscribed n-gons get closer and closer to the perimeter of the circle.

If we let L_n denote the perimeter of a regular n-gon circumscribed about the above circle, and recall that the perimeter of the circle is $2\pi r$, we can summarize what we have said as follows:

The numbers L_n get closer and closer to $2\pi r$ as n gets larger and larger.

In the following exercises we will see how this observation enables us to find a formula for the area enclosed by a circle in terms of its radius.

44. Given a circle of radius r, let L_n denote the perimeter of a regular n-gon circumscribed about the circle. Discuss with your classmates why the following is plausible:

The numbers $\frac{1}{2}rL_n$ get closer and closer to πr^2 as n gets larger and larger.

[*Hint:* What do the numbers L_n get closer and closer to? If each number L_n is multiplied by $\frac{1}{2}r$, what do the resulting numbers get closer and closer to?] $\frac{1}{2}r(2\pi r) = \pi r^2$

45. Suppose L_n is the perimeter of a regular n-gon circumscribed about a circle of radius r, and A_n is the area of the polygon. Show that

$$A_n = \frac{1}{2}rL_n.$$

[*Suggestion:* The following picture illustrates how to proceed in the case of a regular 5-gon. Cut the circumscribed pentagon into 5 triangles by drawing lines from the center to the vertices. Draw the radii to the points of

The derivation is subtle. Do not belabor it if students have difficulty.

contact of the sides of the 5-gon with the circle (why are the radii perpendicular to the sides?). Each side of the pentagon has length s. Each triangle has area $\frac{1}{2}rs$, and there are 5 triangles. Hence $A_5 = 5(\frac{1}{2}rs)$. What is L_5? Why is $A_5 = \frac{1}{2}rL_5$?]

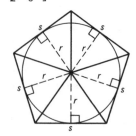

46. Let A denote the area enclosed by a circle of radius r. Let A_n denote the area of a regular n-gon circumscribed about the circle. Take for granted the following plausible statement:

The numbers A_n get closer and closer to A as n gets larger and larger.

Discuss with your classmates why this and the preceding two exercises show that $A = \pi r^2$. [*Hint:* The numbers A_n get closer and closer to A. But $A_n = \frac{1}{2}rL_n$. What do the numbers $\frac{1}{2}rL_n$ get closer and closer to?]

The preceding exercises derived the following very useful formula for the area enclosed by a circle:

The area enclosed by a circle of radius r is πr^2.

This will be needed in the following exercises.

$\pi r^2 = 400, r = \dfrac{20}{\sqrt{\pi}}$

$C = 2\pi r = 40\sqrt{\pi} \doteq 40(\sqrt{3.14}) > 70.8$

47. A farmer wants to put a fence around a circular field of area 400 square feet. Can he do this with 70 feet of fencing material? (π is greater than 3.14.)

no

48. If the radius of a circle is doubled, what happens to

(a) the perimeter?   (b) the area?
doubled   *quadrupled*

Justify each answer in two *different* ways. (Use the formula for the area of a circle, and also use the fact that any two circles are similar.)

424   10: *Circles*

49. A square of area 100 is inscribed in a circle. What is the area of the circle?

Side of the square is 10, diagonal is $10\sqrt{2}$, radius of the circle is $5\sqrt{2}$, area is 50π.

50. Suppose the cost of ingredients for a 10″ diameter pizza is 75¢. What would the cost be for a 12″ diameter pizza of the same thickness?

$$75¢ \; \frac{\text{area (12″ pizza)}}{\text{area (10″ pizza)}} = \$1.08$$

Area of a Sector of a Circle

DEFINITION

A *sector* of a circle is that portion of the interior enclosed by an arc of the circle and the two radii from the center to the endpoints of the arc.

A sector is a "piece of pie."

sector of a circle

It will help to examine simple special cases. For example, if we cut a circle in half ($\theta = 180°$) we get two sectors, each having half the area of the whole circle. If we cut a circle into 4 equal sectors, each sector has central angle $90°$ and area one-fourth that of the whole circle. For a sector of central angle $60°$, we have $60°$ = one-sixth of $360°$ and the area of the sector = one-sixth the area of the whole circle. It may help to make a verbal statement: "The fraction of the area of the whole circle taken up by the sector is equal to the fraction of the whole central angle taken up by the angle of the sector."

It is easy to compute the area of a sector if we know the radius and the central angle. Consider the shaded sector of the circle with center P below.

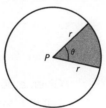

We have

$$\frac{\text{area of sector}}{\text{area of whole circle}} = \frac{\text{central angle at } P}{\text{whole angle around } P}.$$

In other words, we can find the area of the sector by using

$$\frac{\text{area of sector}}{\pi r^2} = \frac{\theta \text{ (degrees)}}{360°}.$$

51. Find the area of a sector of a circle of radius r, and central angle θ, if
 (a) $r = 4$ and $\theta = 90°$ (b) $r = 5$ and $\theta = 60°$
  4π $\dfrac{25\pi}{6}$

(c) $r = 1$ and $\theta = 1°$ $\dfrac{\pi}{360}$ (d) $r = 3$ and $\theta = 200°$. 5π

52. What is the area of the shaded region?

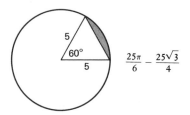

[*Hint:* The area of the shaded region is the area of the sector minus the area of the triangle.]

The application of π to a piece of pizza is always of interest.

53. The Sultan of Polygonia has a liking for pizza. The royal cook baked a 10″ diameter circular pizza which had 10 calories per square inch. The Sultan, being on a diet, could eat only a 75-calorie slice. What should be the central angle of the (sector-shaped) slice to contain exactly 75 calories?

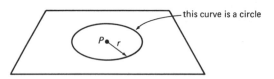

The Ball and its Surface, The Sphere

Given a point P in a plane and a number $r > 0$, those points *in the plane* at a distance r from P form a circle of radius r centered at P.

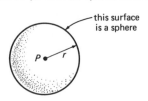

DEFINITION

Emphasize that the points on the surface of the ball are the points at distance r from P. A sphere is the surface of a ball.

Given a point P in space and a number $r > 0$, the *sphere* of radius r and center P consists of all those points in space at a distance r from P.

Exercise 54: Remind students of the definition of a circle.

54. When you cut an orange in half with a knife, you obtain a circular cross section. Explain why this happens. In other words, explain why the intersection of a sphere of radius r with a plane through its center is a circle of radius r.

Exercise 55 may be difficult for some students to visualize, even with pictures.

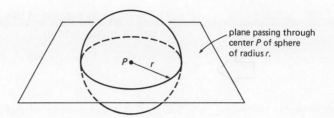
plane passing through center P of sphere of radius r.

55. When you cut an orange with a straight cut of a knife, the cross section is circular, even if the cut is not through the center.

Suppose a sphere of radius r is cut by a plane at distance $k < r$ from the center P. Explain why the intersection is a circle of radius $\sqrt{r^2 - k^2}$. [*Hint:* The right triangle, $\triangle PQR$, has $PQ = k$ and $PR = r$.]

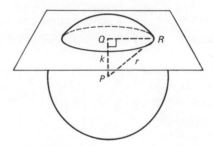

A good model of a sphere is the *surface* of an ordinary rubber ball. The ball itself is a *solid* object.

DEFINITION Given a point P and a number $r > 0$, the *ball* of radius r and center P consists of all those points inside and on the sphere of radius r centered at P. In other words, the ball consists of all points at distance *less than or equal to* r from P.

56. A ball of radius r is cut by a plane through the center. What is the perimeter of the cross section? What is the area of the cross section?
$2\pi r$ πr^2

427 Central

In the Projects section of this chapter it is proved that if V is the volume of a ball of radius r, then

$$V = \frac{4}{3}\pi r^3.$$

Exercise 57 is similar to the statue problem in Chapter 9. The answer should be a surprise. You might ask, "Could you lift a steel ball of radius 6 inches?"

57. (a) A steel marble of radius $\frac{1}{2}$ inch weighs 2 ounces. How much would a steel wrecking ball of radius 2 feet weigh? (Give your answer in pounds, 1 pound is equal to 16 ounces.) *13,824 pounds*

(b) Discuss with your classmates how you can arrive at the correct answer in *two* different ways.

An amusing digression — what is the better deal, buying a cylindrical tin can (radius of base = r) with a spherical tomato in it or buying a spherical tin can (of radius r) with a cylindrical tomato in it?

58. In the Projects section it is shown that a right circular cylinder (imagine an ordinary tin can) of height h and base radius r has volume

$$V = \pi r^2 h.$$

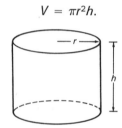

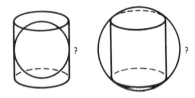

Suppose such a tin can (with $h = 2r$) is completely filled with water and a perfectly spherical tomato. What fraction of the volume is taken up by the water? $\frac{1}{3}$

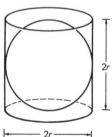

This raises the nontrivial problem of finding the cylinder of maximum volume inscribed in a sphere, which turns out to be a cylinder of height $2r/\sqrt{3}$, where r is the radius of the sphere. The fraction of the volume of the sphere that the cylinder takes up is $1/\sqrt{3}$.

Important Ideas in This Chapter

Definitions

Central angle determined by an arc of a circle

Bend of an arc of a circle

Inscribed angle

Arc *intercepted* by an inscribed angle

Sector of a circle

Semicircle

Secant line of a circle

Tangent line of a circle

Inscribed circle of a triangle

Perimeter of a closed curve

The number π

Circumference of a circle

Sphere

Ball

Theorems

Inscribed Angle Theorem   The size of any inscribed angle is half the bend of its intercepted arc. (Hence, in any circle, all inscribed angles which intercept the same arc are the same size.)

The Theorem of Thales   Any angle inscribed in a semicircle is a right angle.

The product of the segments is the same for all chords through a given point inside a circle.

The product of the external segment and the whole segment is the same for all extended chords through a given point outside a circle.

Any tangent of a circle is perpendicular to the radius drawn to the point of contact. Conversely, a line that intersects a circle at a point and is perpendicular to the radius at that point is a tangent line of the circle.

Tangent segments drawn to a circle from an external point are of equal length.

The angle bisectors of any triangle all pass through the center of the inscribed circle.

The ratio of the circumference to the diameter is the same for all circles, and this number is denoted by the symbol π. Hence, if C is the circumference and D the diameter of a circle, then

$$C = \pi D.$$

If a circle has radius r, circumference C, and area A, then

$$C = 2\pi r \quad \text{and} \quad A = \pi r^2.$$

The length of an arc of a circle of radius r with central angle θ is given by

$$\frac{\text{length of arc}}{2\pi r} = \frac{\theta}{360°}.$$

The area of a sector of a circle of radius r with central angle θ is given by

$$\frac{\text{area of sector}}{\pi r^2} = \frac{\theta}{360°}.$$

The volume of a ball of radius r is $V = \frac{4}{3}\pi r^3$.

Words circle, chord, diameter, arc, bend, semicircle, tangent, secant, sector, perimeter, circumference, sphere, ball, inscribed, circumscribed

Symbols notation for angles: $\angle APB$

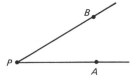

π (pi), $\overset{\frown}{AB}$ (arc of a circle from A to B)

Summary

In this chapter we studied some useful and interesting properties of circles. Some of these properties, such as the Inscribed Angle Theorem and the theorems about intersecting chords, were rather unexpected. The formulas for computing perimeter and area of a circle with given radius are of great importance in applications. We also looked briefly at the sphere, the 3-dimensional analogue of the circle.

Historical Note

Archimedes of Syracuse (287–212 B.C.) was the greatest mathematician of ancient times, and one of the greatest of all time. His fundamental discoveries ranged over many fields, including mechanics and hydraulics. In his computations of areas and volumes he anticipated methods that were used two thousand years later when Newton and Leibniz invented the calculus. Of all his discoveries, Archimedes was especially proud of his proof that the volume of a sphere is two-thirds that of the smallest

This traditional story of Archimedes's death has been passed down by various authors and is probably true in outline. His actual words to the Roman soldier are, of course, open to dispute.

cylinder enclosing it, and he requested that the diagram of a sphere in a cylinder be engraved on his tombstone. Archimedes died after the siege of Syracuse by the Romans. A Roman soldier had found him drawing geometric figures in the sand; when Archimedes exclaimed, "Do not disturb my circles," the soldier killed him. The Roman commander Marcellus was incensed by the soldier's act and gave Archimedes an honorable burial. The location of his tomb was forgotten for two hundred years, until it was discovered by Cicero, then governor of Sicily, who recognized the tombstone engraved with the figure of a sphere in a cylinder.

Central Self Quiz

1. Find x (O is the center of the circle).

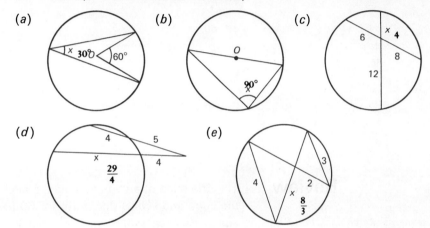

(e): $\dfrac{x}{2} = \dfrac{4}{3}$

2. A circle passes through the vertices of a square of side 3. What is the area of the circle? *radius of circle* $= \dfrac{3}{2}\sqrt{2}$

3. A circle has circumference 12 cm. Find the bend of an arc of length 2 cm.

4. If O is the center of the circle, and $OA = 6$, find AB.

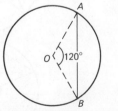

[*Hint:* Draw the altitude from O to $\overline{AB}$.]

5. Find x.

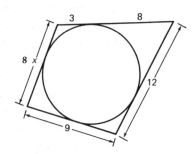

Answers
1. (a) 30° (b) 90° (c) 4 (d) $\frac{29}{4}$ (e) $\frac{8}{3}$

2. $\frac{9\pi}{2}$

3. 60°

4. $6\sqrt{3}$

5. 8

Review

Have students sketch a picture. Remind them of the Theorem of Thales.

1. The edge of a certain pond is a circle. A goldfish starts at a point on the edge and swims due north for 60 feet, which takes him to the edge again. He then swims due east, reaching the edge after going 80 feet. What is the diameter of the pool?

2. What relationship exists between x and y? (O is the center.)

(a) 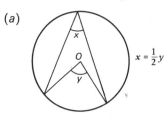 $x = \frac{1}{2}y$

(b) 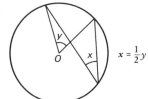 $x = \frac{1}{2}y$

(c) 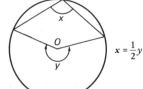 $x = \frac{1}{2}y$

(d) 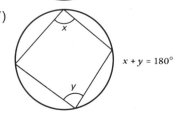 $x + y = 180°$

3. Which must be larger, ∡APB or ∡AQB? Prove your answer. [*Hint:* Draw the chord containing P and Q.]

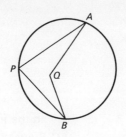

4. Find x.

(a)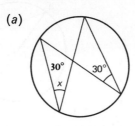

(b)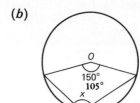

(c)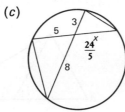

(d)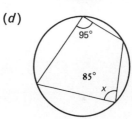

(e)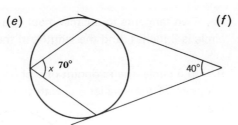

(f)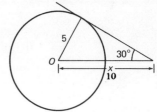

(g)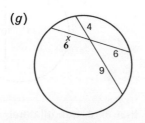

(h)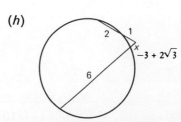

($\frac{1}{3}$ and 3 are both wrong answers!)

433   Review

Exercise 5 is not easy! A solution alternate to that given in the Review Answers follows quickly from Projects Exercises 27-31. You may wish to do these exercises along with the Central section.

5. Prove that ∡1 = ∡2.

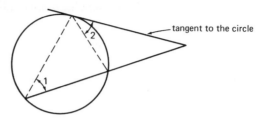

6. In the figure below, O is the center of the circle, and $\overline{OD}$ is parallel to $\overline{AC}$. Prove that the arcs $\overset{\frown}{BD}$ and $\overset{\frown}{DC}$ have the same length. [*Hint:* Draw $\overline{OC}$.]

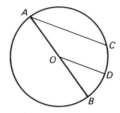

7. Use what you have learned in this chapter to show that △*ABC* is a right triangle.

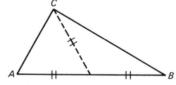

8. Two tangents to a circle meet at an angle of 60°. If the radius of the circle is 8 inches, find the lengths of the tangents.

9. In a circle, the midpoint of a chord 10 inches long is 1 inch from the midpoint of the smaller arc with the same endpoints. Find the radius of the circle.

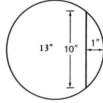

10. (a) Prove that if a quadrilateral *ABCD* is circumscribed about a circle, then $AB + CD = AD + BC$.

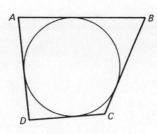

(b) Can you draw a quadrilateral whose sides have lengths 3, 4, 7, and 9 cm., and such that the quadrilateral is circumscribed about some circle? If you think so, do it. If not, explain why not.

11. (a) Prove that if a quadrilateral is inscribed in a circle, then the sum of its opposite angles is 180°.

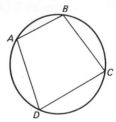

In other words, prove: $\angle A + \angle C = \angle B + \angle D = 180°$.

(b) Alex Smart states that he found a quadrilateral with angles of size 70°, 130°, 95°, and 60° which is inscribed in a circle. Explain how part (a) shows he must be wrong, and then find another (and worse) mistake he has made.

(c) What kinds of parallelograms can be inscribed in circles? Explain.

12. (a) What is the definition of the number π?

(b) If a circle has circumference C and diameter D, then why is $C = \pi D$?

(c) If a circle has radius r and circumference C, then why is $C = 2\pi r$?

(d) Explain why a circle of radius 5 cm. has perimeter greater than 30 cm. but less than $31\frac{1}{2}$ cm. What is the exact perimeter?

13. The radius of a circle is 10 cm.

(a) What is the circumference of this circle?

(b) What is the bend (in degrees) of an arc of this circle which is 2 cm. long?

14. Suppose the radius of the circle in the figure below is 5 cm.

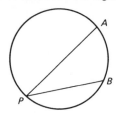

Find the size of ∡APB if the length of the smaller arc of the circle from A to B is

(a) 2π cm.   (b) 5π cm.   (c) x cm.

15. A railroad company wants to build a circular track which passes through 3 given locations A, B, and C.

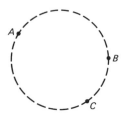

It is possible only if A, B, and C are not on the same line.

Will this always be possible? If it is possible, is there more than one circle which will work? If there is only one such circle, how could they locate where its center must be?

The center is at the intersection of the perpendicular bisectors of the sides. The radius is determined once the center is found.

16. Using only straightedge and compass, construct a circle which passes through the vertices of this triangle.

The center is at the intersection of the angle bisectors. The radius must be found by constructing a line through the center perpendicular to one of the sides.

17. (a) Using only straightedge and compass, construct a circle such that the sides of the triangle below are tangent to the circle.

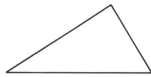

(How do you find the center? the *radius*?)

436   10: Circles

(b) Explain why your method works. *See Central Exercise 33.*

(c) Find a formula which relates the area and perimeter of the triangle with the radius of this circle.

18. Draw a segment 6 cm. long and mark its endpoints A and B. Now, using straightedge and compass, construct a pair of circles having the same center, such that segment $\overline{AB}$ is a chord of the larger circle and is tangent to the smaller circle.

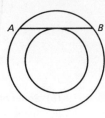

19. Using only straightedge and compass, find all points P on ℓ which make $\angle APB = 90°$

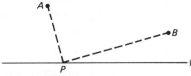

Review Answers

1. 100 ft. (Use the Theorem of Thales.)

2. (a), (b), (c) $x = \frac{1}{2}y$ (d) $x + y = 180°$

3. $\angle AQB$. Apply the Exterior Angle Theorem to $\triangle PAQ$ and $\triangle PBQ$ after drawing the chord through P and Q.

4. (a) 30°   (b) $x = \frac{1}{2}(360° - 150°) = 105°$ (c) $\frac{24}{5}$ (d) 85°

   (e) 70°   (f) 10   (g) 6   (h) $x(x + 6) = 1(1 + 2), x = -3 + 2\sqrt{3}$

5. $\angle 2 = 90° - \angle 3 = \frac{1}{2}(180° - 2\angle 3) = \frac{1}{2}\angle BOC = \angle 1$

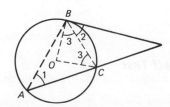

6. $\angle CAB = \angle ACO = \angle COD$ (explain). Also $\angle CAB = \frac{1}{2}\angle COB$ (why?). It follows that $\angle DOB = \angle COD$. Hence $\overarc{BD}$ and $\overarc{DC}$ have the same bend and also the same length.

7. $\overline{AB}$ is a diameter of a circle through A, B, and C (why?). $\angle ACB = 90°$ by the Theorem of Thales.

8. $8\sqrt{3}$ inches

9. 13 inches

10. (a)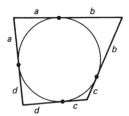

(b) Some pair of numbers would have to add to the sum of the other pair by part (a), but that is not possible with the numbers given.

11. (a) Use the Inscribed Angle Theorem.
(b) Some pair of angles would have to add up to the sum of the other pair by part (a), but that is not possible. But, in fact, the sum of the angles is not 360°; hence there cannot be *any* quadrilateral with these angles!
(c) only rectangles, why?

12. (d) $30 < 10\pi < 31.5$, because $3 < \pi < 3.15$.

13. (a) 20π cm. (b) $\left(\dfrac{2}{20\pi}\right)(360°) = \dfrac{36°}{\pi}$

14. (a) $36°$ (b) $90°$ (c) $\left(\dfrac{1}{2}\right)\left(\dfrac{x}{10\pi}\right)(360°) = \dfrac{18x°}{\pi}$

15. Center is at the point of intersection of the perpendicular bisectors of $\overline{AB}$ and $\overline{BC}$. This is not possible if A, B, and C are on the same line.

17. (c) If A is the area, L the perimeter, and r the radius of the inscribed circle, then $A = \tfrac{1}{2}rL$.

18. Construct the perpendicular bisector of $\overline{AB}$. Pick any point on it as the center of the two circles.

19. Construct the circle having $\overline{AB}$ as a diameter.

Review Self Test

1. Find the area of a circle whose circumference is 10 cm.

2. The picture shows a *regular pentagon* inscribed in a circle.

(a) Find the bend of $\widehat{AB}$.
$\tfrac{1}{5}(360°)$

(b) Find $\angle DAC$.
$\tfrac{1}{2}(72°)$

(c) Find $\angle ADC$.

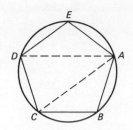

radius of large circle

$= \sqrt{(4)^2 + \left(\frac{AB}{2}\right)^2} = \sqrt{52}$

area of large circle = 52π
area of small circle = 16π

3. If $AB = 12$, find the area *between* the two circles.

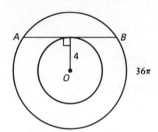

36π

4. Express the area of $\triangle ABC$ below in terms of r, if r is the radius of the inscribed circle.

area $= \frac{1}{2}r(AC) + \frac{1}{2}r(AB) + \frac{1}{2}r(BC)$

$= \frac{1}{2}r(5+3) + \frac{1}{2}r(5+8) + \frac{1}{2}r(3+8)$

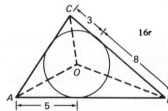

$16r$

5. If AB is equal to the radius of the circle, find $\angle APB$.

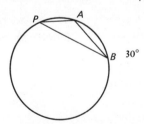

30°

6. The area of the intersection of a ball with a plane through its center is 9π. Find the volume of the ball.   *The radius is 3.*

Answers

1. $\dfrac{25}{\pi}$

2. (a) 72°   (b) 36°   (c) 72°

3. 36π

4. $16r$

5. $30°$, because $\overarc{AB}$ has bend $60°$ (why?). *If O is the center of the circle, $\triangle AOB$ is equilateral.*

6. 36π

Projects

Constructible Segments

Given a segment of unit length, what lengths are possible for segments that can be constructed using straightedge, compass, and the given unit segment? The impossibility of "squaring a circle" is equivalent to the fact that a segment of length π cannot be so constructed.

Suppose we are given two line segments of lengths a and b respectively, with $a > b$. Then it is easy to construct segments of lengths $a + b$ and $a - b$ using only straightedge and compass: draw a line, mark off a segment of length a using a compass, and then mark a segment of length b from one endpoint.

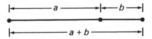 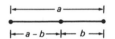

Suppose we are given a segment of length 1 (unit length), a segment of length a, and a segment of length b. In the following exercises we learn how to construct segments of length ab and certain other combinations, using only straightedge and compass.

It must be stressed that the lengths $1, a, b$ are given on paper first and then copied. Students tend to construct an arbitrary triangle and then label the sides, completely missing the point of the problem.

1. Show how to construct with straightedge and compass a segment of length ab, given segments of lengths 1, a, and b.

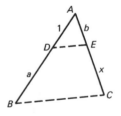

[*Hint:* Construct $\triangle ADE$ with $AD = 1$ and $AE = b$, construct segment $\overline{AB}$ with $DB = a$, and then construct $\overline{BC}$ parallel to $\overline{DE}$. Why is $\dfrac{b}{x} = \dfrac{1}{a}$?]

2. Given segments of lengths 1, a, and b, construct a segment of length $\dfrac{a}{b}$, using only straightedge and compass. $\dfrac{x}{1} = \dfrac{a}{b}$

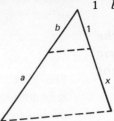

The Geometric Mean

The *geometric mean* of two nonnegative numbers a and b is defined to be $\sqrt{ab}$, the square root of their product. The following exercise gives a simple method for constructing, with straightedge and compass, a line segment whose length is the geometric mean of the lengths of two given line segments.

3. (a) Using similar triangles, show that $x = \sqrt{ab}$.

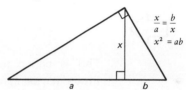

$\dfrac{x}{a} = \dfrac{b}{x}$

$x^2 = ab$

(b) In the figure below, P and Q are the endpoints of a diameter of the circle. Explain why $x = \sqrt{ab}$. (Recall the Theorem of Thales.)
∡PRQ is a right angle, use (a).

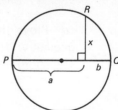

(c) Given segments of lengths a and b, give a method for constructing with straightedge and compass alone a segment of length $\sqrt{ab}$ (base your method on part (b)). *Construct a circle with diameter $a+b$, erect a perpendicular where segments a and b meet.*

Exercise 4 contains some good construction problems.

4. Given segments of lengths 1, a, and b, show how to construct, using only straightedge and compass, a segment of length

(a) $\sqrt{a}$ (b) a^2 (c) $\sqrt{a + \sqrt{b}}$ (d) $\sqrt{\sqrt{a}}$.

(a): *Use Exercise 3(c) with a circle of diameter $a+1$.*

5. Draw a segment of length 8 cm. Now, using only straightedge and compass, construct a segment of length $\sqrt{6}$ cm. When you are finished, check the accuracy of your construction by measurement. *Bisect the 8 cm. length twice to get a length of 6 cm, use Exercise 4(a).*

The Inequality of the Arithmetic and Geometric Means

The idea of a "mean" or "average" is very important in science. We encountered the *geometric mean* $\sqrt{ab}$ of positive numbers a and b in a previous exercise. A more familiar mean is the *arithmetic mean* of a and b, defined as $\dfrac{a+b}{2}$.

Exercise 6 contains one of the most useful inequalities in mathematics — the derivation in (d) is particularly elegant.

6. (a) Take some specific pairs of positive numbers, a and b, and check that their arithmetic mean is always larger than or equal to their geometric mean.

(b) What you observed in part (a) holds for any nonnegative numbers a and b. In other words, it is always true that

$$\frac{a+b}{2} \geq \sqrt{ab}.$$

This is called the *inequality of the arithmetic and geometric means*. Prove it, using algebra, by examining $(\sqrt{a} - \sqrt{b})^2$. [*Hint:* When is $(\sqrt{a} - \sqrt{b})^2 \geq 0$ a true statement?]

(c) Under what conditions is the arithmetic mean equal to the geometric mean? *when a = b*

(d) Prove the inequality of the arithmetic and geometric means using geometry. [*Hint:* In the picture below the circle has diameter $PQ = a + b$. The chord $\overline{RS}$ is perpendicular to $\overline{PQ}$.

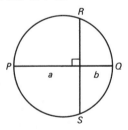

$RS = \underline{\quad 2\sqrt{ab} \quad}$ (in terms of a and b). Recall Projects Exercise 3 of this chapter. Compare RS and $a + b$; $a + b$ is the $\underline{\quad diameter \quad}$ of the circle.]

7. (*a*) Prove that the area of any rectangle is always less than that of a square of the same perimeter. [*Hint:* What is the perimeter of an $a \times b$ rectangle? What is the length of a side of a square of the same perimeter? What is the area of the square? Is the area of the square larger than that of the rectangle? See the preceding exercise.]

(*b*) A farmer wants to build a pen in the shape of a rectangle. He has enough fencing material to build a fence of perimeter 200 feet. What is the pen of greatest area he can build and what is that area? Prove that your answers are correct.   2500 *sq. ft.*

(*c*) Suppose you have a closed loop of string 12 inches long. What is the maximum area you can enclose by laying the string down (on a table) in the shape of a rectangle?   *9 sq. inches, a square*

The Isoperimetric Theorem

Possible subjects for discussion related to the Isoperimetric Theorem: Why did the early American pioneers arrange their wagons in a circle? How can the size of an enemy army be estimated by reconnaissance of the perimeter of its camp? Why are soap bubbles spherical (Isoperimetric Theorem in 3 dimensions)?

8. (*a*) Suppose you have a closed loop of string 12 inches long. What is the area enclosed if you lay the string down in the shape of an equilateral triangle? 8.9   square? 9   regular hexagon? 10.4   circle? 11.5

(*b*) Suppose an equilateral triangle, a square, a regular hexagon, and a circle all have the same perimeter. Which has largest area? *circle*

(*c*) Of all closed curves having a given fixed perimeter, which curve do you think encloses the largest area? *circle*

(*d*) One of the famous theorems of geometry is the *Isoperimetric Theorem* (iso- means "same" in Greek). Complete the statement of the theorem:

Of all closed curves of the same perimeter in a plane, the *circle* **encloses the largest area.**

(*e*) A closed curve has perimeter L and area A. Using the Isoperimetric Theorem, show that

$$L^2 \geq 4\pi A.$$

[*Hint:* What is the area of a circle with the same perimeter as the curve?]

Mathematics and Plausible Reasoning, vol. 1, *by G. Polya (Princeton: Princeton University Press, 1954), chapter 10, contains a stimulating treatment of this subject.*

(*f*) Read about the Isoperimetric Theorem (and how it is proved) in the book *What Is Mathematics?* by R. Courant and H. Robbins (New York: Oxford University Press, 1941), pp. 373–376.

The Golden Section

The ancient Greeks were interested in geometric shapes which are pleasing to behold—those shapes which are "symmetrical" or "pleasantly proportioned." In particular, they felt that aside from the square a certain shape of a rectangle was the best proportioned of all rectangles. A rectangle with these proportions they called a "golden rectangle," and this is the subject of Exercises 9–14.

9. Suppose the point P is chosen on the line segment $\overline{AB}$ in such a way that

$$\frac{AB}{AP} = \frac{AP}{PB}.$$

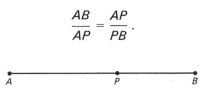

The Greeks then said that $\overline{AB}$ was divided in the ratio of the *Golden Section*. Show that this ratio is exactly

$$\frac{AB}{AP} = \frac{1+\sqrt{5}}{2} \doteq 1.618\ldots. \qquad (\sqrt{5} \doteq 2.236)$$

This number is called the *Golden Ratio*. [*Hint:* Let $AP = 1$ and $PB = x$.

Solve for x:

$$\frac{x+1}{1} = \frac{1}{x}$$

Then $x = \dfrac{-1+\sqrt{5}}{2}$. What is $\dfrac{AB}{AP}$?] $\dfrac{AB}{AP} = \dfrac{x+1}{1} = \dfrac{1+\sqrt{5}}{2}$

10. (a) According to the Greeks, a *golden rectangle* is one such that the ratio of the length of the longer side to the length of the shorter side is the golden ratio. The rectangle on the next page is a golden rectangle if

$$\frac{AB}{BC} = \frac{1+\sqrt{5}}{2}.$$

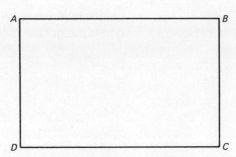

Show that any two golden rectangles are similar.

(b) In the figure below, *ABCD* is a golden rectangle, and by cutting along the dotted line we cut off a square *APQD*. Show that the remaining rectangle *PBCQ* is also a golden rectangle.

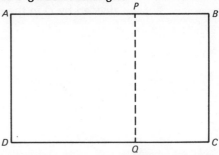

11. (a) Prove that in the triangle below, $\dfrac{AB}{BC}$ is the golden ratio.

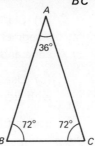

[*Hint:* Look at the similar triangles which are formed when you draw the line bisecting the angle at *C*; call the new triangle $\triangle DCB$; show $\dfrac{AB}{BC} = \dfrac{BC}{BD}$ and $BC = CD = AD$; then show $\dfrac{AB}{AD} = \dfrac{AD}{BD}$.]

(b) If the polygon *ABCDE* is a regular pentagon, show that $\dfrac{AC}{AB}$ is the

445 Projects

golden ratio (concisely: the ratio of a diagonal to a side of a regular pentagon is the golden ratio).

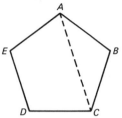

[*Hint:* △ADC is a 72°–72°–36° triangle.]

12. (*a*) In the figure below, *ABCD* is a square of side length 1, *P* is the midpoint of $\overline{CD}$, and *Q* is located so that *PQ* = *PA*. Show that *CQ* is divided by *D* in the ratio of the golden section.

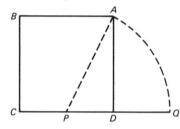

[*Hint:* Show that $PA = \dfrac{\sqrt{5}}{2}$, then $\dfrac{CQ}{CD} = \underline{}$.]

(*b*) Starting with a segment of length 1, describe how one can construct a segment whose length is equal to the golden ratio, using only straightedge and compass as tools. *Construct CQ as in (a).*

(*c*) Use compass and straightedge only to construct a regular pentagon with sides of length 1. [*Hint:* Starting with a given segment of length 1, use part (*a*) to construct a segment $\overline{AC}$ of length $\dfrac{1 + \sqrt{5}}{2}$. Then construct △*ABC* with *AB* = *BC* = 1. Then ∡*ABC* is one angle of the required pentagon (see Exercise 11(*b*)).]

The Fibonacci Numbers and the Golden Section

The *Fibonacci sequence* is the sequence of integers

$$1, 1, 2, 3, 5, 8, 13, 21, 34, \ldots$$

where each integer in the sequence is the sum of the previous two integers.

This sequence has many remarkable properties, and some surprising applications in geometry. You can read more about this in the book *Fibonacci and Lucas Numbers*, by V. E. Hoggatt, Houghton Mifflin Mathematics Enrichment Series (Boston: Houghton Mifflin Co., 1969).

13. Using the Fibonacci sequence, one can obtain a new sequence of numbers

$$\frac{1}{1}, \frac{2}{1}, \frac{3}{2}, \frac{5}{3}, \frac{8}{5}, \frac{13}{8}, \frac{21}{13}, \ldots$$

by forming the ratio of each Fibonacci number to the preceding Fibonacci number.

Note that the terms oscillate about the Golden Ratio: $1, 2, 1.5, 1.67, 1.6, 1.63, 1.615, 1.619, \ldots$
$\frac{1 + \sqrt{5}}{2} \doteq 1.618$

(*a*) Johannes Kepler observed that as one goes farther and farther along this new sequence, the numbers get closer and closer to a certain interesting number. Write down a few of the numbers in the new sequence in decimal notation and make a guess as to what number they are approaching.

(*b*) The "*n*th Fibonacci number" is the *n*th number in the Fibonacci sequence. Let us use the symbol F_n for the *n*th Fibonacci number:

$$F_1 = 1, \quad F_2 = 1, \quad F_3 = 2, \quad F_4 = 3, \quad F_5 = 5, \quad \ldots$$

Then for each *n* we have

$$F_{n+2} = F_n + F_{n+1}.$$

Let R_n denote the *n*th fraction in the sequence obtained by forming the ratio of each Fibonacci number to the preceding Fibonacci number:

$$R_1 = \frac{1}{1}, \quad R_2 = \frac{2}{1}, \quad R_3 = \frac{3}{2}, \quad \ldots$$

We have for each *n*

$$R_n = \frac{F_{n+1}}{F_n}.$$

Show that for each *n*

$$R_{n+1} = \frac{1}{R_n} + 1.$$

[*Hint:* Divide both sides of the equation $F_{n+2} = F_n + F_{n+1}$ by F_{n+1}.]

(*c*) As Kepler observed, the ratios R_n get closer and closer to some number L as we choose larger and larger values of *n*. Why would you expect

For large values of n, R_{n+1} and R_n are very close to L.

that this number L satisfies the equation

$$L = \frac{1}{L} + 1\,?$$

[*Hint:* part (*b*)]

(*d*) Use part (*c*) to prove that $L = \dfrac{1 + \sqrt{5}}{2}$ (The Golden Ratio).
$L^2 - L - 1 = 0$

The Golden Section and a Certain Regular Polyhedron

14. The picture below shows 3 congruent golden rectangles which have their centers at one point and which are mutually perpendicular (the 3 rectangles "pass through" each other).

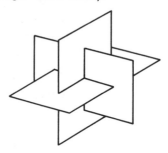

(*a*) The twelve corners of the rectangles are the vertices of a certain regular polyhedron. What is it? See the Projects section of Chapter 4. Build a model and connect the corners with threads to see the polyhedron. *icosahedron*

(*b*) Read about the golden section and its applications in the book of H. S. M. Coxeter, *Introduction to Geometry,* (New York: John Wiley & Sons, 1961), chapter 11. (It is pointed out there how one can make a fairly accurate model of the three golden rectangles using postcards with slits. Measure an ordinary postcard to see if it is nearly a golden rectangle.)

Some Exercises With Circles

15. The shaded region between the large semicircle and the two smaller semicircles has the shape of an arbelos, or cobbler's knife.

Archimedes showed that the shaded area is the same as that of a circle with diameter *x*. Can you show it? (The vertical segment is assumed to be perpendicular to the diameter of the large semicircle. Recall Projects Exercise 3(*b*) of this chapter.)

16. In the figure below is a right triangle with semicircles sitting on its sides. Prove that the area enclosed by the semicircle on the hypotenuse is equal to the sum of the areas enclosed by the semicircles on the legs.

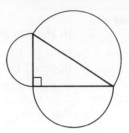

This can be posed as a problem of dividing a circular field into 3 regions of equal area using 2 fences of equal length. The generalization, of dividing into n regions of equal area with n−1 fences of equal length, for various values of n, is an excellent problem.

17. In the design below, the diameter *AB* of the large circle is divided into three equal segments, and semicircles are drawn as indicated.

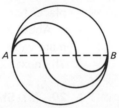

(*a*) Show that the circle is divided into three pieces which have equal area.

(*b*) Show that the perimeters of the pieces are all equal.

18. If $AB = 7$, what is the area between the circles?

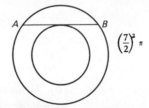

$\left(\frac{7}{2}\right)^2 \pi$

(Yes, you *do* have enough information to work the problem!)

19. You are floating in a balloon, $\frac{1}{2}$ mile above Davisville in Loco County. Davisville is 60 miles from the ocean. Can you see the ocean? (Assume it is a very clear day! See the picture on the next page.)

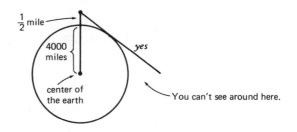

Another Helping of π

In the Central section of this chapter it was noted that for all circles the ratio of the circumference to the diameter is always the same number, denoted by the symbol π. It was proved in 1767 (by J. H. Lambert) that π is not a rational number, that is, there do *not* exist integers m and n such that $\pi = \dfrac{m}{n}$. However, since ancient times much interest has centered on finding rational numbers (fractions) which are good *approximations* to π. A commonly used approximation for π is $\frac{22}{7}$. Around A.D. 500 the Chinese used the remarkable approximation $\frac{355}{113}$. This is correct to six decimal places!

About 250 B.C. the great mathematician Archimedes obtained approximate values for π by using inscribed and circumscribed regular polygons. The next exercise deals with a simple example of his method.

20. (*a*) Show that a regular hexagon *inscribed* in a circle of radius r has perimeter $6r$, and that a regular hexagon *circumscribed* about the circle has perimeter $(4\sqrt{3})r$.

(*b*) Assume that the perimeter of the inscribed hexagon is less than the perimeter of the circle and the perimeter of the circumscribed hexagon is greater than the perimeter of the circle. Use this to show that

$$3 < \pi < 3.47.$$

(*c*) Consider a regular octagon with side length s. Let x be the distance from its center to the midpoint of a side, and let y be the distance from the center to a vertex.

This tests a student's facility with algebra.

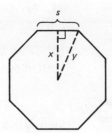

Show that $s = 2(\sqrt{2} - 1)x$ and that $s = y\sqrt{2 - \sqrt{2}}$. [*Hint:* First note that the octagon can be obtained by appropriately cutting off the corners of a square of side $2x$. Use this to deduce that

$$2x = \frac{s}{\sqrt{2}} + s + \frac{s}{\sqrt{2}}.$$

Simplification gives $s = 2(\sqrt{2} - 1)x$. Use the Pythagorean Theorem.]

(*d*) Show that a regular octagon *inscribed* in a circle of radius r has perimeter $8r\sqrt{2 - \sqrt{2}}$, and a regular octagon *circumscribed* about a circle has perimeter $16r(\sqrt{2} - 1)$. [*Hint:* Use part (*c*)!]

(*e*) Use the preceding to show that

$$3.06 < \pi < 3.32.$$

(*f*) Discuss how to obtain estimates for π from regular polygons.

(*g*) Archimedes used regular 96-gons to prove that

$$\frac{223}{71} < \pi < \frac{22}{7}.$$

Explain why this shows that the estimate 3.14 for π is correct to two decimal places.

(*h*) The following poem by A. C. Orr pays tribute to Archimedes, the "Immortal Syracusan." It also gives the decimal expansion of π, correct to 30 decimal places! Can you see how?

> Now I, even I, would celebrate
> In rhymes unapt, the great
> Immortal Syracusan, rivaled nevermore,
> Who in his wondrous lore,
> Passed on before,
> Left men his guidance
> How to circles mensurate.
>
> Adam C. Orr, *The Literary Digest,* Jan. 20, 1906

Chapter 5 of Famous Problems of Mathematics, by H. Tietze (New York: Graylock Press, 1965), contains interesting material on π and "squaring the circle."

21. (a) Does there exist any circle whose diameter and circumference are both integers? Explain. *no*

(b) Does there exist any square whose diagonal length and perimeter are both integers? Explain. *no*

Volume of a Right Circular Cone

Start with a circle lying in a horizontal plane. Draw a line through the center of the circle perpendicular to the plane, and choose a point P on that line. Now connect P to each point inside and on the circle with a line segment.

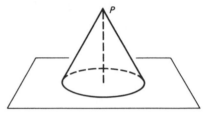

The solid we obtain this way is called a *right circular cone*. The point P is the *apex* of the cone, and the line through P and the center of the base circle is the *axis* of the cone. The distance from the apex to the center of the base is the *altitude*.

Analogy with a pyramid leads us to believe that the volume of a right circular cone is equal to one-third the altitude times the base area. Exercises 22 and 23 prove that this is indeed true.

22. A right circular cone has altitude h and base circle of radius r. The cone is cut by a plane parallel to the base at distance k from the apex P.

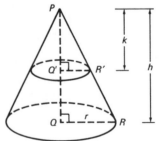

(a) Prove that the cross section is a circle of radius $\dfrac{kr}{h}$. [*Hint:* The cutting plane intersects the axis of the cone at Q'. Use similar right triangles to show that $Q'R' = \dfrac{kr}{h}$ for each point R' where the plane cuts the cone.]

452   10 : Circles

(b) Prove that the area of the cross section is $\pi \dfrac{k^2}{h^2} r^2$.

23. Consider a right circular cone of altitude h and base circle of radius r. Choose a triangular pyramid with same base plane as the cone, with base a triangle of the same base area as the cone (namely, πr^2), and same altitude h.

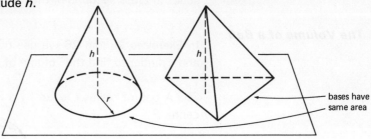

bases have same area

(a) Explain why each plane parallel to the base plane cuts the cone and the pyramid in cross sections of the same area. [*Hint:* Use the preceding exercise and Projects Exercise 12, Chapter 9.]

(b) Prove that the volume V of the cone is equal to one-third base area times altitude:

$$V = \frac{1}{3} \pi r^2 h$$

[*Hint:* By Cavalieri's Principle (Projects section, Chapter 9), the cone and the pyramid have the same volume.]

Volume of a Right Circular Cylinder

An ordinary tomato can is a good model of a right circular cylinder.

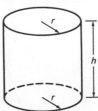

In the picture above, the cylinder has altitude h and base circle of radius r. Each plane parallel to the base cuts the cylinder in a circle of radius r.

24. Prove that the volume V of a right circular cylinder is equal to base area times altitude:

$$V = \pi r^2 h$$

[*Hint:* Choose a box with same base plane as the cylinder, same altitude h, and same base area. Then each plane parallel to the base cuts both solids in cross sections of the same area. Apply Cavalieri's Principle.]

The Volume of a Ball

In Exercises 25 and 26 we use our knowledge of the volume of a cone and cylinder to find the volume of a ball.

25. A ball of radius r is cut by a horizontal plane at distance k from the center P.

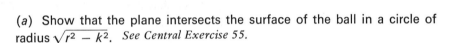

(a) Show that the plane intersects the surface of the ball in a circle of radius $\sqrt{r^2 - k^2}$. *See Central Exercise 55.*

(b) Show that the area of the cross section of a ball of radius r, cut by a plane at distance k from the center, is equal to $\pi(r^2 - k^2)$.

$area = \pi(\sqrt{r^2 - k^2})^2$

26. In the diagram below, the solid A is a half-ball of radius r. The solid B is obtained by removing a right circular cone from a right circular cylinder of altitude r and base radius r. The cone which is removed is upside down, with apex at the center of the base of the cylinder.

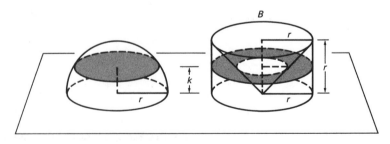

Both solids are cut by a plane at height k and the cross sections are shaded.

(a) The cross section of B is an *annulus,* a region bounded by two concentric circles. Show that the area of this cross section is

$$\pi(r^2 - k^2).$$

(b) Prove that A and B have the same volume. [*Hint:* Show that each plane parallel to the base plane cuts both solids in cross sections of the same area. Then apply Cavalieri's Principle.]

(c) Show that the volume of B is $\frac{2}{3}\pi r^3$.
[*Hint:* The volume of B is $(\pi r^2)(r) - \frac{1}{3}(\pi r^2)(r)$. Why?]

(d) Using parts (b) and (c), conclude that the volume V of a ball of radius r is given by

$$V = \frac{4}{3}\pi r^3.$$

The General Theorem for Intercepted Arcs

Students should not be required to go through these proofs unless they really want to. However, you may wish to give a special presentation, since they tie together several theorems of the Central section.

In Exercises 27–31 we will prove several theorems about the bends of the arcs of a circle determined by intersecting lines. We will see that these theorems are all a special case of *one* general theorem, the *General Theorem for Intercepted Arcs.* The main tools we will use in our proofs are the *Inscribed Angle Theorem:* Any inscribed angle is half the bend of its intercepted arc.

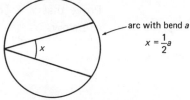

and the *Exterior Angle Theorem:* An exterior angle of a triangle equals the sum of the two opposite interior angles.

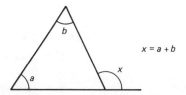

27. (a) In the following figure we have two secants of a circle intercepting arcs with *bends a* and *b* (keep in mind that a letter next to an arc in the following exercises indicates the *bend* of that arc, that is, the size of the central angle corresponding to that arc).

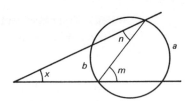

Prove that $x = \frac{1}{2}(a - b)$. [*Hint:* Use the Inscribed Angle Theorem and the Exterior Angle Theorem; $x = \underline{m-n}$ (in terms of m and n).]

(b) The angle between two secants is _____ the intercepted arcs. *equal to $\frac{1}{2}$ the difference of the bends of*

28. (a) In the figure below we have a tangent to a circle and a chord drawn from the point of contact.

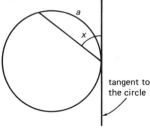

Prove that $x = \frac{1}{2}a$. [*Hint:* Draw the diameter from the point of contact as below.

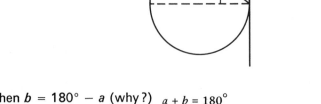

Then $b = 180° - a$ (why?) $a + b = 180°$
$y = $ _____ (in terms of b) $\frac{1}{2}b$ (*Inscribed Angle Theorem*)
$x = 90° - y$ (why?) $x + y = 90°$
$x = \frac{1}{2}a$ (why?).]

(b) The angle between a tangent and a chord drawn from the point of contact is _____ the intercepted arc. *$\frac{1}{2}$ the bend of*

29. (a) In the following figure we have a tangent and a secant of a circle intercepting arcs of bends *a* and *b*.

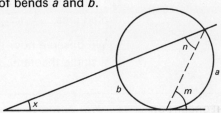

Prove that $x = \frac{1}{2}(a - b)$. [*Hint:* Use the previous exercise to express *m* in terms of *a*.] $m = x+n, \ m = \frac{1}{2}a, \ n = \frac{1}{2}b$

(*b*) The angle between a tangent and a secant is _____ the intercepted arcs. $\frac{1}{2}$ the difference of the bends of

30. (*a*) In the figure below we have two tangents to a circle intercepting arcs of bends *a* and *b*.

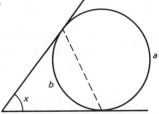

Label *m* and *n* as in Ex. 29.
$x = m - n$
$= \frac{1}{2}a - \frac{1}{2}b$

Prove that $x = \frac{1}{2}(a - b)$. [*Hint:* Exercise 28]

(*b*) The angle between two tangents is _____ the intercepted arcs. $\frac{1}{2}$ the difference of the bends of

31. In Exercises 27–30 we considered arcs intercepted by lines which meet outside a circle. Let us now consider the situation for lines which meet *inside* a circle.

(*a*) In the figure below we have two chords of a circle intercepting arcs of bends *a* and *b*.

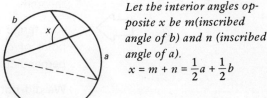

Let the interior angles opposite *x* be *m*(inscribed angle of *b*) and *n* (inscribed angle of *a*).
$x = m + n = \frac{1}{2}a + \frac{1}{2}b$

Prove that $x = \frac{1}{2}(a + b)$. [*Hint:* Express *x* as an exterior angle.]

(b) The angle between two chords is _____ the intercepted arcs. $\frac{1}{2}$ *the sum of the bends of*

We observe now that the results of Exercises 27–31 can be expressed in a single theorem.

GENERAL THEOREM FOR INTERCEPTED ARCS

I The angle between two lines meeting *inside* or *on* a circle is one-half the *sum* of the bends of the intercepted arcs.

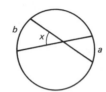   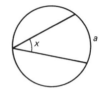

$x = \frac{1}{2}(a + b)$ $x = \frac{1}{2}(a + b)(b = 0,$ so $x = \frac{1}{2}a)$

II The angle between two lines (which intersect a given circle) meeting *outside* the circle is one-half the *difference* of the bends of the intercepted arcs.

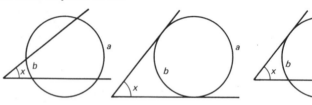

$x = \frac{1}{2}(a - b)$ in each case

Tangent Lines of Circles

In the Central section of this chapter we stated and used the following theorem:

Any tangent line of a circle is perpendicular to the radius drawn to the point of contact. Conversely, a line intersecting a circle at a point and perpendicular to the radius drawn to that point is a tangent line of the circle.

We did not prove this theorem. The proof is now given in Exercises 32 and 33.

32. (*a*) The following figure shows a line ℓ intersecting a circle in two different points *P* and *Q*.

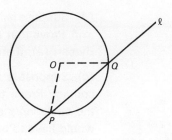

Prove that ℓ cannot be perpendicular to $\overline{OP}$. [*Hint:* Why is $\measuredangle OPQ = \measuredangle OQP$? Is it possible that $\measuredangle OPQ = 90°$? Why?]

(*b*) Suppose a line ℓ passing through a point P on a circle is perpendicular to the radius $\overline{OP}$. Why must ℓ be tangent to the circle? [*Hint:* Read the definition of a tangent line. If ℓ also intersected the circle in another point Q, then what does part (*a*) tell us?]

(*c*) Part (*b*) proves that a line intersecting a circle at a point and perpendicular to the radius drawn to that point is a _____ line of the circle.

33. (*a*) In the figure below we have a line ℓ, a point O not on ℓ, and a point P on ℓ such that $\overline{OP}$ is *not* perpendicular to ℓ.

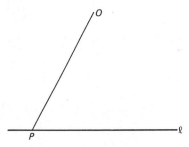

Prove that there is another point Q on ℓ such that $OP = OQ$. [*Hint:* Let F be the point on ℓ such that $\overline{OF}$ is perpendicular to ℓ. Then choose Q so that $FQ = FP$, as below. Then $OQ = OP$ (why?).]

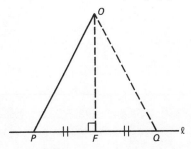

(b) Suppose a line ℓ meets a circle in a point P and $\overline{OP}$ is *not* perpendicular to ℓ. Prove that ℓ must also meet the circle at some other point Q. [*Hint:* By part (a), there is another point O on ℓ such that $OQ = OP$.]

(c) Suppose a line ℓ is a tangent line to a circle and P is the point of contact. Prove that ℓ is perpendicular to the radius $\overline{OP}$. [*Hint:* ℓ meets the circle *only* at P (why?). If ℓ were not perpendicular to $\overline{OP}$, what would have to be true, by part (b)?]

(d) Part (c) proves the *converse* of Exercise 32(c): Any tangent line of a circle is ———— to the radius drawn to the point of contact.

Chapter 11 Coordinates

This chapter is a straightforward introduction to the fundamentals of analytic geometry. We have not presented the usual coordinate geometry with proofs of geometric properties using coordinates. Instead, we have limited the presentation to what will be most useful to students later.

Students interested in investigating properties of polyhedra will find the exercises on convex polyhedra in the Projects stimulating.

Central

A bit of a struggle with the opening problem will provide motivation for coordinates in space.

A fly is flying inside a room. At a certain instant his distance from the left wall is 3 feet, from the right wall is 4 feet, and from the floor is 7 feet. Exactly one second later his distance from the left wall is 2 feet, from the right wall is 6 feet, and from the floor is 5 feet. Is this particular fly able to fly at least 3 feet per second? *yes*

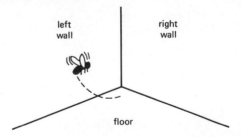

Coordinate Systems

René Descartes (dā cart′) (1596–1650) was the inventor of *analytic geometry,* which makes it possible to change geometric problems into algebraic problems and vice versa. In order to apply the methods of algebra to geometry, Descartes introduced the idea of a *coordinate system.* By

463

using a coordinate system, geometric objects (points and lines) can be expressed in algebraic terms (numbers and equations). Thus problems dealing with geometric objects can be solved algebraically.

A *coordinate system* for a plane is constructed as follows. First, choose a pair of *perpendicular* lines in that plane. One of the lines is named the *x-axis* (or *horizontal axis*) and the other line is called the *y-axis* (or *vertical axis*). The point where these lines intersect is called the *origin*.

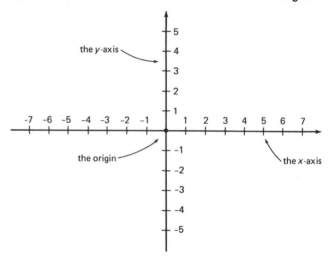

In our picture, the marks on the x-axis and y-axis are equally spaced with unit distance between adjacent points.

We can now assign to each point in the plane a pair of numbers, called the *coordinates* of that point, describing the exact position of that point. For example, the picture below shows four points and their coordinates.

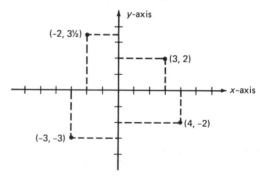

Point out to students that if we start at the origin and move x units horizontally and y units vertically, then we arrive at the point having coordinates (x, y).

To obtain the first number of the pair of coordinates, called the *x-coordinate,* find where the vertical line through the point intersects the x-axis. To obtain the second number of the pair, called the *y-coordinate,* find where the horizontal line through the point intersects the *y*-axis.

Notice that the *x*-coordinate of a point is

> *positive* if the point is to the *right* of the *y*-axis
> *negative* if the point is to the *left* of the *y*-axis,

and the *y*-coordinate is

> *positive* if the point is *above* the *x*-axis
> *negative* if the point is *below* the *x*-axis.

1. Draw a coordinate system on your paper and plot (mark) the points (2, 3) and (3, 2).

Note. When we write the coordinates (*x, y*) of a point, the *first* number always represents the *x-coordinate* and the *second* number the *y-coordinate*.

2. (*a*) In the figure below, the equally spaced points on each axis are a distance of 1 unit apart. Write the coordinates of each of the points *A, B, C, D, E, F*, and *G*. For example, *D* has coordinates $(-5\frac{1}{2}, 3)$.

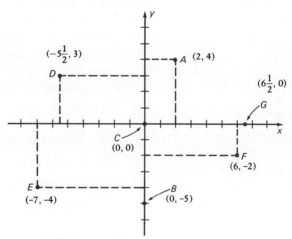

(*b*) In a coordinate system drawn on your paper, plot the points with the following coordinates:

$(1, -1)$, $(2, 2)$, $(-3\frac{1}{2}, -2\frac{1}{2})$, $(0, \frac{1}{3})$, $(-1\frac{1}{2}, 2\frac{1}{4})$, $(-7, 0)$.

3. (*a*) What are the coordinates of the origin? (0, 0)

(*b*) Plot (3, 0), (−4, 0), (5, 0).

(c) What can you say about the second coordinate (y-coordinate) of any point on the x-axis? *It equals zero.*

(d) Plot (0, −1), (0, 2), (0, 4).

(e) What can you say about the first coordinate (x-coordinate) of any point on the y-axis? *It equals zero.*

4. (a) Which points have a positive x-coordinate? a negative x-coordinate? a positive y-coordinate? a negative y-coordinate? *right of y-axis, left of y-axis, above x-axis, below x-axis*

(b) Draw a picture showing all those points for which

   (i) both coordinates are positive

   (ii) both coordinates are negative

   (iii) the x-coordinate is positive and the y-coordinate is negative

   (iv) the x-coordinate is negative and the y-coordinate is positive.

It is convenient to use graph paper when working with coordinates. When using graph paper, draw the x-axis and y-axis, and choose a unit to represent the distance 1 (the units on the graph paper are usually convenient). Generally, you should use the same unit on both the x-axis and the y-axis (occasionally you may find a graph where different scales may be preferred).

After we have chosen a coordinate system in a plane (by choosing an x-axis and a y-axis) we refer to the plane as "the (x, y)-plane." Then, instead of speaking of "the point whose coordinates are (x, y)," we simply say "the point (x, y)."

Do not rush through these exercises, which call for graphing sets of points satisfying given conditions.

5. Using a sheet of graph paper, plot and label all those points (x, y) such that

(a) the x-coordinate is 1

(b) the y-coordinate is $-1\frac{1}{4}$

(c) the x-coordinate is <0

(d) the x-coordinate is <1.

6. Make a sketch which shows all those points in the (x, y)-plane such that

(a) $x = 0$ (b) $y = 0$ (c) $y = -1$ (d) $y > 0$

(e) $x > 0$ (f) $y < 0$.

7. (a) Find numbers a and b such that the points (0, 0), (1, 0), and (a, b) are the vertices of an equilateral triangle in the (x, y)-plane. $\left(\frac{1}{2}, \frac{\sqrt{3}}{2}\right)$ or $\left(\frac{1}{2}, -\frac{\sqrt{3}}{2}\right)$

(b) How many possible answers are there to part (a)? Compare with your classmates. *two*

8. In each case, plot the three points on graph paper. Find the area of the triangle with these vertices.

(a) (0, 0), (3, 0), and (1, 2) 3

(b) (0, 1), (0, 4), and (3, 0) $\frac{9}{2}$

(c) (−2, 0), (0, −1), and ($\frac{1}{2}$, 0) $\frac{5}{4}$

9. (a) List examples where something similar to a coordinate system is used in daily life. *coordinates of places on a map, intersections of streets*

(b) Has your home address anything to do with coordinate systems? Explain.

Absolute Value of a Number

Consider the numbers on the number line:

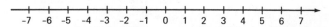

Make sure the definition of absolute value is understood.

The *absolute value* of a number is its distance from 0 on the number line. For example, the absolute value of −2 is 2, and the absolute value of 5 is 5. Observe that if x is a positive number, then the absolute value of x is simply x, but if x is a negative number, then the absolute value of x is $-x$. For example, the absolute value of −3 is −(−3) = 3.

We denote the absolute value of a number x by the symbol $|x|$.

$$|x| = x, \text{ if } x \geq 0$$

but

$$|x| = -x, \text{ if } x < 0.$$

Note. If $x < 0$, $-x > 0$. Thus the absolute value of a number is always nonnegative.

Exercises 10-11 review the use of absolute value while preparing for the distance formula. If students have trouble here, give other specific examples.

10. Find

(a) $|-7|$ (b) $|14|$ 14 (c) $|.5|$ (d) $|-\frac{1}{4}|$ $\frac{1}{4}$

(e) $|2 - 9|$ (f) $|3\frac{1}{3}|$ $3\frac{1}{3}$ (g) $|-6 - (-4)|$ (h) $|\frac{1}{2} - 1|$ $\frac{1}{2}$

(i) $|3 - 3|$.

[Ans: (a) 7   (c) .5   (e) 7   (g) 2   (i) 0]

11. Show that if *x* is any number (positive or negative), then the square of *x* is equal to the square of its absolute value, that is,

$$(x)^2 = |x|^2.$$

(Try some specific numbers for *x*. Discuss with your classmates.)

The following result gives the distance between two numbers on the number line in terms of absolute value.

Actually a special case of the distance formula!

DISTANCE BETWEEN TWO NUMBERS ON THE NUMBER LINE	If *a* and *b* are any numbers, then $|a - b|$ is the distance between *a* and *b* on the number line.

Example 1. The distance between 4 and $-2 = |4 - (-2)| = |4 + 2| = |6| = 6$.

Example 2. The distance between -1 and $-5 = |(-1) - (-5)| = |-1 + 5| = |4| = 4$.

Example 3. The distance between 0 and $-3 = |0 - (-3)| = |3| = 3$.

12. Find the distance on the number line between

   (a) 5 and 8   3
   (b) 3 and -2 5
   (c) $6\frac{1}{2}$ and -3 $9\frac{1}{2}$
   (d) $-1\frac{3}{4}$ and -2
   (e) -12 and 12
   (f) 8.3 and -7.2
   (g) 6.5 and 2.3   4.2
   (h) 10 and -7.   17
   (i) 33 and 40   7
   (j) -4 and 6   10
   (k) -1 and 0   1

   [Ans: (d)$\frac{1}{4}$ (e) 24   (f) 15.5]

The Distance Formula

We will derive a formula which enables us to compute the distance between any two points of the (*x, y*)-plane in terms of the coordinates of the points.

If two points have the same *y*-coordinate, then it is easy to find the distance between the points. For example, $(-2, 2)$ and $(3, 2)$ have the same *y*-coordinate, 2. The distance between the points (see picture on the next page) is $|3 - (-2)| = 5$.

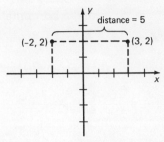

Similarly, if two points have the same x-coordinate, then the distance between them is easily computed. For example, (−2, 2) and (−2, −3) have the same x-coordinate, −2. The distance between the points (see picture below) is $|2 - (-3)| = 5$.

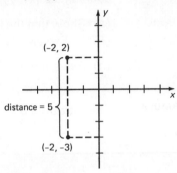

Finding distance between points on a horizontal segment or a vertical segment prepares students for the derivation of the distance formula.

13. Find the distance between

(a) (2, −2) and (5, −2)   3

(b) (35, 3) and (35, −4)   7

(c) (7, y) and (−3, y)   10

(d) (x, −3) and (x, −2)   1

(e) (x_1, y_1) and (x_1, y_2)

(f) (x_2, y_2) and (x_1, y_2).

[Ans: (e) $|y_1 - y_2|$ (f) $|x_1 - x_2|$]

14. (a) Plot the two points (3, 3) and (−1, −3).

(b) Form a right triangle having (3, 3) and (−1, −3) as two of its vertices, by drawing a vertical line through (3, 3) and a horizontal line through (−1, −3). What are the coordinates of the vertex with the right angle?   (3, −3)

(c) Find the lengths of the legs of the triangle.   6, 4

(d) Find the distance from (3, 3) to (−1, −3). $2\sqrt{13}$

15. Suppose we are given two points (x_1, y_1) and (x_2, y_2) in the (x, y)-plane. We form a right triangle by drawing a vertical line through (x_1, y_1)

and a horizontal line through (x_2, y_2).

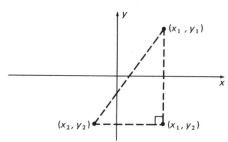

Observe that the horizontal and vertical lines intersect in the point with coordinates (x_1, y_2).

(a) Show that the legs of the triangle have lengths $|x_1 - x_2|$ and $|y_1 - y_2|$. [*Hint:* Exercise 13(e), (f)]

(b) Prove the following, recalling that $|x_1 - x_2|^2 = (x_1 - x_2)^2$.

THE DISTANCE FORMULA	The distance d between the points (x_1, y_1) and (x_2, y_2) is given by $$d = \sqrt{(x_1 - x_2)^2 + (y_1 - y_2)^2}.$$

Observe that the distance formula enables us to find the distance between two points (x_1, y_1) and (x_2, y_2) without drawing any figures. For example, to find the distance between $(3, 2)$ and $(1, -1)$, we let $x_1 = 3$, $y_1 = 2$ and $x_2 = 1$, $y_2 = -1$ and substitute in the distance formula:
$$d = \sqrt{(3 - 1)^2 + (2 - (-1))^2}$$
$$= \sqrt{4 + 9}$$
$$= \sqrt{13}.$$

Although students should simplify radicals whenever possible, the chief goal in Exercise 16 is, of course, skill in using the distance formula, not in manipulation of radicals.

16. Find the distance between the points

(a) $(2, 1)$ and $(-3, 4)$

(b) $(6, -3)$ and $(4, -5)$ $2\sqrt{2}$

(c) $(1, 8)$ and $(-3, -7)$

(d) the origin and $(-5, -6)$ $\sqrt{61}$

(e) $(5, 0)$ and $(-3, 0)$ 8

(f) $(2, 3)$ and $(-2, -3)$. $2\sqrt{13}$

[Ans: (a) $\sqrt{34}$ (c) $\sqrt{241}$ (f) $\sqrt{52}$]

17. (a) Draw the triangle with vertices $(1, 1)$, $(5, 1)$, and $(3, 1 - 2\sqrt{3})$.

In Exercise 17(b), be sure students compute the lengths of all three sides.

Does it look equilateral? *yes*

(b) Using the distance formula, prove that the triangle is equilateral.

18. (a) Draw the triangle with vertices $(-9, 1)$, $(4, 1)$, and $(0, -5)$. Does it look like a right triangle? *yes*

(b) Using the distance formula, prove it is a right triangle. [*Hint:* Use the *converse* of the Pythagorean Theorem.]

19. Find the perimeter of the triangle with vertices $(-1, -1)$, $(2, 3)$, and $(-1, 2)$. $8 + \sqrt{10}$

Exercise 20(a): Tell students to use the distance formula, not to plot the points.

20. (a) Show that each of the following points lies on the circle with center $(1, -2)$ and radius 3:

$(1, 1)$, $(1, -5)$, $(4, -2)$, $(-2, -2)$, $(0, 2\sqrt{2} - 2)$, $(-1, \sqrt{5} - 2)$.

[*Hint:* Find the distance from each point to the point $(1, -2)$.]

(b) Suppose the numbers x and y satisfy

$$\sqrt{(x-1)^2 + (y+2)^2} = 3.$$

Explain why the point (x, y) must lie on the circle in part (a).

(c) Suppose (x, y) is a point such that

$$(x-1)^2 + (y+2)^2 = 9.$$

Explain why (x, y) must lie on the circle in part (a).

In Exercise 20(d), point out the usefulness of the distance formula: It tells precisely whether or not the point is inside the circle. We could not be certain from plotting the point.

(d) Is the point $(0, 0.83)$ inside the circle in part (a)? How could you be absolutely certain that your answer is correct? Discuss this with your classmates. *no,* $\sqrt{(1-0)^2 + (-2-0.83)^2} = \sqrt{9.0089} > 3$

Graphs of Equations

Suppose we have an *equation,* such as

(*) $\qquad\qquad x^2 + y = 5.$

We say that a pair of numbers (x, y) *satisfies* the equation if we get a true statement when we substitute these numbers (in the correct order!) into the equation. For example, the pair of numbers $(2, 1)$ does satisfy the equation, since (substituting $x = 2$, $y = 1$)

$$2^2 + 1 = 5.$$

But the pair of numbers $(2, 3)$ does *not* satisfy the equation, since (sub-

471 Central

stituting $x = 2$, $y = 3$)

$$2^2 + 3 \neq 5.$$

The pair of numbers $(4, -11)$ also satisfies the equation, since

$$4^2 + (-11) = 5.$$

The *graph* of an equation consists of all those points (x, y) which satisfy that equation. For example, the graph of the equation (*) on p. 471 consists of all those points (x, y), in the (x, y)-plane, such that $x^2 + y = 5$.

A good way to find points (x, y) which satisfy an equation is to first substitute some arbitrary number for x and then see how you must choose y in order to get a true statement. For example, we might substitute $x = 3$ in equation (*) to get

$$(3)^2 + y = 5$$

or

$$9 + y = 5.$$

Then we see that

$$y = -4.$$

Hence $(3, -4)$ satisfies the equation $x^2 + y = 5$. Sometimes you might find it easier to first substitute some number for y and then find the corresponding number for x.

Warn students that even one point plotted incorrectly can confuse the graph. If one of the points seems out of place, it should be checked. Emphasize that plotting extra points helps in drawing the graph.

21. In each of the following cases, find at least 6 points (x, y) which satisfy the given equation and plot the corresponding points in the (x, y)-plane.

(a) $y - x = 0$ (b) $x + y = 6$ (c) $x^2 + y^2 = 4$

(d) $y = x^2$ (e) $y = 3$ (f) $x = 0$

22. (a) Show that $(0, 1)$ and $(1, 3)$ both satisfy the equation $y - 2x = 1$.

$1 - 2(0) = 1$

(b) Plot the points $(0, 1)$ and $(1, 3)$.

$3 - 2(1) = 1$

Students often make algebraic errors in substituting for x and then solving for y in order to find a point which satisfies an equation. Hence it should be emphasized that they check their (x, y) by direct substitution in the given equation.

(c) Find three more points (x, y) which satisfy the equation in part (a) and plot them. (You should always check that the coordinates of a point satisfy the equation by substituting the numbers into the equation.)

(d) What do you think is true about the graph of the equation $y - 2x = 1$? (What kind of geometric figure is it?) Discuss with your classmates.
straight line

23. In each case, find some points (x, y) which satisfy the equation,

Exercise 23: While we are concerned primarily with lines in this chapter, we do throw a curve or two at the student to keep him honest!

Check the work on Exercise 23, (d) and (f).

and use them to make a rough sketch of the graph of the equation. (In some cases you will need to find only a few points before seeing what the graph looks like; in other cases, you may need many points.)

(a) $y = x$ (b) $y = 2x + 1$ (c) $x = \frac{1}{3}$

(d) $xy = 1$ (e) $x^2 + y^2 = 0$ (f) $y = x^2$

Linear Equations

A *linear equation* is an equation which can be put into the form

$$ax + by + c = 0,$$

where a, b, and c are some fixed numbers, and not both a and b are zero.

24. Which of the following equations are linear?

(a) $3x + 5y + 2 = 0$ (b) $x - 3y + 1 = 0$ (c) $y = 2x + 1$

(d) $2x^2 + y^2 = 1$ *no*   (e) $x = 1$ (f) $y - 2 = x$

(g) $xy + 2y + 3 = 0$ *no*   (h) $-x - y = 3x + 2y$

Emphasize plotting a third point as a check when graphing a line.

25. (a) Devise two linear equations and draw their graphs. Compare your results with those of your classmates.

(b) The graph of any linear equation is a _____. *straight line*

As you might have discovered in the preceding exercise, the graph of any linear equation is a straight line. Therefore you can find the entire graph of the equation by finding two points on the graph and drawing the straight line through them. You should plot a third point as a check.

26. (a) Find three points that satisfy the linear equation

$$3x + 2y + 4 = 0.$$

(b) Plot the points you found in part (a) and draw the straight line through them.

(c) Find three more points that satisfy the equation. Plot these points and check that they lie on the line you drew in part (b).

Slope of a Line

The *slope* of a straight line in the (x, y)-plane is the ratio

$$\frac{\text{rise}}{\text{run}}, \text{ or } \frac{\text{fall}}{\text{run}},$$

Face the class; hold a yardstick at various angles with the horizontal. Have the class determine whether the slope is positive, negative, zero, or undefined (uphill, downhill, horizontal, or vertical. Keep in mind that "uphill" for you will be "downhill" for them!)

depending on whether the line, *left to right,* goes "uphill" (positive slope)

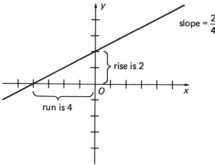

or "downhill" (negative slope).

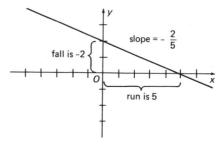

A horizontal line has a *slope of zero,*

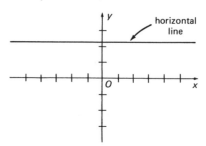

while a vertical line has an *undefined slope.*

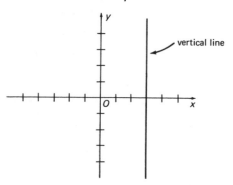

In the figure below, the two right triangles (dotted legs) are similar, because their corresponding angles are equal. Therefore $\frac{a}{b} = \frac{c}{d}$.

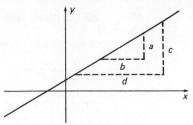

Therefore, we can calculate the ratio

$$\frac{\text{rise}}{\text{run}}$$

to find the slope of the line by computing either $\frac{a}{b}$ or $\frac{c}{d}$.

Any two points, P and Q, on a line determine a rise-and-run triangle, where the hypotenuse is the line segment $\overline{PQ}$. Thus, *any* two points on the line may be used to compute the slope.

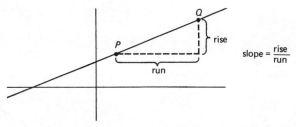

27. As we have seen, the graph of a linear equation is always a straight line. In each of the following cases, draw the graph and find its slope.

(a) $y = x - 1$ (b) $y = x + 1$ (c) $y = 2x - 3$

(d) $y = -\frac{2}{3}x - 2$ (e) $y = -2$ (f) $y + 2x + 3 = 0$ -2

[Ans: (a) 1   (c) 2   (d) $-\frac{2}{3}$ (e) 0]

28. (a) Draw the graph of $y = 2x - 3$.

(b) Find a pair of points on the graph and calculate $\frac{\text{rise}}{\text{run}}$ using an appropriate right triangle.

(c) Do part (b) for a different pair of points on the graph and check that you get the same value of $\dfrac{\text{rise}}{\text{run}}$.

29. Explain why two parallel lines in the (x, y)-plane always have the same slope. [*Hint:* Calculate $\dfrac{\text{rise}}{\text{run}}$ for each line using appropriate right triangles. Are the triangles similar?]

30. In each case, find the slope of the line.

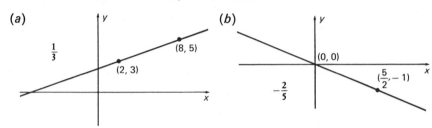

The Slope Formula

The picture below shows a line in the (x, y)-plane with two points (x_1, y_1) and (x_2, y_2) on the line.

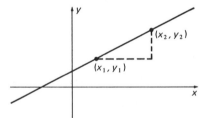

Using the right triangle (dotted legs) we see that

$$\text{slope} = \dfrac{\text{rise}}{\text{run}} = \dfrac{y_2 - y_1}{x_2 - x_1}.$$

A short lecture on the slope formula may be helpful.

THE SLOPE FORMULA   If m denotes the slope of a line, and (x_1, y_1) and (x_2, y_2) are two points on the line, then the slope is given by

$$m = \dfrac{y_2 - y_1}{x_2 - x_1}.$$

---

31. In each case, draw the straight line which is the graph of the given equation, choose a pair of points on the line, and use the slope formula to find the slope.

(a) $y = 2x + 1$ 2   (b) $2y + 3x = 4$ $-\frac{3}{2}$ (c) $y = -3$ 0

32. Find the slope of the straight line passing through the two points $(1, 3)$ and $(-4, 2)$, without drawing the graph. Compare your answer with those of your classmates. $\frac{1}{5}$

33. In obtaining the slope formula, we used a picture of a "rising" line. Discuss with your classmates why the same formula also gives the correct slope for any "falling" line.

Exercise 34: A short presentation is in order here.

34. Is the following true? yes
$$m = \frac{y_1 - y_2}{x_1 - x_2} = \frac{(y_1 - y_2)(-1)}{(x_1 - x_2)(-1)} = \frac{y_2 - y_1}{x_2 - x_1}$$

Discuss with your classmates.

35. (a) Draw the graph of the line passing through $(-3, 2)$ and $(2, 2)$. Use the slope formula to find the slope of this line. *zero*

(b) The slope of any horizontal line in the (x, y)-plane is ___0___.

(c) Discuss with your classmates why the slope formula gives the correct answer in part (b). $y_2 = y_1$ and $m = \frac{y_2 - y_1}{x_2 - x_1} = \frac{0}{x_2 - x_1} = 0$

You may wish to present the $y = mx + b$ form of the equation of a straight line, where m is the slope and b the y-intercept.

36. (a) Try to compute the slope of the straight line passing through $(3, 5)$ and $(3, 7)$ using the slope formula. *Division by zero is undefined.*

(b) What happens when you try to compute the slope of any vertical line using the slope formula? Discuss with your classmates.

(c) When does the slope formula fail? Discuss with your classmates. *for vertical lines*

Exercise 37: Emphasize that it is useful to check answers by plotting the given points to see if the slope of the resulting line appears to agree with that given by the formula.

37. In each of the following cases, compute the slope of the line passing through the two given points by using the slope formula, then check your answer by graphing the line. (For example, you can be sure you have made a mistake if your computation gives a positive number but the line you have drawn is "falling" from left to right.)

(a) $(1, 2)$ and $(3, 4)$ 1   (b) $(2, 1)$ and $(4, \frac{1}{2})$ $-\frac{1}{4}$

(c) $(-2, -3)$ and $(1, 0)$ 1   (d) $(-\frac{1}{2}, -\frac{1}{2})$ and $(-1, -2)$ 3

(e) (a, a) and (b, b) 1   (f) $(2, 5)$ and $(2, 7)$ $(??)$ *undefined*

[Ans: (a) 1   (c) 1   (d) 3]

Simultaneous Linear Equations

Suppose you are asked to find a single pair of numbers x and y such that
$$3x + 2y = 4$$
and also
$$2x - 6y = 10.$$

Could you find such a pair of numbers? The pair of numbers $x = 2$ and $y = -1$ will work, because
$$3(2) + 2(-1) = 4$$
and also
$$2(2) - 6(-1) = 10.$$

We shall see that there are systematic methods for finding x and y in such problems. This particular example is typical of a problem in which one must solve a pair of *simultaneous linear equations*:

Using a brace to indicate a pair of simultaneous equations helps the student relate them when he solves them algebraically.

$$\begin{cases} a_1x + b_1y = c_1 \\ a_2x + b_2y = c_2 \end{cases}$$

where $a_1, b_1, c_1, a_2, b_2, c_2$ are given numbers. To *solve* these equations means to find a single pair of numbers x and y which satisfies *both* equations simultaneously.

If a pair of numbers (x, y) satisfies *both* equations, then the point (x, y) lies on the graphs of both equations in the (x, y)-plane. This forms the basis for the *graphical method* of solving simultaneous linear equations: Draw the graphs—the solution is the intersection of graphs. This method has the great drawback that an inaccurate drawing will lead to an incorrect "solution."

38. Consider the two lines in the (x, y)-plane:
$$\begin{cases} x + y = 3 \\ 2x + 4y = 10 \end{cases}$$

(a) By drawing the graphs, estimate the coordinates (x, y) of the point where the lines meet. (1, 2)

(b) Find numbers x and y which simultaneously satisfy both of the given equations. How many such pairs of numbers x and y can you find?
$x = 1, y = 2$: *Two nonparallel lines cross in only one point.*

39. Solve each pair of equations graphically. In each case, check that you have actually found a solution by substituting the *x*-coordinate and the *y*-coordinate of the point of intersection of the two lines into both equations to see if you get a true statement.

(a) $\begin{cases} x + y = 1 \\ 2x - y = 2 \end{cases}$ (b) $\begin{cases} 2x + 4y = 3 \\ x - y = 0 \end{cases}$ $\left(\frac{1}{2}, \frac{1}{2}\right)$

(c) $\begin{cases} 2x - 6y = 2 \\ -x + 3y = 1 \end{cases}$ (d) $\begin{cases} x + y = 1 \\ 3x + 3y = 3 \end{cases}$ same line

[Ans: (a) (1, 0)   (c) no solution (why?)]
parallel lines

Algebraic Solution to Simultaneous Equations

It is usually best to solve simultaneous linear equations algebraically, and there are two algebraic methods.

The following example illustrates the *addition-subtraction method* for doing this. Consider the pair of linear equations

$$\begin{cases} 2x + 3y = 17 \\ 5x - 2y = 14. \end{cases}$$

We want to find numbers *x* and *y* which satisfy both equations simultaneously. We can "eliminate *y*" by multiplying both sides of the first equation by 2 and both sides of the second equation by 3 and then adding the resulting equations, as the following scheme indicates.

Demonstrate the method to the class with another example.

$\begin{cases} 2(2x + 3y) = 2(17) \\ 3(5x - 2y) = 3(14) \end{cases}$ Multiply each side by 2.
Multiply each side by 3.

$\begin{cases} 4x + 6y = 34 \\ 15x - 6y = 42 \end{cases}$
$19x = 76$ Add.
$x = 4$

We substitute $x = 4$ into the first equation of our original pair of equations to find *y*:

$2(4) + 3y = 17$
$3y = 17 - 8$
$3y = 9$
$y = 3$

You should check that $x = 4$ and $y = 3$ satisfy both equations; that is, (4, 3) is a point on the graph of each equation:

$$\begin{cases} 2(4) + 3(3) \stackrel{?}{=} 17 \\ 5(4) - 2(3) \stackrel{?}{=} 14 \end{cases}$$

$$\begin{cases} 8 + 9 = 17 \\ 20 - 6 = 14 \end{cases} \quad \text{Yes, they both check.}$$

40. In each case, solve using the addition-subtraction method.

(a) $\begin{cases} x + 2y = 5 \\ 2x - y = 1 \end{cases}$ (b) $\begin{cases} 5x - 2y = 11 \\ 2x + 7y = 20 \end{cases}$ $x = 3$, $y = 2$

(c) $\begin{cases} (.5)x + (.6)y = 3 \\ x + (.7)y = 2 \end{cases}$ (d) $\begin{cases} \frac{2}{3}x + \frac{1}{6}y = 4 \\ 3x + y = 3 \end{cases}$ $x = 21$, $y = -60$

[*Hint:* (c) Clear of decimals. (d) Clear of fractions.]

[Ans: (a) $x = \frac{7}{5}$, $y = \frac{9}{5}$ (c) $x = -3.6$, $y = 8$]

We can also solve simultaneous linear equations algebraically by the *substitution method*. To illustrate this, we again use the equations

$$\begin{cases} 2x + 3y = 17 \\ 5x - 2y = 14. \end{cases}$$

Using the first equation, we *express y in terms of x:*

$$2x + 3y = 17$$
$$3y = -2x + 17$$
$$y = -\frac{2}{3}x + \frac{17}{3}$$

Substituting this expression for *y* into the second equation, we solve for *x:*

$$5x - 2\left(-\frac{2}{3}x + \frac{17}{3}\right) = 14$$
$$5x + \frac{4}{3}x - \frac{34}{3} = 14$$
$$3\left(5x + \frac{4}{3}x - \frac{34}{3}\right) = (14)3$$
$$15x + 4x - 34 = 42$$
$$19x = 76$$
$$x = 4$$

Substituting $x = 4$ into one of the original pair of equations, we get

$$y = 3.$$

Point out that the substitution method works best when both x-coefficients (or both y-coefficients) are the same: If

$$\begin{cases} y = 2x - 3 \\ y = -7x + 5 \end{cases}$$

then

$2x - 3 = -7x + 5$, etc.

Again, you should check that the solution satisfies both of the original equations.

41. Solve each pair of linear equations algebraically, using the substitution method. Check that your answer is correct in each case.

(a) $\begin{cases} 2x + y = 21 \\ x - 3y = 14 \end{cases}$ (b) $\begin{cases} 3x + 2y = 3 \\ \frac{1}{2}x + y = \frac{5}{6} \end{cases}$

(c) $\begin{cases} 3x - y = 7 \\ x + 4y = 1 \end{cases}$ (d) $\begin{cases} 2y = \frac{1}{3}x + 2 \\ 2x = 3y - 1 \end{cases}$

[*Hint:* (b), (d) Clear of fractions.]

[Ans: (a) $x = 11, y = -1$ (b) $x = \frac{2}{3}, y = \frac{1}{2}$
(c) $x = \frac{29}{13}, y = -\frac{4}{13}$ (d) $x = \frac{4}{3}, y = \frac{11}{9}$]

The Graph of a Condition

We often describe a set (that is, a collection) of points by stating a certain condition satisfied by those points. For example, consider all those points in the (x, y)-plane that satisfy the condition

$$y > 0.$$

Observe that this describes all points above the x-axis (shaded in the picture below).

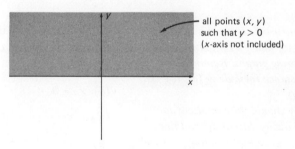

We define "condition" and "graph of a condition" informally, because nothing really new is involved here. We are simply saying, "Find all those points (x, y) such that ...," and there is no ambiguity.

We have already encountered this idea in graphing equations. For example, all those points in the (x, y)-plane that satisfy the condition

$$y + 2x = 4$$

form a certain straight line in the (x, y)-plane (see the figure on the next page). The *graph* of a condition consists of all those points in the (x, y)-plane that satisfy the condition. For example, suppose we want to find the graph of all those points (x, y) such that

$$\sqrt{x^2 + y^2} < 3.$$

From the distance formula, we see that $\sqrt{x^2 + y^2}$ is simply the distance from (x, y) to $(0, 0)$. Hence, we are looking for all those points (x, y) whose distance from $(0, 0)$ is less than 3. These are the points inside a circle of radius 3 centered at $(0, 0)$.

Point out that the points on the circle do not satisfy the condition, only points inside the circle do.

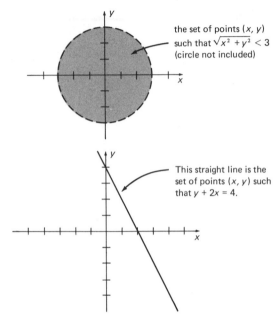

the set of points (x, y) such that $\sqrt{x^2 + y^2} < 3$ (circle not included)

This straight line is the set of points (x, y) such that $y + 2x = 4$.

Do not rush Exercise 42. Students will learn much by discussing and comparing graphs. Remind them of the distance formula in (d), (e), (f), and (g).

They should note in discussion and show by dotted or solid lines the reference figure (a line or circle) for each graph.

42. In each case, draw the graph of those points in the (x, y)-plane that satisfy the given condition. Compare your results with those of your classmates.

(a) $x > 0$ (b) $y < -1$ (c) $x + 2y = 1$

(d) $\sqrt{x^2 + y^2} < 4$ (e) $\sqrt{x^2 + y^2} = 4$ (f) $\sqrt{x^2 + y^2} > 4$

(g) $\sqrt{(x - 1)^2 + (y + 2)^2} < 3$

43. (a) Plot five different points in the (x, y)-plane that satisfy $x + y > 1$.

(b) Do the points you plotted in part (a) all lie on the same side of the line with equation $x + y = 1$? Compare your results with those of your classmates. *yes, above the line*

(c) Consider the set of all (x, y) that satisfy the condition

$$x + y > 1.$$

What do you think this set of points looks like? *half-plane*

If you plot a few points that satisfy the condition

$$y - 2x - 3 > 0,$$

you will find that they all lie on one side of the line with equation $y - 2x - 3 = 0$. The graph of those points in the (x, y)-plane that satisfy the above condition is a certain *half-plane*, shaded below:

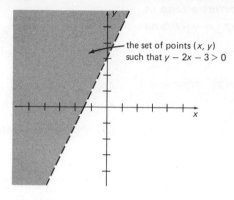

the set of points (x, y) such that $y - 2x - 3 > 0$

Alternate approach:

$$y - 2x - 3 > 0$$

means

$$y > 2x + 3.$$

This implies that the y-value for any x is above *the line $y = 2x + 3$ for the same x. Similarly,*

$$y < 2x + 3$$

implies that the y-value for any x is below *$y = 2x + 3$ for the same x.*

The line in the picture is dotted to indicate that points on this line do *not* satisfy the condition $y - 2x - 3 > 0$. In fact, $y - 2x - 3 = 0$ for points on the line. The preceding condition is an example of a *linear inequality*. More generally we have the following definition.

DEFINITION If a, b, and c are given numbers, and a and b are not both zero, then the conditions

$$ax + by + c > 0 \quad \text{and} \quad ax + by + c < 0$$

are called *linear inequalities*.

The following result holds for linear inequalities.

The graph of a linear inequality

$$ax + by + c > 0$$

or

$$ax + by + c < 0$$

is a half-plane. The *edge* of the half-plane is the line with equation

$$ax + by + c = 0.$$

The graph is one of the two half-planes determined by a certain line. To determine which of the two, it suffices to check one point on either side of the line. Note that the line is not included in any of these graphs.

44. In each case, find the half-plane that is the graph of the given linear inequality. Compare your results with those of your classmates.

(a) $x < 0$ (b) $x + y < 1$ (c) $x - y + 1 > 0$

(d) $x - y + 1 < 0$ (e) $2x - y > 0$ (f) $2x - y < 0$

Reflection Across a Line in the (x, y)-Plane

In Chapter 1 we studied the reflection of a point across a line. Now we will see how this reflection is described in terms of coordinates.

Remind students of the definition of reflection across a line.

45. (a) What are the coordinates of the reflection of (5, 2) across the x-axis? across the y-axis? (Plot on graph paper.)   (5, −2), (−5, 2)

(b) Do part (a) with (5, 2) replaced by

     (i) $(-1, 3)$ (ii) $(2, -3)$ (iii) $(-3, -4)$.

(c) Discuss with your classmates why the following is true:

The reflection of (x, y) across the x-axis is $(x, -y)$. The reflection of (x, y) across the y-axis is $(-x, y)$.

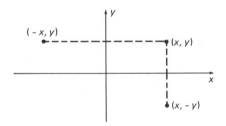

46. Find the length of the shortest path, in the (x, y)-plane, from (0, 2) to (5, 6) which touches the x-axis.
$\sqrt{89}$

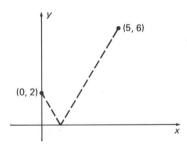

[*Hint:* Reflect one of the points across the x-axis.]

Coordinates in Space

Exercise 47: Given a line $ax + by + c = 0$ and a point (x, y), how does one find the coordinates (x^*, y^*) of the reflection of (x, y) across the line? The midpoint of the segment joining (x, y) to (x^*, y^*) lies on the line; hence

$$a\left(\frac{x+x^*}{2}\right) + b\left(\frac{y+y^*}{2}\right) + c = 0.$$

This segment is perpendicular to the line, implying that

$$b(x-x^*) - a(y-y^*) = 0.$$

Solving the two equations gives

$$x^* = \frac{(b^2-a^2)x - 2aby - 2ac}{a^2 + b^2}$$

$$y^* = \frac{(a^2-b^2)y - 2abx - 2bc}{a^2 + b^2}.$$

Lecture on 3-dimensional coordinates. Students have difficulty visualizing 3-dimensional space from a representation on 2-dimensional paper.

A corner cut from a cardboard box is a good model.

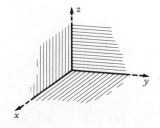

47. Suppose a point with coordinates (a, b) is reflected across the line $y = x$. What are the coordinates of the reflection? (Draw the graph of the line $y = x$ on graph paper and reflect a few "nice" points across the line.)
(b, a)

By introducing a pair of perpendicular lines in the plane we have seen that we can represent points in the plane by pairs of numbers (x, y). Similarly, we can represent points in *space* by *triples* of numbers (x, y, z), if we introduce a coordinate system in space as follows.

Imagine three mutually perpendicular lines in space. These lines will be referred to as the *x-axis*, the *y-axis*, and the *z-axis*.

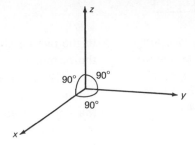

Imagine you are looking into the corner of a room. The *x*-axis is where the left hand wall intersects the floor, the *y*-axis is where the right hand wall intersects the floor, and the *z*-axis is where the left and right walls intersect each other.

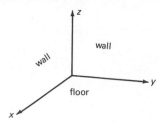

From now on, we will refer to the left-hand wall as the (x, z)-*plane*, the right-hand wall as the (y, z)-*plane*, and the floor as the (x, y)-*plane*. These planes are called the *coordinate planes*.

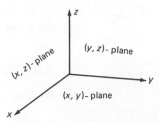

485 Central

Graph paper should be used here to make plotting easy (the x scale will be slightly off).

The *coordinates* of a point indicate how far the point is from each of these planes.

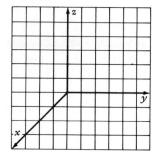

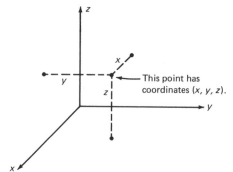

The z-coordinate is the "height" of the point above the (x, y)-plane. The x- and y-coordinates give the location of the projection of the point on the (x, y)-plane.

In the picture above, the distance from the point to the (y, z)-plane is x, the distance to the (x, z)-plane is y, and the distance to the (x, y)-plane is z. (Thus the x-coordinate is "distance to the right wall," the y-coordinate is "distance to the left wall," and the z-coordinate is "distance to the floor." If you have trouble keeping track of symbols, it will help to remember that the x-coordinate measures distance parallel to the x-axis, the y-coordinate measures distance parallel to the y-axis, and the z-coordinate measures distance parallel to the z-axis.)

A cardboard box with yarn or string may be helpful in Exercise 48.

48. The picture shows a rectangular box (dotted lines) with one vertex (F) at the origin. The marks on each axis represent unit distances.

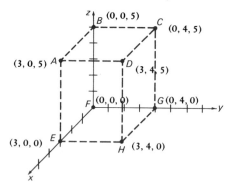

(a) Give the coordinates of each of the eight vertices of the box.

[Ans: $A = (3, 0, 5)$, $F = (0, 0, 0)$, $G = (0, 4, 0)$]

Part (b) prepares for the distance formula in space.

(b) Compute the distances.

(i) EH	(ii) DG	(iii) BD	(iv) FC	(v) FD	(vi) AG
4	$\sqrt{34}$	5	$\sqrt{41}$	$\sqrt{50}$	$\sqrt{50}$

[*Hint:* Find right triangles and use the Pythagorean Theorem. For example, △FHD is a right triangle.]

[Ans: (*ii*) $\sqrt{34}$ (*v*) $\sqrt{50}$]

Exercise 49 is more preparation for the distance formula in space.

49. The picture shows a box with one vertex at the origin *A* and "opposite" vertex at *B*. The line segment joining *A* to *B* is called a *major diagonal* of the box.

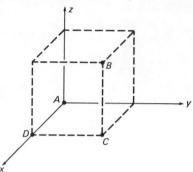

(*a*) If *B* has coordinates (*x*, *y*, *z*), then find the coordinates of *C* and of *D*.   (*x,y*,0) (*x*, 0,0)

(*b*) Using part (*a*), express the distance *AC* in terms of *x* and *y*. [*Hint:* △*ADC* is a right triangle.] $AC = \sqrt{x^2 + y^2}$

$AB = \sqrt{(AC)^2 + (BC)^2}$
$= \sqrt{(x^2+y^2) + z^2}$

(*c*) Using part (*b*), express the distance *AB*, the length of the major diagonal, in terms of *x*, *y*, and *z*. [*Hint:* △*ACB* is a right triangle.]

(*d*) Complete:

The distance from (*x*, *y*, *z*) to (0, 0, 0) is _____ . $\sqrt{x^2+y^2+z^2}$

Compare your answer with those of your classmates.

50. The picture shows a box with edges parallel to the coordinate axes.

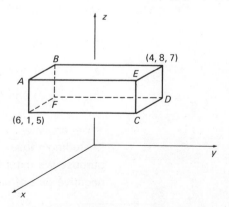

(a) Find the lengths of the edges of the box. (For example, $EC = 2$, since the z-coordinate of E is 7 (why?) and the z-coordinate of C is 5 (why?).)

(b) Find the distance from (6, 1, 5) to (4, 8, 7). Compare your answer with those of your classmates. [*Hint:* Apply the Pythagorean Theorem to an appropriate right triangle.] $\sqrt{57}$

51. The picture shows a box *with edges parallel to the coordinate axes.*

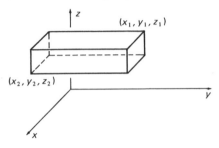

Point out the analogy with the distance formula in the plane. Recall to the class the derivation of the length of the diagonal of a box (Chapter 7).

(a) Explain why the edges parallel to the z-axis have length $|z_1 - z_2|$, those parallel to the y-axis have length $|y_1 - y_2|$, and those parallel to the x-axis have length $|x_1 - x_2|$. (Keep in mind that the coordinates represent distances to various planes.)

(b) Prove the following formula.

THE DISTANCE FORMULA IN SPACE

The distance d from (x_1, y_1, z_1) to (x_2, y_2, z_2) is

$$d = \sqrt{(x_1 - x_2)^2 + (y_1 - y_2)^2 + (z_1 - z_2)^2}$$

[*Hint:* The length of the hypotenuse of a certain right triangle is d.]

52. Using the distance formula, find the distance between

(a) (2, 1, 3) and (5, 2, 1) $\sqrt{14}$ (b) (0, 0, 0) and (2, 1, 3) $\sqrt{14}$

(c) (0, 0, 0) and (x, y, z)   (d) (1, 1, 0) and (0, 1, 1). $\sqrt{2}$

$\sqrt{x^2 + y^2 + z^2}$

Generally we want to represent *all* points in space using coordinates (including those "behind the walls" and "beneath the floor"). For this purpose we extend our three mutually perpendicular lines. We show the negative part of each line in the figure on the following page.

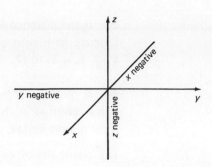

The coordinates of a point still represent the distances to the various coordinate planes, except for a minus sign if the point is on the negative side of the corresponding coordinate plane. For example, points *below* the (x, y)-plane have a *negative* z-coordinate.

53. (*a*) The eight points $(1, 1, 1)$, and $(1, -1, 1)$, and $(-1, 1, 1)$, $(-1, -1, 1)$, $(1, 1, -1)$, $(1, -1, -1)$, $(-1, 1, -1)$, $(-1, -1, -1)$, are the vertices of a certain well-known solid. What is it? Discuss with your classmates. *cube*

(*b*) Make a sketch of the solid using coordinates in space.

Students may have difficulty visualizing these sets of points. It is helpful to think of the walls and floor of the classroom as coordinate planes. Part (a) asks, "Where are all those points that are at height 1 above the floor?" Part (b) asks, "Where are all those points that are at distance 1 from two adjacent walls?" (This is a good place to mention that the intersection of two nonparallel planes is a straight line.) Part (c) asks, "Where are all points within distance 1 of a corner?"

54. In each case, consider all those points (x, y, z) which satisfy the given condition, and make a sketch showing all such points.

(*a*) all points (x, y, z) such that $z = 1$

(*b*) all points (x, y, z) such that $x = 1$ and $y = 1$

(*c*) all points (x, y, z) such that $x^2 + y^2 + z^2 \leq 1$

(*d*) all points (x, y, z) such that $x^2 + y^2 + z^2 \leq 4$ and $z \geq 0$

[Ans: (*a*) a plane parallel to the (x, y)-plane and passing through $(0, 0, 1)$

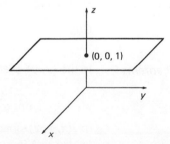

(*b*) a straight line parallel to the z-axis and passing through $(1, 1, 0)$

(*c*) a ball of radius 1 with center at $(0, 0, 0)$]

55. Is the distance formula valid if some of the coordinates are negative? Discuss with your classmates. (For example, consider the distance from $(-2, 1, -3)$ to $(1, -4, 5)$. *yes*

56. A rectangular box has a vertex at the origin, one side in the (x, y)-plane, another in the (x, z)-plane, and another in the (y, z)-plane. The most remote vertex, with reference to the one at the origin, is at $(-3, 4, 5)$.

(a) Name the other vertices. $(0, 0, 0), (0, 0, 5), (0, 4, 5), (0, 4, 0), (-3, 0, 0),$
$(-3, 4, 0), (-3, 0, 5)$

(b) Sketch the box in 3-dimensional coordinates.

57. Solve the problem of the fly at the beginning of this chapter. Compare your solution with those of your classmates.

Important Ideas in This Chapter

Summarize the chapter.

Definitions

Coordinate system in a plane and in space

x-axis, y-axis

Absolute value of a number

Graph of a condition

Graph of a linear inequality

Linear equations

Simultaneous linear equations

Slope of a straight line in the (x, y)-plane

Linear inequality

Half-plane

Origin

Theorems

In the (x, y)-plane, the distance between two points (x_1, y_1) and (x_2, y_2) is
$$d = \sqrt{(x_2 - x_1)^2 + (y_2 - y_1)^2}.$$

In (x, y, z)-space, the distance between two points (x_1, y_1, z_1) and (x_2, y_2, z_2) is
$$d = \sqrt{(x_2 - x_1)^2 + (y_2 - y_1)^2 + (z_2 - z_1)^2}.$$

If (x_1, y_1) and (x_2, y_2) are two points on a line, then the slope of the line is

$$m = \frac{y_2 - y_1}{x_2 - x_1}.$$

The graph of a linear equation

$$ax + by + c = 0 \qquad (a \text{ and } b \text{ not both zero})$$

is a straight line.

The graph of a linear inequality

$$ax + by + c > 0$$

or

$$ax + by + c < 0$$

where a and b are not both zero, is a half-plane.

Words   coordinates, Descartes, analytic geometry, absolute value, graph, origin, slope, linear, half-plane

Symbols (x, y), (x, y, z)

Summary

In this chapter we learned how to represent points in the plane by pairs of numbers (x, y). We extended this idea to three dimensions and represented points in space by triples of numbers (x, y, z). We found formulas giving the distance between points in terms of their coordinates. In the plane we found a formula for the slope of a line in terms of the coordinates of two points on that line. We also studied linear equations and methods for solving simultaneous linear equations.

Historical Note

The reason that analytic geometry is such a powerful mathematical tool was expressed eloquently by the eighteenth century mathematician J. L. Lagrange, who said:

As long as algebra and geometry proceeded along separate paths, their advance was slow and their applications limited. But when these sciences joined company, they drew from each other vitality and thenceforward marched on at a rapid pace toward perfection. [Morris Kline, *Mathematics in Western Culture* (New York: Oxford University Press, 1953), p. 159]

Although René Descartes is given credit for the discovery of analytic geometry, most of the same ideas were found independently by Pierre de Fermat (1601–1665) at about the same time. It is remarkable that Fermat, perhaps the greatest French mathematician of the seventeenth century, was a lawyer by profession who devoted only his spare time to mathematical research.

Central Self Quiz

1. What is the slope of the line that goes through
   (a) the origin and $(-7, 5)$?   (b) $(6, -2)$ and $(-1, 3)$?

2. Use the distance formula to show that the triangle with vertices $(1, 2)$, $(4, 6)$, and $(1, 7)$ is isosceles.

3. (a) Graph on the same axes:
$$\begin{cases} 2x + 3y = 7 \\ x - 2y = 5 \end{cases}$$

   (b) Use the graph in (a) to estimate the point where they intersect.

   (c) Use algebra to find the precise intersection point.

4. One corner of a rectangular box is at the origin, another at $(-3, -2, -1)$, and another at $(0, 0, -1)$. Name the remaining corners using 3-dimensional coordinates.

5. Find the graph, in the (x, y)-plane, of the condition
$$2x + y - 4 < 0.$$

Answers

1. (a) $-\frac{5}{7}$ (b) $-\frac{5}{7}$

2. The distance from $(1, 2)$ to $(1, 7)$ equals the distance from $(1, 2)$ to $(4, 6)$. $= \sqrt{25}$

3. (c) $x = \frac{29}{7}, y = -\frac{3}{7}$

4. $(-3, 0, 0), (0, -2, 0), (-3, -2, 0), (0, -2, -1), (-3, 0, -1)$

5. the shaded half-plane

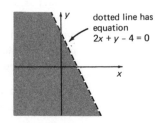

dotted line has equation $2x + y - 4 = 0$

Review

1. Using the distance formula, find the distance between
   (a) (2, 2) and (6, 4)
   (b) (−1, 4) and (5, −4)
   (c) (−3, 0) and (0, 5)
   (d) (a, b) and (−a, −b).

2. Prove that the triangle with vertices (0, 0), (6, 4), and (4, 7) is a right triangle.

3. The picture shows a rectangle with vertices at (0, 0), (a, 0), (a, b), and (0, b). Use the distance formula to prove that the diagonals of the rectangle have equal length.
   Both diagonals have length $\sqrt{a^2 + b^2}$.

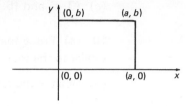

4. Prove that the quadrilateral with vertices (−1, −3), (2, 1), (1, 5) and (−2, 1) is a parallelogram.

5. Using graph paper, plot and label all those points (x, y) such that
   (a) the x-coordinate is 0
   (b) the x-coordinate is 1
   (c) the y-coordinate is 2
   (d) $xy = 0$
   (e) $x + y = 1$
   (f) $x^2 + y^2 = 1$.

6. (a) Plot the points (1, 4), (2, 6), and (4, 10). Do they look as though they lie in a straight line? *yes*

   Let $A = (1, 4)$, $B = (2, 6)$, and $C = (4, 10)$. $AB = \sqrt{5}, BC = 2\sqrt{5}$, $AC = 3\sqrt{5}$. Since $AB + BC = AC$, the points A, B, and C do not form a triangle.

   (b) Using the distance formula, prove that they lie on a straight line. [*Hint:* What does the Triangle Inequality state about the lengths of the sides of a triangle formed by three points?]

7. Start with any pair of points in the (x, y)-plane. Add the same number h to both x-coordinates and the same number k to both y-coordinates.

   (a) Prove that the new pair of points are the same distance apart as the original pair.

   (b) Prove that the line through the new pair has the same slope as the line through the original pair.

8. Is the graph of the given equation a straight line? If so, what is the slope?

(a) $y = 2x + 3$ (b) $2x - 3y = 4$ (c) $x^2 + y^2 = 4$

(d) $x + 2 = 0$ (e) $y - 1 = 0$ (f) $xy = y^2$

9. In each case, compute the slope of the line passing through the two given points by using the slope formula, then check your answer by graphing the line.

(a) $(2, 1)$ and $(4, \frac{1}{2})$ $-\frac{1}{4}$ (b) $(-1, 1)$ and $(2, 2)$ $-\frac{1}{3}$

(c) $(3, a)$ and $(5, a)$ 0 (d) $(a, -a)$ and $(-b, b)$ -1

10. (a) Prove that the triangle with vertices $(0, 0)$, $(2, 0)$ and $(1, 1)$ is similar to the triangle with vertices $(3, 2)$, $(4, 7)$, and $(1, 5)$.

(b) What is the ratio of similarity of the large triangle to the smaller?

11. What are the coordinates of the reflection of $(3, 4)$

(a) across the x-axis?

(b) across the y-axis? (Plot on graph paper.)

12. (a) Given the point $(2, 5)$, find that point (a, b) such that the origin bisects the segment joining (a, b) to $(2, 5)$.

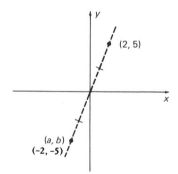

(b) Do part (a) with $(2, 5)$ replaced by

(i) $(1, -2)$ (ii) $(-1, -1)$ (iii) $(-3, 4)$

13. (a) Find the reflection of $(2, 3)$ across the line $y = x$. (Drawing a graph will help.)

(b) Find the reflection of each of the following points across the line $y = x$: $(1, 0), (2, 4), (-1, 1), (0, 1), (-4, -3), (a, b)$.

14. (a) Find the shortest path, in the (x, y)-plane, from $(0, 6)$ to $(12, 3)$ which touches the x-axis.

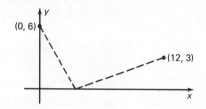

(b) What are the coordinates of the point where the shortest path meets the x-axis?

(c) What is the length of the shortest path?

15. Find x and y in the picture below.

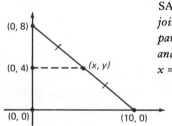

The triangles are similar by SAS. Hence the line segment joining $(0, 4)$ and (x, y) is parallel to that joining $(0, 0)$ and $(10, 0)$. Thus $y = 4$ and $x = 5$.

16. Solve simultaneously for x and y.

(a) $\begin{cases} 3x - 5y = 8 \\ 2x + 3y = 18 \end{cases}$

(b) $\begin{cases} 9x + 2y = 27 \\ 2x - 3y = 6 \end{cases}$

(c) $\begin{cases} (2.5)x + y = 3.1 \\ x - .4y = 1.22 \end{cases}$

17. Using the distance formula in space, find the distance between

(a) $(1, 2, 1)$ and $(3, 2, 5)$ $2\sqrt{5}$

(b) $(2, -1, 3)$ and $(-5, 2, 2)$ $\sqrt{59}$

(c) $(-2, 1, -3)$ and $(-1, 2, 0)$ $\sqrt{11}$

(d) $(\frac{1}{2}, 1, \frac{1}{3})$ and $(1, \frac{1}{2}, 0)$. $\sqrt{\frac{11}{18}}$

18. (a) Sketch a polyhedron with vertices at $(1, 0, 0), (0, 1, 0), (-1, 0, 0), (0, -1, 0),$ and $(0, 0, 1)$.

(b) Show that the polyhedron is a square pyramid, with a square base and four other faces which are equilateral triangles.

(c) Find the volume of the pyramid ($V = \frac{1}{3}$ (altitude) (base area)).

19. Prove that the triangle with vertices (2, 0, 2), (2, 2, 0), and (0, 2, 2) is equilateral.

20. Consider a pair of points $(x_1, y_1, 0)$ and $(x_2, y_2, 0)$. What does the distance formula give for the distance between the two points? Compare the result with the distance formula in a plane. Discuss why one should not be surprised by this result.

Review Answers

1. (a) $\sqrt{20}$ (b) 10   (c) $\sqrt{34}$ (d) $2\sqrt{a^2 + b^2}$

2. The distance from (0, 0) to (4, 7) is $\sqrt{65}$, from (6, 4) to (4, 7) is $\sqrt{13}$, and from (0, 0) to (6, 4) is $\sqrt{52}$. Since $(\sqrt{52})^2 + (\sqrt{13})^2 = (\sqrt{65})^2$, the triangle is a right triangle by the converse of the Pythagorean Theorem.

4. Use the slope formula to show opposite sides have the same slope, hence are parallel.

5. (a) the y-axis   (b) vertical line through (1, 0)

(c) horizontal line through (0, 2)

(d) all points (x, y) such that either $x = 0$ or $y = 0$, in other words, the x-axis together with the y-axis

(e) straight line through (1, 0) and (0, 1)

(f) circle of radius 1 with center (0, 0)

7. (a) $\sqrt{[(x_1 + h) - (x_2 + h)]^2 + [(y_1 + k) - (y_2 + k)]^2}$
$= \sqrt{(x_1 - x_2)^2 + (y_1 - y_2)^2}$

(b) $\dfrac{(y_1 + k) - (y_2 + k)}{(x_1 + h) - (x_2 + h)} = \dfrac{y_1 - y_2}{x_1 - x_2}$

8. (a) yes, $m = 2$ (b) yes, $m = \frac{2}{3}$ (c) no   (d) yes, slope undefined

(e) yes, $m = 0$ (f) No, the graph consists of two lines, $y = x$ and $y = 0$.

10. (a) Corresponding sides are proportional (they are both isosceles right triangles).

(b) $\sqrt{\frac{13}{2}}$

11. (a) (3, −4)   (b) (−3, 4)

12. (a) (−2, −5)   (b) (i) (−1, 2)   (ii) (1, 1)   (iii) (3, −4)

13. (a) (3, 2) (b) (0, 1), (4, 2), (1, −1), (1, 0), (−3, −4), (b, a)

14. (a) Join (0, 6) to (12, −3), the reflection of (12, 3) across the x-axis. The point where this line segment meets the x-axis is the point we are looking for.

(b) (8, 0)   (c) 15.

15. $x = 5, y = 4$.

16. (a) $x = 6, y = 2$ (b) $x = 3, y = 0$ (c) $x = 1.23, y = 0.025$

17. (a) $\sqrt{20}$ (c) $\sqrt{11}$

18. (c) $\frac{2}{3}$

Review Self Test

1. (a) Draw on graph paper the quadrilateral with vertices (5, 2), (9, 5), (6, 9), and (2, 6).

   (b) Prove that the quadrilateral is a square.

2. Find the slope of the straight line through

   (a) (1, 4) and (3, 2)   (b) $(\frac{3}{2}, -1)$ and $(-\frac{1}{2}, \frac{2}{3})$

3. (a) Find the point of intersection of the graphs of
$$\begin{cases} 3x - 4y = 2 \\ 5x + 2y = 7. \end{cases}$$

   (b) Find (x, y) which satisfies both equations simultaneously by solving algebraically.

4. Find y such that $(0, y)$ is the same distance from $(0, 0)$ and $(4, 3)$.

5. Prove that if $A = (0, 0, 4)$, $B = (3, 1, 0)$, $C = (1, 3, 0)$, then $\triangle ABC$ is an isosceles triangle in (x, y, z)-space.

Answers

1. (b) Each side has length 5; each diagonal has length $5\sqrt{2}$. The converse of the Pythagorean Theorem shows that each angle is 90°.

2. (a) −1   (b) $-\frac{5}{6}$

3. (b) $x = \frac{16}{13}, y = \frac{11}{26}$

4. $|y| = \sqrt{(0-4)^2 + (y-3)^2} = \sqrt{16 + y^2 - 6y + 9}$. Squaring gives $y^2 = 16 + y^2 - 6y + 9$, or $y = \frac{25}{6}$.

5. Using the distance formula in space, $AB = AC = \sqrt{26}$.

Projects

Reflection Across a Plane

Given a plane in space and a point P not on that plane, we define the *reflection* of P across the plane to be that point P* such that the plane is the perpendicular bisector of the line segment $\overline{PP*}$.

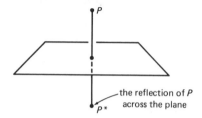

When we say the plane is the "perpendicular bisector" of $\overline{PP*}$, we mean

(i) it passes through the midpoint of $\overline{PP*}$, and

(ii) it is perpendicular to $\overline{PP*}$.

(To understand what it means for a line to be perpendicular to a plane, imagine the plane as "horizontal." Then the line is "vertical.")

1. Sketch a 3-dimensional coordinate system and in each case plot the given point and its reflection across the (x, y)-plane.

  (a) (0, 0, 1)   (b) (1, 1, 1)   (c) (2, 3, −5)   (d) (a, b, c)

2. Find the coordinates of the reflection of (2, 1, 3) across

  (a) the (x, y)-plane   (b) the (x, z)-plane   (c) the (y, z)-plane.
  (2, 1, −3)   (2, −1, 3)   (−2, 1, 3)

3. Consider the vertical plane containing the z-axis and also passing through (1, 1, 0). What is the reflection of (1, 0, 2) across this plane?
(0, 1, 2)

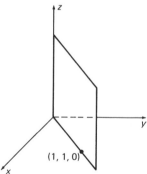

498 11 : Coordinates

Reflection Through the Origin

The reflection of a point *P through the origin* in (*x, y, z*)-space is defined to be that point *P** such that the origin bisects $\overline{PP^*}$.

4. Find the coordinates of the reflection through the origin of

   (a) (1, 0, 0)   (−1,0,0)
   (b) (0, 0, 1)   (0,0,−1)
   (c) (1, 1, 0)   (−1,−1,0)
   (d) (1, 1, 1)   (−1,−1,−1)
   (e) (1, 3, 4)   (−1,−3,−4)
   (f) (−3, 4, 0) (3,−4,0)
   (g) (*a, b, c*). (−*a*,−*b*,−*c*)

A Shortest Path Problem in Space

5. A butterfly leaves the point (0, 0, 6), lands on the (*x, y*)-plane, and then flies to (6, 8, 4).

   (a) What is the length of the shortest path he can travel in doing this? [*Hint:* Consider the vertical plane which contains (0, 0, 6) and (6, 8, 4).] $10\sqrt{2}$

   (b) What are the coordinates of the landing point in the (*x, y*)-plane for the shortest path? $(\frac{18}{5}, \frac{24}{5}, 0)$

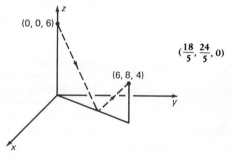

Convex Polyhedra

DEFINITION   A polyhedron is *convex* if the line segment joining any pair of points in the polyhedron is contained in the polyhedron.

Students will find the Projects on polyhedra enjoyable and useful. Exercises 6-9 are especially recommended.

For example, a cube is a convex polyhedron. It is easy to see that no matter how you choose a pair of points inside a cube, the line segment joining these points is contained in the cube. The three polyhedra in the picture below are convex:

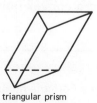
triangular prism

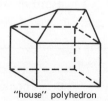
"house" polyhedron

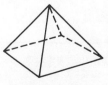
square pyramid

The regular polyhedra (Projects section, Chapter 4) are all convex.

Recall that a *face* of a polyhedron is one of the polygons making up its surface. An *edge* is a line segment along which two faces meet, and a *vertex* is a point where three or more edges meet (a "corner").

6. In the following table, V represents the number of vertices, E the number of edges, and F the number of faces of the polyhedron. Complete the table by counting the vertices, edges, and faces in pictures of the polyhedra in the Projects section of Chapter 4.

Polyhedron	V	E	F
triangular prism	6	9	5
"house"	10	17	9
square pyramid	5	8	5
regular tetrahedron	4	6	4
cube	8	12	6
regular octahedron	6	12	8
regular dodecahedron	20	30	12
regular icosahedron	12	30	20

The relationship was known to Descartes. An excellent account of Euler's Formula is given in Mathematics and Plausible Reasoning, *vol. 1, by G. Polya, chapter 3. Descartes's work is described in the exercises in that chapter.*

7. For each polyhedron in the preceding table, compute the number $V - E + F$. What do you notice?   *It always equals* 2.

The relationship you observed in the preceding exercise is known as Euler's Formula, in honor of the mathematician Leonhard Euler (oi' ler), who proved it in 1752.

EULER'S FORMULA **If V is the number of vertices, E the number of edges, and F the number of faces of a convex polyhedron, then**

$$V - E + F = 2.$$

---

8. Check that Euler's Formula holds for the following convex polyhedra.

(a) a pyramid whose base is a pentagon $V = 6, E = 10, F = 6$

(b) a pyramid whose base is a hexagon $V = 7, E = 12, F = 7$

(c) a pyramid whose base is an *n*-gon $V = n+1, E = 2n, F = n+1$

(d) a prism whose base is a pentagon $V = 10, E = 15, F = 7$

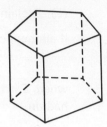

(e) a prism whose base is an *n*-gon $V = 2n, E = 3n, F = n+2$

(f) a "truncated" cube $V = 10, E = 15, F = 7$

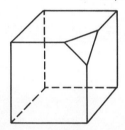

Exercise 9: Encourage students to report to the class.

9. Read about Euler's Formula in the book *Mathematics, The Man-made Universe*, 2nd edition, by S. K. Stein (San Francisco: W. H. Freeman, 1969), chapter 14.

10. Suppose we have a polyhedron *whose faces are all triangles*. Imagine making three marks on each face, near the edges. For example, we have marked the faces of a triangular pyramid; the marks on the back faces are not visible.

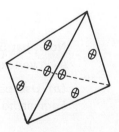

(a) Explain why the total number of marks is 2E, if E is the number of edges of the polyhedron. (Do this generally, not only for the pyramid in the picture!)

(b) Explain why the total number of marks is 3F, if F is the number of faces of the polyhedron.

(c) Explain why the following is true:

If all the faces of a polyhedron are triangles, then

$$2E = 3F,$$

where E is the number of edges and F is the number of faces of the polyhedron.

(d) Check that the relationship in part (c) holds for a tetrahedron, regular octahedron, regular icosahedron, and the convex polyhedron in the following picture.

11. Suppose all the faces of a convex polyhedron are triangles, with 5 triangles meeting at each vertex.

(a) Prove that $2E = 5V$. [*Hint:* Place two marks on each edge, near the ends of the edges.

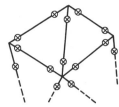

Note that 5 edges meet at each vertex; hence there are 5 marks around each vertex. But also note that there are 2 marks on each edge.]

(b) Use part (a) of this exercise, part (c) of Exercise 10, and Euler's Formula, to show that for our polyhedron

$$V = 12, \quad E = 30, \quad F = 20.$$

[*Hint:* Write V and F in terms of E and substitute into Euler's Formula.]

(c) Part (b) shows that if all the faces of a convex polyhedron are triangles, and if 5 triangles meet at each vertex, then the polyhedron must have 12 vertices, 30 edges, and 20 faces. Can you find a specific example of such a polyhedron? [*Hint:* Look at the regular polyhedra!]

12. Suppose all the faces of a convex polyhedron are triangles, with 4 triangles meeting at each vertex.

(a) Prove that $2E = 4V$. (See Exercise 11(a).)

(b) Prove that $V = 6, E = 12, F = 8$. (See Exercise 11(b).)

(c) Part (b) shows that if all the faces of a convex polyhedron are triangles, with 4 triangles meeting at each vertex, then the polyhedron must have 6 vertices, 12 edges, and 8 faces. Can you find a specific example of such a polyhedron?

13. Suppose all the faces of a convex polyhedron are triangles, with 3 triangles meeting at each vertex.

(a) Prove that $2E = 3V$.

(b) Prove that $V = 4, E = 6, F = 4$.

(c) Part (b) shows that if all the faces of a convex polyhedron are triangles, with 3 triangles meeting at each vertex, then the polyhedron has 4 vertices, 6 edges, and 4 faces. Can you find a specific example of such a polyhedron?

Why There Are Only Five Regular Polyhedra

A highly recommended reference is What is Mathematics? *by Courant and Robbins, Chapter 5. A proof of Euler's Formula is also given there.*

14. Suppose all the faces of a convex polyhedron are *n*-gons (same *n* for all faces!) and that there are *m* faces meeting at each vertex (same *m* for all vertices!)

(a) Prove that $2E = mV$. [*Hint:* Place two marks on each edge, one near each end. Note that there will then be *m* marks around each vertex.]

(b) Prove that $2E = nF$. [*Hint:* Place *n* marks on each face, near the edges. Note that each edge will then have two marks "near" it.]

(c) Prove that

$$\frac{2E}{m} - E + \frac{2E}{n} = 2.$$

[*Hint:* Substitute into Euler's Formula using (*a*) and (*b*).]

(*d*) Prove that
$$\frac{1}{m} + \frac{1}{n} > \frac{1}{2}.$$

[*Hint:* Deduce from (*c*) that $\frac{1}{m} - \frac{1}{2} + \frac{1}{n} = \frac{1}{E} > 0.$]

(*e*) Check that the only positive integers $m > 2$ and $n > 2$ which can satisfy the inequality in part (*d*) are

(*i*) $m = 3 \quad n = 5$

(*ii*) $m = 3 \quad n = 4$

(*iii*) $m = 3 \quad n = 3$

(*iv*) $m = 4 \quad n = 3$

(*v*) $m = 5 \quad n = 3$

(*f*) In view of part (*e*), discuss with your classmates the possible types of convex polyhedra, all of whose faces are *n*-gons with *m* faces meeting at each vertex.

[*Hint:* Once we are given *m* and *n*, *E* is determined by the equation in part (*c*), and *V* and *F* are determined by parts (*a*) and (*b*).]

(*g*) Notice that we did not require the *n*-gons to be *regular*, yet we still ended up with only five possible types. Discuss this with your classmates.

A proof that there are only five regular polyhedra may also be found in Chapter 13 of The Enjoyment of Mathematics, *by Rademacher and Toeplitz (Princeton: Princeton University Press, 1957).*

Chapter 12 The Conic Sections

This chapter uses once again some ideas from Chapter 1, namely, the shortest path, reflection, and distance, which are used in describing some interesting properties of the ellipse and other conics.

Central

The chapter emphasizes the interesting properties of the ellipse; the other conics are only briefly treated.

All students should perform this experiment; it is essential to an understanding of the definition of an ellipse.

In this chapter we shall study certain curves, called the *conic sections*, that are of interest not only in geometry but also in physics, astronomy, and other sciences. In our study we will meet an old friend from Chapter 1, the Problem of the Shortest Path, in a surprising new context.

The first exercise of the chapter deals with the "pins and string" construction of the most important of the conic sections, the *ellipse*.

1. Place a sheet of paper over a heavy piece of cardboard (or even better, a wooden board). Stick in two pins, or thumbtacks, about 4 inches apart. Next take a piece of string, about 6 inches long, and tie the ends to the tacks. Now stretch the string with the point of your pencil and draw the curve as you let the pencil move around with the string stretched taut (you will have to lift the string over a tack in certain places to keep it from becoming tangled).

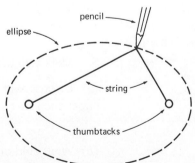

The curve traced by the pencil is called an *ellipse.* Each of the points where the thumbtacks are placed is called a *focus* (fō′kus) of the ellipse (pl. *foci* (fō′sī)).

2. Using the pins and string method, and a piece of string 6 inches long, draw an ellipse whose foci are

(*a*) 3 inches apart (*b*) 2 inches apart (*c*) 1 inch apart

(*d*) zero inches apart (that is, the foci are the same point).

3. (*a*) Discuss with your classmates how the shape of an ellipse changes if we keep the string length fixed and move one focus closer and closer to the other. *The ellipse gets closer to the shape of a circle.*

(*b*) What happens to the shape if we keep the string length fixed and move one focus farther and farther from the other? *The ellipse flattens to a line segment.*
(*c*) What curve is obtained when the foci come together at the same point? *circle*

Exercises 4 and 5 prepare the student for the formal definition of an ellipse.

4. (*a*) Draw an ellipse with the pins and string method using a string 10″ long, label one focus of the ellipse *A* and the other *B*, and let *P* be any point on the ellipse.

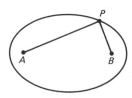

What is true about $AP + PB$? Compare your answer with those of your classmates. *$AP + PB$ is always 10″, because the length of the string is fixed.*

You may wish to introduce the terminology: The ellipse is the locus *of points P such that $AP + PB = 10″$, where A and B are fixed points.*

(*b*) Where are all those points *P* on the sheet of paper such that $AP + PB = 10″$? *on the ellipse*

5. Mark two points *A* and *B* four inches apart on a sheet of paper.

(*a*) Discuss with your classmates how to find all those points *P* on the paper such that
$$AP + PB = 8″.$$
Make an ellipse with foci A and B and string length 8″.

(*b*) What kind of curve is formed by all those points *P* on the paper such that $AP + PB = 8″$? *ellipse*

DEFINITION Let A and B be two points in a plane. Then the curve formed by all those points P such that the sum of the distances, $AP + PB$, is constant is an *ellipse*. Each of the points A and B is a *focus* (pl. *foci*) of the ellipse.

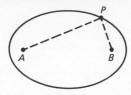

A very readable account of Kepler's prodigious labors is found in The Watershed: A Biography of Johannes Kepler *by A. Koestler (Garden City, N.Y.: Anchor Books, Doubleday, 1960).*

One of the highpoints in the history of science was the discovery by Johannes Kepler in 1605 (after years of incredible calculation) that each planet follows a path which is an ellipse with the sun at one focus.

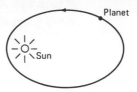

Nearly two thousand years before Kepler, the ancient Greeks studied the geometric properties of the ellipse without suspecting the applications in astronomy. They knew the ellipse as one of the *conic sections*, the curves obtained by cutting a right circular cone with a plane. If a right circular cone is cut with a plane in the manner shown below, the curve obtained is an ellipse. (This is proved in the Projects section.)

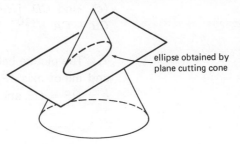

The Major Axis and Minor Axis of an Ellipse

Consider an ellipse with foci A and B. The line drawn through the foci intersects the ellipse in two points C and D. The line segment $\overline{CD}$ is called the *major axis* of the ellipse.

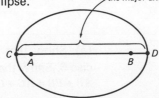

509 *Central*

You may wish to point out that the major and minor axes of the ellipse are the axes of symmetry.

The perpendicular bisector of the major axis intersects the ellipse in points E and F. The line segment $\overline{EF}$ is the *minor axis* of the ellipse. The point O of intersection of the major axis and minor axis is the *center* of the ellipse.

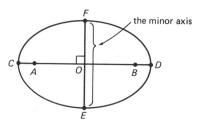

6. Suppose the ellipse in the picture below consists of all those points P such that $AP + BP = 10$, and suppose $AB = 8$. $\overline{CD}$ and $\overline{EF}$ are respectively the major axis and minor axis.

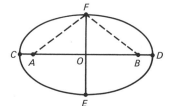

AF = BF = 5
AO = OB = 4
By the
Pythagorean
Theorem,
OF = 3

(a) Find OF. [*Hint:* $AF = BF$, what is true about $AF + BF$?]

(b) Find EF. $EF = 2(OF) = 6$

(c) Find CD. [*Hint:* Use the fact that $CA = BD$. $CD = CB + BD = CB + CA = \underline{\ 10\ }$]

7. The picture below shows an ellipse with foci A and B, major axis $\overline{CD}$ and minor axis $\overline{EF}$ intersecting at the center O. We let $a = OD = OC$, $b = OF = OE$, and $c = OB = OA$.

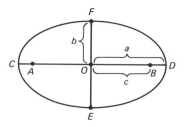

(a) Show that $AF + FB = 2a$. [*Hint:* Proceed as in part (c) of Exercise 6. Show that $AF + FB = AC + CB = BD + CB = CD = 2a$.]
$AF + FB = AC + CB$ because $AP + PB$ is constant for any point P on the ellipse.

510   12: The Conic Sections

Since AP + PB is constant for every point P on the ellipse, AP + PB = AF + FB = 2a.
length of major axis = 2a
length of minor axis = 2b
$= 2\sqrt{a^2 - c^2}$

Students are better off memorizing this picture rather than formulas.

Enclosing the ellipse in a rectangle emphasizes the above relationships.

Cutting a Cylinder to Obtain an Ellipse

(b) Using $AF = FB$, show that
$$a^2 = b^2 + c^2.$$
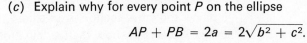
$AF = FB = a$
Use the Pythagorean Theorem.

(c) Explain why for every point P on the ellipse
$$AP + PB = 2a = 2\sqrt{b^2 + c^2}.$$

(d) An ellipse is constructed using a string of length $2a$ tied to tacks at distance $2c$ apart. What is the length of the major axis of the ellipse (in terms of a and c)? What is the length of the minor axis (in terms of a and c)? [*Hint:* Use parts (a), (b), (c) of this exercise.]

The relationships in the previous exercises are easily remembered using the following picture of an ellipse with major axis length $2a$, minor axis length $2b$, and distance between foci $2c$.

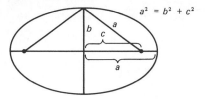

8. Suppose you want to use the pins and string method to draw an ellipse which fits exactly in a 24″ by 30″ rectangle, as indicated below.

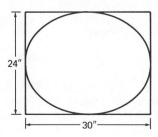

How far apart should the tacks be and how long the string? Discuss with your classmates. [*Hint:* Use the preceding exercise.]   18″, 30″

If a cylindrical stick, such as a broomstick, is sawed in two at a slant, the resulting cross section appears to have the shape of an ellipse.

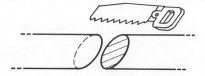

511 Central

Experiments are very important aids to understanding this chapter.

9. By cutting a cardboard tube at a slant, demonstrate how the shape of an ellipse is obtained.

A later exercise will prove that when a right circular cylinder is cut by a plane, the cross section is an ellipse. (An almost identical proof is used in the Projects to show that certain cross sections of a cone are ellipses.) But some preparation, dealing with properties of tangent planes and tangent lines of spheres, is needed before the proof.

Tangent Planes and Tangent Lines of Spheres

We digress for a moment to consider some useful properties of tangent planes and tangent lines of spheres.

DEFINITION A plane is said to be *tangent* to a sphere if it intersects the sphere in exactly one point. In this case it is called a *tangent plane* of the sphere.

Hold a piece of cardboard against a globe to illustrate tangent plane of a sphere.

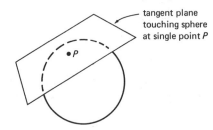

tangent plane touching sphere at single point P

The point *P* is called the *point of contact,* or *point of tangency,* of the plane and the sphere.

10. (*a*) Place a smooth ball (such as a Ping-Pong ball) on a sheet of paper. The plane of the paper is _____ to the ball at the point where the ball rests. *tangent*

Exercise 10(b): To give students an intuitive idea of a line perpendicular to a plane, have them imagine a horizontal plane and a vertical line (for example, a flagpole and the ground). Let them imagine tilting the entire configuration of vertical line and horizontal plane to see the general case.

(*b*) If *P* is the point where the ball touches the paper, and *O* is the center of the ball, then $\overline{OP}$ is _____ to the plane of the paper. Discuss your answer with your classmates. *perpendicular*

(*c*) Guess how to complete the following statement, based on your observation in part (*b*).

The radius drawn to the point of contact of a tangent plane of a sphere is _____ to the tangent plane.
perpendicular

DEFINITION   A straight line that lies in a tangent plane of a sphere and passes through the point of contact is called a *tangent line* of the sphere.

An equivalent definition: A tangent line of a sphere is a line meeting the sphere in exactly one point.

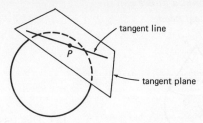

11. (a) Mark a point P on a sheet of paper and draw a number of lines through P. Now place a smooth ball on P.

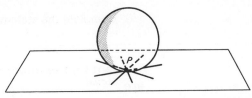

Contrast this with the case of a tangent line to a circle in a plane. There is only one tangent line through a given point on the circle.

Based on your observations in this experiment, what would you say about the number of tangent lines through a point P on a sphere? *an infinite number*

(b) Using a sphere (a ball, or a globe of the world) and a very thin rod, or the edge of a ruler, demonstrate some of the different tangent lines passing through a point on the sphere.

12. The picture below shows a sphere cut by a plane. As we saw in Chapter 10, the intersection is a circle. The point P is in the plane, and lines through P tangent to the circle at Q and R respectively are shown.

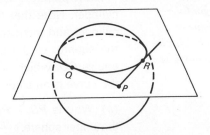

Discuss with your classmates why $PQ = PR$. [*Hint:* Tangent segments drawn to a circle from a point outside a circle *are equal* .]

13. The following picture shows a sphere and two tangent lines drawn

513 Central

If all *the tangent line segments to a sphere from an exterior point P are drawn, one obtains a dunce cap sitting on a spherical head.*

All the tangent line segments have the same length.

to the sphere from a point P outside the sphere.

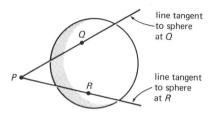

(a) Discuss with your classmates why $PQ = PR$. [*Hint:* If you think of "filling in" the angle formed by the two rays in the picture, you will see that the given lines determine a plane. What does the preceding exercise now tell us?]

(b) Complete the statement:

> **Tangent line segments drawn to a sphere from an exterior point are ___equal___ .**

A Proof That a Cross Section of a Cylinder is an Ellipse

Students should study the figure carefully before they begin Exercise 14(a)-(g). There is value in having the students sketch this picture freehand (without looking at the model in the book).

14. The picture on the following page shows a right circular cylinder cut by a plane X. A sphere with the same diameter as the cylinder fits snugly inside the cylinder above the plane and tangent to the plane at B. Similarly, a sphere fits snugly below the plane and tangent to the plane at A. (Think of the cylinder as a can for tennis balls, cut at a slant with a plane, with a tennis ball touching above and a tennis ball touching below.)

In the picture, P is any point on the curve where the plane intersects the cylinder. Note that the vertical line through P lies in the surface of the cylinder and intersects the equator of the upper sphere in a point C and the equator of the lower sphere in point D. We will see that $AP + PB$ equals CD, which is the same for every choice of P on the curve; hence the curve is an ellipse with foci A and B.

(a) Why is $PB = PC$? [*Hint:* Observe that $\overline{PC}$ and $\overline{PB}$ are tangent to the upper sphere.] *Exercise 13*

(b) Why is $PA = PD$? *similar to (a)*

(c) Why is $PA + PB = PD + PC$?

(d) Why is $PA + PB = CD$? *PD + PC = CD*

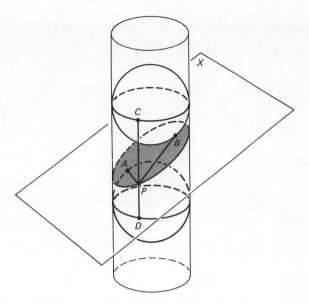

(e) Does the length CD change for different choices of P on the curve of intersection of the plane with the cylinder? *No, CD is constant for all choices of P.*

(f) Does PA + PB change for different choices of P on the curve? Discuss with your classmates. *no, since PA + PB = CD*

(g) Discuss with your classmates why the curve of intersection of the plane X with the cylinder is an ellipse with foci A and B. *Since PA + PB is constant for any point P in the intersection, the intersection is an ellipse.*

The Reflection Property of an Ellipse

The picture below shows an ellipse and a line ℓ tangent to the ellipse. The line meets the ellipse at a single point P.

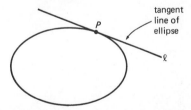

The experiment in the next exercise introduces a surprising property of a tangent line to an ellipse.

Exercise 15 should be done carefully; it prepares for Exercises 16, 17, and 18.

15. Use the pins and string method to make a careful drawing of an ellipse (or you may use the ellipse constructed in Exercise 1, if it was drawn carefully with a sharp pencil). Choose a point P on the ellipse and draw, as accurately as you can, the tangent line t of the ellipse at P (this will have to be done by trial and error).

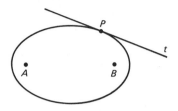

(a) Choose a point Q on the tangent line in your drawing, and find $AQ + QB$ by measuring with a ruler (where A and B are the foci).

(b) Do part (a) for at least five more choices of Q on the tangent line. Choose points on both sides of P on the tangent line, some near P and some far from P.

(c) Measure $AP + PB$. How does it compare with $AQ + QB$ for the choices of Q you made in part (b)? Compare your results with those of your classmates. $AP + PB \leqslant AQ + QB$ *if measurements are accurate.*

(d) On the basis of your measurements, how do you think Q should be chosen on the tangent line in order to make $AQ + QB$ as small as possible? Discuss with your classmates. *Choose $Q = P$.*

The next exercise proves the result you discovered experimentally in the preceding exercise.

16. The picture below shows an ellipse with foci A and B. The line ℓ is tangent to the ellipse, meeting it at the point P, and Q is any other point on ℓ.

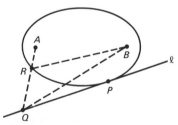

(a) If R is the point where $\overline{AQ}$ intersects the ellipse, why is
$$RQ + QB > RB ?$$ *Triangle Inequality*

(b) Why is
$$AQ + QB > AR + RB ?$$
$AQ + QB = AR + (RQ + QB) > AR + RB$

[*Hint:* $AQ + QB = AR + RQ + QB$]

(c) Why is
$$AQ + QB > AP + PB?$$
$AR + RB = AP + PB$, since R and P are on the ellipse.

[*Hint:* Since P and R are on the ellipse, what is true about $AP + PB$ and $AR + RB$?]

(d) Discuss with your classmates why part (c) shows the following.
If Q is not chosen at the point of tangency, the distance $AQ + QB$ is larger.

If ℓ is a tangent line of an ellipse with foci A and B, then $AQ + QB$, with Q on ℓ, has its least value when Q is chosen to be the point where ℓ meets the ellipse.

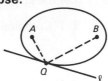

---

17. For this exercise you will again need to carefully draw an ellipse by the pins and string method, with foci A and B.

(a) Choose a point P on your ellipse and draw as accurately as you can the tangent line at P. Let $\angle 1$ and $\angle 2$ be the angles indicated below:

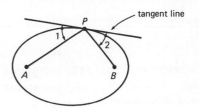

Measure $\angle 1$ and $\angle 2$ with a protractor. What do you notice? $\angle 1 = \angle 2$

(b) Do the experiment in part (a) using three other choices of P on the ellipse. Compare your results with those of your classmates. *The angles are equal.*

The result you observed experimentally in the preceding exercise is proved in the next exercise.

18. The following picture shows an ellipse with foci A and B, and a tangent line ℓ meeting the ellipse at P.

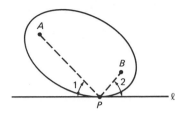

(a) Discuss with your classmates why the length of the shortest path from A to B meeting the line ℓ is given by $AP + PB$. [*Hint:* Exercise 16(d)!] *By 16(d), any other path is longer.*

(b) Why must $\angle 1 = \angle 2$? Discuss with your classmates. [*Hint:* Remember the shortest path in Chapter 1.] *When $AP + PB$ is the shortest path, $\angle 1 = \angle 2$.*

You may wish to summarize the ideas that lead to this statement.

(c) Complete the statement:

THE REFLECTION PROPERTY OF AN ELLIPSE

A line tangent to an ellipse at a point P makes ____equal____ angles with the lines drawn from the foci to P.

$\angle 1$ ____ $\angle 2$

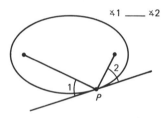

Exercise 19: A ball driven from either focus in any direction will pass through the other focus after bouncing once. A ball that misses a focus between bounces will thereafter always miss the foci (Review Exercise 14).

19. Imagine a billiards table with the shape of an ellipse and having a red ball at one focus. A cue ball is placed at the other focus.

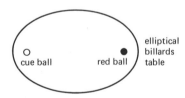

In what directions can you shoot the cue ball so it will hit the red ball after bouncing once off the side? Discuss with your classmates. *any direction*

20. Imagine an "elliptical mirror," in which someone standing "inside"

Exercise 20: The light ray will always pass through the foci, and its path will converge toward the major axis.

The path of a ray missing the foci has interesting properties. See the discussion in Mathematical Snapshots, *2nd ed., by H. Steinhaus (London: Oxford University Press, 1969), chapter 10. These and other interesting properties of the ellipse are discussed in chapter 15 of* New Mathematical Diversions from Scientific American, *by Martin Gardner (New York: Simon and Schuster, 1966).*

could see his reflection in the ellipse. A light ray starts at one focus, hits the mirror and is reflected. What is the path of the light ray as it is reflected repeatedly by the "inside" of the ellipse? Discuss with your classmates.

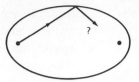

(Draw a careful model with string, then show the path of a ray on its successive reflections.)

21. Some "whisper chambers" (where a whisper can be heard across the room) such as the Mormon Tabernacle in Salt Lake City, are based on the ellipse. How does this help explain their "whisper" characteristics? Discuss this with your classmates.

Other Conic Sections

The old House of Representatives chamber in the Capitol in Washington, D.C., is a well-known whispering gallery. According to an often repeated story John Quincy Adams, while serving in the House, was able to overhear important conversations of the opposition party because his desk was at a focal point of sound.

It is proved in the Projects section that an ellipse is obtained when a cone is cut appropriately with a plane. The other conic sections are obtained by cutting a *double cone;* the two halves are called the *nappes* of the double cone. The picture below shows three types of conic sections obtained as intersections of a double cone with a plane.

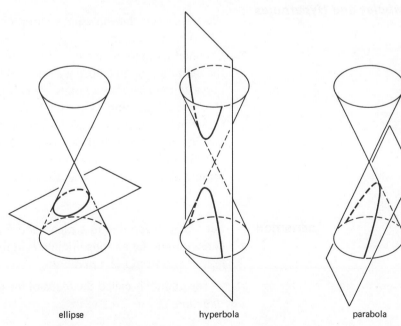

ellipse   hyperbola   parabola

Encourage students to devise models if there are no commercial ones in the classroom.

For a hyperbola, the plane need not *be parallel to the axis of the cone.*

Emphasize that the hyperbola and the parabola are "unbounded" curves.

A *hyperbola* is obtained when the plane cuts both nappes of the cone. Note that a hyperbola is in two "pieces," called the *branches* of the hyperbola. A *parabola* is obtained when the plane is parallel to one of the straight lines lying in the surface of the cone. In the case of the hyperbola and parabola, you must imagine that the double cone and the intersecting plane extend indefinitely upward and downward, so these curves also extend indefinitely.

22. Discuss with your classmates and make sketches that show how to intersect a double cone with a plane to obtain the following "degenerate" conic sections.

(*a*) a single point   (*b*) a straight line

(*c*) two intersecting straight lines

There will be some difficulty in distinguishing between a parabola and a branch of a hyperbola.

23. A flashlight produces a "cone of light." Bring a flashlight to class and demonstrate how the following shapes can be produced by shining it against a wall at appropriate angles.

(*a*) circle   (*b*) ellipse   (*c*) parabola

(*d*) branch of a hyperbola

Other Ways to Obtain Parabolas and Hyperbolas

We have seen that an ellipse can be obtained by intersecting a cylinder with a plane. It is shown in the Projects section that an ellipse can also be obtained as the intersection of a cone with a plane. Recall however that the definition of an ellipse was quite different. An ellipse was *defined* as a curve formed by those points P such that the sum of the distances from P to two fixed points is constant.

A similar situation exists for the parabola and hyperbola. These curves can be defined as consisting of those points P satisfying a certain condition. Then it can be shown how such curves may be obtained by intersecting a cone with a plane. We now give these definitions.

DEFINITION   Let ℓ be a given line in a plane and F a point not on ℓ in the plane. The curve formed by all those points P in the plane that are an equal distance from F and from ℓ is a *parabola*.

The point F is called the *focus* of the parabola, and the line ℓ is called the *directrix*.

Remind students of the definition of distance from a point to a line.

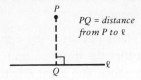

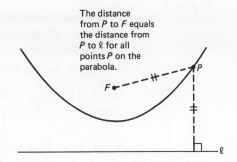

DEFINITION

Equivalently one could state that a hyperbola consists of those points P such that |AP − BP| is constant.

Let A and B be fixed points in a plane. The curve formed by all those points P such that the positive *difference* of the distances, $AP - BP$ or $BP - AP$, is constant is a *hyperbola*.

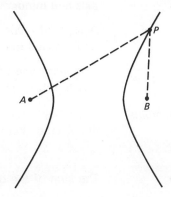

Note that a branch of a hyperbola is not similar to a parabola.

In the picture, $AP - BP$ is constant for P on the right-hand *branch* of the hyperbola, and $BP - AP$ is equal to the same constant for P on the left-hand branch. The points A and B are the *foci* of the hyperbola.

Some properties of the parabola are treated in the Projects section. It is shown there that the parabola, like the ellipse, has an interesting and useful reflection property. An equation for the graph of a parabola in the (x, y)-plane is also derived.

Important Ideas in This Chapter

Definitions

Given two points A and B in a plane, the curve formed by all those points P in the plane such that the sum of the distances, $AP + PB$, is constant is an *ellipse*. The points A and B are the *foci* of the ellipse.

Given two points A and B in a plane, the curve formed by all those points P in the plane such that the positive difference of the distances, $AP - BP$ or $BP - AP$, is constant is a *hyperbola*. The points A and B are the *foci* of the hyperbola. A hyperbola has two *branches*.

Given a line ℓ in a plane and a point F in the plane not on ℓ, the curve formed by all those points P in the plane such that the distance from P to F is equal to the distance from P to ℓ is a *parabola*. The point F is the *focus* of the parabola, and ℓ is the *directrix*.

The *major axis* of an ellipse is the line segment with endpoints on the ellipse and passing through the foci. The *minor axis* is the line segment with endpoints on the ellipse, which is the perpendicular bisector of the major axis. The *center* of an ellipse is the point of intersection of the major axis and minor axis.

A *tangent plane* of a sphere is a plane touching the sphere in a single point, called the *point of contact*, or *point of tangency*.

A *tangent line* of a sphere is a straight line lying in a tangent plane and passing through the point of contact.

Theorems

If the major axis of an ellipse has length $2a$, the minor axis length $2b$, and the distance between the foci is $2c$, then

$$a^2 = b^2 + c^2.$$

The sum of the distances from the foci to any point on the ellipse is $2a$.

The radius drawn to the point of contact of a tangent plane of a sphere is perpendicular to the tangent plane.

Tangent line segments drawn to a sphere from an exterior point are equal.

An ellipse can be obtained by intersecting a right circular cylinder with a plane.

If ℓ is a tangent line of an ellipse with foci A and B, then for points Q on ℓ, $AQ + QB$ has its least value when Q is the point of contact.

The Reflection Property of an Ellipse A line tangent to an ellipse makes equal angles with the lines drawn from the foci to the point of contact.

Words

conic section, ellipse, hyperbola, parabola, focus, foci, directrix, branch, major axis, minor axis, center of an ellipse, tangent, Johannes Kepler, whisper chamber, nappe, double cone.

Summary

This chapter provided an introduction to some of the most interesting curves in mathematics, namely, the conic sections. The bulk of our study was concerned with the ellipse. An ellipse can be obtained either by the "pins and string" construction or by intersecting a cylinder or cone appropriately with a plane. The reflection property of the ellipse recalled our study of the shortest path in Chapter 1.

Historical Note

The quote is from a letter written by Newton in 1676 to Robert Hooke.

Sir Isaac Newton (1642–1727), considered by many the greatest scientist who ever lived, once said, "If I have seen farther, it is by standing on the shoulders of giants." One of those giants of science whose discoveries made possible Newton's achievements was Johannes Kepler (1571–1630). Kepler's most important scientific contribution was the discovery of three laws of planetary motion. Basing his computations on the observations of the astronomer Tycho Brahe, he found:

I   Each planet moves along an ellipse with the sun at one focus.

II   The line drawn from the sun to the planet sweeps out equal areas in equal times.

III   The square of the time required to revolve once around the sun, divided by the cube of the major axis of the ellipse along which the planet moves, is the same for all the planets.

Johannes Kepler, The Six-Cornered Snowflake, edited and translated by Colin Hardie (Oxford: Oxford University Press, 1966). This is a very interesting example of the quality of Kepler's thought. Among so many other things, one finds in this essay the remark that the ratio of successive Fibonacci numbers tends to the Golden Ratio (Chapter 10, Projects Exercise 13).

Kepler also applied his talents to geometry. In an essay titled *The Six-Cornered Snowflake,* Kepler's speculations as to why snowflakes have a hexagonal shape led him to a number of interesting results about tiling space with polyhedra. Motivated by the crude methods used in those days to estimate the volumes of wine casks, Kepler developed formulas for the volumes of various solids. This work was published under the title *The Solid Geometry of Wine Barrels.*

Central Self Quiz

1. As you tip a (cylindrical) glass of water to your lips to take a sip, what shape does the edge of the water surface take? Explain.

2. Suppose you want to use the pins and string method to draw an ellipse with major axis 20" long and minor axis 12" long. How long should the string be, and how far apart the tacks?

3. The next picture shows an ellipse with foci A and B, and a line t tangent to the ellipse at P. The tangent line intersects the extension of the major axis at Q.

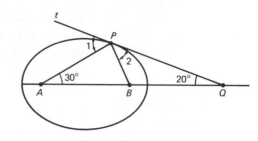

(a) What must be true about ∡1 and ∡2?

(b) Using the fact that ∡1 is an exterior angle of △AQP, find the size of ∡1.

(c) Find the size of ∡ABP.

Answers

1. ellipse *(intersection of cylinder and plane)*
2. length of string = 20", distance between tacks = 16"
3. (a) ∡1 = ∡2   (b) 50°   (c) 70°

Review

1. It is late afternoon at the beach. A beachball lies in the sun on the flat surface of the sand. Explain why its shadow has the shape of an ellipse with one focus where the ball touches the sand.

If students have trouble, they should refer back to Central Exercise 14.

Since the ball corresponds to one of the spheres of Central Ex. 14, it touches the ellipse at a focus.

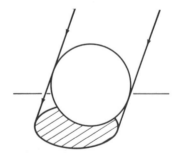

(Assume the beach is a plane and the rays of the sun are parallel to each other. Tilt the picture in Central Exercise 14 so that the plane X is horizontal.)

An acceptable answer for Exercise 2: A circle can be obtained by cutting a right circular cylinder or a cone with a plane.

2. Explain why a circle can be considered a special case of an ellipse.

Let P be any point on the curve. Then the length of string is AB + BP + PA which is constant. Since AB is also constant, BP + PA must be constant and the curve is an ellipse.

3. Suppose there are tacks at fixed points A and B and that you have a *loop* of string whose length is greater than twice the distance between A and B. Stretch the string taut with a pencil. What curve do you get when the pencil moves along the string? Explain.

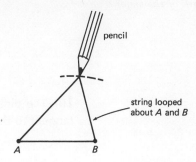

4. An ellipse is constructed using a piece of string 26″ long tied to a pair of thumbtacks 24″ apart. What is the length of the major axis of the resulting ellipse? the length of the minor axis? $2a = \text{major axis} = 26″$
$2b = \text{minor axis} = 10″$

5. If you want to use the tacks and string method to draw an ellipse with major axis 16″ long and minor axis 8″ long, how long should the string be and how far apart the tacks? $2a = \text{length of string} = 16″$
$2c = \text{distance between tacks} = 8\sqrt{3}″$

6. Suppose you want to use a *loop* of string, as in Review Exercise 3, to draw an ellipse with major axis 10″ long and minor axis 8″ long. How long should the loop be, and how far apart the tacks?

7. The royal gardener of Polygonia is told to inscribe an ellipse in the royal garden, which is a 20 ft. by 30 ft. rectangle.

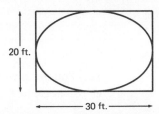

Exercise 8: The ratio of the minor axis to the major axis is about 1 to 15, so the ellipse is quite thin and elongated. But it is still surprising that in a careful drawing the foci would be indistinguishable from the endpoints of the major axis!

Describe how he should place stakes, and how long a string he should use, to construct the ellipse.

8. A certain ellipse has major axis of length 842 and minor axis of length 58. What is the distance from a focus to the nearest endpoint of the major axis? [*Hint:* $\sqrt{1764} = 42$]

Construct the minor axis (it lies on the perpendicular bisector of the major axis). The distance from an endpoint of the minor axis to a focus is half the length of the major axis.

9. The picture below shows an ellipse and its major axis. Using only straightedge and compass, find the foci of the ellipse. [*Hint:* What is the distance from a focus to an endpoint of the minor axis?]

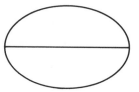

10. The picture below shows an ellipse with foci *A* and *B*, and a line *t* tangent to the ellipse at *P*. Explain why $\angle 1 = \angle 2$.

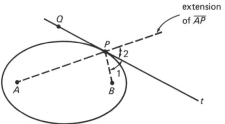

11. For this exercise, use an ellipse you have constructed with the pins and string method. Choose a point *P* on your ellipse, and construct, using only straightedge and compass, the tangent line through *P* (no trial and error allowed!). [*Hint:* If *A* and *B* are the foci, draw $\overline{AP}$ and $\overline{BP}$. Extend $\overline{AP}$ through *P* and apply the preceding exercise.]

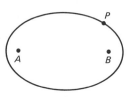

12. You are given an ellipse with foci *A* and *B*, and a line *t* tangent to the ellipse at point *P*. If *s* is perpendicular to *t* at *P*, explain why $\angle 1 = \angle 2$.

$$\angle 1 = 90° - \angle APQ$$
$$= 90° - \angle BPR$$
$$= \angle 2$$

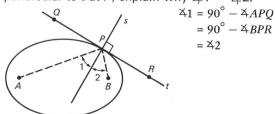

13. A ball is driven over one focus of an elliptical billiards table. Discuss the path of the ball if it keeps rebounding forever (no friction and no holes in the table).

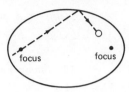

14. You are playing billiards on the elliptical billiards table in the preceding exercise. A red ball rests on one focus. You drive a cue ball which bounces once off the side and *misses* the red ball.

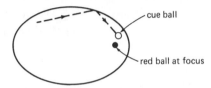

Will the cue ball, after a number of rebounds, *ever* hit the red ball? Explain. (Imagine both balls as points.)

15. The line t is tangent at P to an ellipse with foci A and B, as below. The tangent line intersects the extension of the major axis at Q. Prove that $\angle 1 + \angle 2 = \angle 3$.

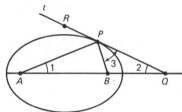

Some students may have difficulty with Exercise 16.

16. The picture below shows an ellipse with foci A and B, and $\overline{PB}$ perpendicular to the major axis.

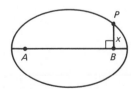

Suppose the major axis has length $2a$ and the minor axis has length $2b$.

Prove that
$$PB = \frac{b^2}{a}.$$

[*Hint:* Let $PB = x$. Consider the right triangle $\triangle ABP$. Why is $AB = 2\sqrt{a^2 - b^2}$? Why is $AP = 2a - x$?]

17. Given below is a line segment $\overline{CD}$ and two points A and B.

(a) Using only straightedge and compass, find at least eight different points P such that
$$AP + PB = CD.$$

[*Hint:* Choose a point Q on $\overline{CD}$. Let P be a point where a circular arc of radius CQ centered at A intersects a circular arc of radius DQ centered at B. Then $AP + PB = \underline{}$.]

(b) Consider the ellipse which consists of all points P such that $AP + PB = CD$. In part (a) you constructed some points on the ellipse. Sketch this ellipse by drawing a smooth curve (freehand) through the points constructed in part (a). (You may need to find a few more points to have enough for a good freehand sketch.)

18. On a piece of graph paper, let ℓ be one of the horizontal lines, and let F be a point 4 units above ℓ.

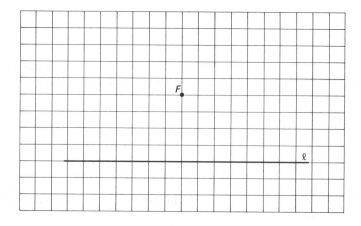

(a) Use your compass to find a point P which is 3 units from F and 3 units from ℓ. [*Hint:* Draw an arc of a circle of radius 3 units centered at F. Let P be a point where this arc intersects the horizontal line 3 units above ℓ.]

(b) Find a second point 3 units from both F and ℓ.

(c) Find at least ten more points which are equidistant from F and ℓ.

(d) By drawing a smooth curve through the points you located in the preceding parts, sketch the parabola with focus F and directrix ℓ.

Some students may need a demonstration of how to find one of the points.

19. In the picture below you are given two points A and B and a line segment $\overline{CD}$.

(a) Using only straightedge and compass, find a point P such that

$$AP - BP = CD.$$

[*Hint:* Draw an arc of a circle of radius r centered at B (r should be larger than $\frac{1}{2}(AB - CD)$). Next draw an arc of a circle of radius $r + CD$ centered at A, and let P be a point where the arcs intersect. Why is P the required point?] $AP - BP = CD + r - r = CD$

(b) Find at least five different points P satisfying the condition in part (a). These points lie on a branch of a hyperbola with foci A and B (why?).

(c) Sketch the branch of the hyperbola by drawing a smooth curve through the points you found in part (b). (You may need to find a few more points to make a good sketch.)

(d) Discuss how you would find points on the *other* branch of this hyperbola.   *Consider points P such that BP − AP = CD.*

Review Answers

1. See Central Exercise 14.

2. A circle is an ellipse with non-distinct foci.

3. ellipse

4. 26", 10"

5. 6", 8√3"

6. 16", 6"

7. Place stakes $15 - 5\sqrt{5}$ ft. from each 20 ft. side and half way down; use a string 30 ft. long.

8. 1

10. $\angle 1 = \angle APQ = \angle 2$

12. Use the fact that $\angle APQ = \angle BPR$.

13. See Central Exercise 20.

14. no

15. $\angle 3 = \angle RPA = \angle 1 + \angle 2$

16. Use the hint and apply the Pythagorean Theorem to $\triangle ABP$.

Review Self Test

1. An ellipse is constructed using a string 8" long with ends tied to tacks 6" apart.

    (a) What is the length of the major axis of the ellipse?

    (b) What is the length of the minor axis?

2. The picture shows an ellipse with foci A and B, and a line t tangent to the ellipse at P. The point $B*$ is the reflection of B across t. Discuss why $AP + PB = AB*$.

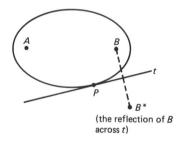

(the reflection of B across t)

3. Describe three different ways to obtain an ellipse.

Answers
1. (a) 8"   (b) $2\sqrt{7}$"

2. $AP + PB = AP + PB* = AB*$

3. *pins and string, shadow of a ball, flashlight shining on a wall, diagonal slice of a can or rod*

Projects

Cutting a Cone to Obtain an Ellipse

Exercise 1 is an excellent subject for general class discussion if there is time. This is a fine exercise for a student to sketch freehand from memory.

The two spheres are called Dandelin spheres after the 19th-century Belgian mathematician G.P. Dandelin, who devised this elegant proof.

This is similar to Review Exercise 1. The shadow cast by a ball on a table under a lightbulb is an ellipse with a focus at the point where the ball rests.

The Reflection Property of a Parabola

1. The picture below shows a right circular cone cut by a plane X. A sphere fits snugly inside the cone above X and tangent to X at a point B. Note that this sphere touches the cone along a circle. Similarly, a sphere fits snugly below X and tangent to X at a point A. This sphere also touches the cone along a circle.

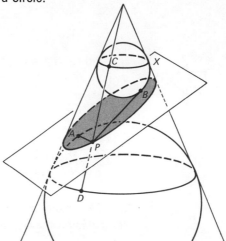

P is any point on the curve of intersection. Note that the line through P and the apex of the cone intersects the circle (along which the upper sphere touches the cone) at point C and the lower circle at point D.

We shall see that the curve of intersection is an ellipse with foci at A and B.

(a) Why is $PB = PC$? *Tangent segments from a point outside a sphere are equal.*

(b) Why is $PA = PD$?

(c) Why is $PA + PB = CD$? *$PA + PB = PC + PD = CD$*

(d) Why is $PA + PB$ the same number for any point P on the curve of intersection? *CD is constant.*

(e) Explain why the curve is an ellipse with foci A and B. *$PA + PB$ is constant for all points P.*

A parabola has an interesting and useful reflection property. Consider a parabola with focus F and directrix ℓ. Let P be a point on the parabola

531 Projects

and t a line tangent to the parabola at P (that is, t touches at the single point P without crossing the parabola).

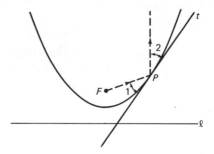

The vertical dotted ray is a ray through P perpendicular to ℓ, and the line segment $\overline{FP}$ is also shown dotted. The *reflection property* of the parabola states that

$$\angle 1 = \angle 2.$$

If you imagine a light ray leaving the focus F and "reflecting off" the parabola, then the ray travels perpendicular to the directrix ℓ after reflection.

The next two exercises prove this reflection property of a parabola.

2. The figure below shows a parabola with focus F and directrix ℓ. Point P is on the parabola, and line t is the *bisector of $\angle FPQ$*.

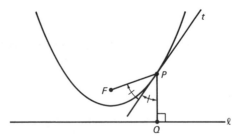

We want to show that t is tangent to the parabola at P. This requires that we show that t does not meet the parabola at any point other than P. In order to see this, let P' be any other point on t.

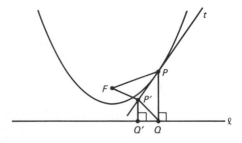

Explain each of the following steps.

(a) $FP' = P'Q$ [Hint: Keep in mind that t bisects $\angle FPQ$, and $FP = QP$, since P is on the parabola. Why is $\triangle FPP' \cong \triangle QPP'$?] SAS

(b) $P'Q > P'Q'$ [Hint: What is the nearest point on ℓ to P'?] Q', hence Q is farther from P.

(c) $FP' \neq P'Q'$ [Hint: parts (a) and (b)] $FP' = P'Q > P'Q'$

(d) P' is not on the parabola. [Hint: Use part (c), and keep in mind the definition of a parabola.] *The parabola consists of all points P such that FP is equal to the distance from P to ℓ, but $FP' \neq P'Q'$.*

3. Suppose P is any point on a parabola with focus F and directrix ℓ. Let Q be the point on ℓ with $\overline{PQ}$ perpendicular to ℓ. Let t be the bisector of $\angle FPQ$.

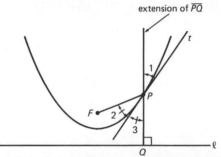

The preceding exercise showed that t is tangent to the parabola. Explain why $\angle 1 = \angle 2$. (In other words, show that the parabola has the reflection property.) *Since t bisects $\angle FPQ$, $\angle 2 = \angle 3$; $\angle 1 = \angle 3$ by vertical angles.*

A parabolic reflector is obtained by revolving a parabola about its axis of symmetry.

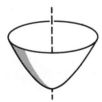

4. Why is a *parabolic* shape good for automobile headlight reflectors? Discuss with your classmates.

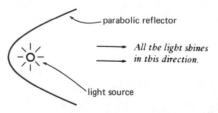

5. A *reflecting telescope*, sketched on the next page, has a curved mirror for focusing the rays of light from distant stars and planets.

(a) Discuss with your classmates why the curve used in telescope mirrors is a parabola. *The light from the stars can be focused at a single point.*

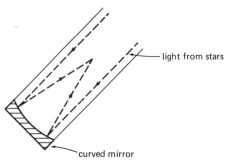

light from stars

curved mirror

(b) At what point are the light rays focused by the parabolic mirror? *focus*

The Equation of a Parabola

6. On a piece of graph paper let F be the point $(0, 2)$ and let ℓ be the line with equation $y = -2$ (horizontal line passing through $(0, -2)$).

(a) Is $(0, 0)$ on the parabola with focus F and directrix ℓ? Is $(2, 2)$ on this parabola? Plot at least five points which are equidistant from F and ℓ. (Recall that the distance from P to ℓ means *perpendicular* distance from P to a point on ℓ.) *yes, no*

(b) Plot a few more points equidistant from F and ℓ and draw a smooth curve through the points you have plotted. This curve is the graph of the parabola with focus $F = (0, 2)$ and directrix ℓ.

This example is an excellent review of algebra as well as geometry.

Let us see how to find the equation of the parabola in the preceding exercise. The picture below shows the parabola with focus $F = (0, 2)$ and directrix ℓ (line with equation $y = -2$). Let $P = (x, y)$ be any point on the parabola.

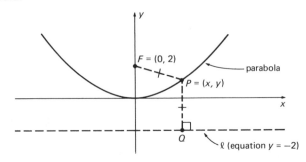

The distance from P to F is given by the distance formula:

$$PF = \sqrt{(x - 0)^2 + (y - 2)^2}$$
$$= \sqrt{x^2 + (y - 2)^2}$$

The distance from P to ℓ is equal to PQ. From the picture you can see that

$PQ = y + 2$, since P is y units above the x-axis, while Q is 2 units below the x-axis. Another way to see this is to observe that Q has coordinates $(x, -2)$. Hence the distance formula gives

$$PQ = \sqrt{(x-x)^2 + (y-(-2))^2}$$
$$= \sqrt{(y+2)^2}$$
$$= y + 2 \quad \text{(since } y + 2 \text{ is positive).}$$

Since for each point P on the parabola

$$PF = \text{distance from } P \text{ to } \ell,$$

we must have

$$PF = PQ$$

or

$$\sqrt{x^2 + (y-2)^2} = y + 2.$$

Squaring both sides to eliminate the radical,

$$x^2 + (y-2)^2 = y^2 + 4y + 4.$$

Simplifying,

$$x^2 + y^2 - 4y + \cancel{4} = \cancel{y^2} + 4y + \cancel{4},$$

so

$$x^2 = 8y.$$

In other words, if a point (x, y) is on the parabola with focus $(0, 2)$ and directrix $y = -2$, then x and y satisfy the equation

$$x^2 = 8y.$$

Some classes may have time to pursue the study further. Let students make up problems choosing their own conditions.

7. Find the equation of the parabola with focus $(0, 1)$ and whose directrix is the line $y = -1$. $x^2 = 4y$

Exercise 8 presents the general case. Point out the nicest case:

$$k = \frac{1}{4}$$

The equation is easy to derive from the focus and directrix.

8. Suppose k is a given positive number. Consider the parabola with focus at $(0, k)$ and whose directrix is the line with equation $y = -k$. Show that the equation of this parabola is

$$x^2 = 4ky.$$

9. On a piece of graph paper sketch the parabola with focus at $(2, 0)$ and directrix $x = -2$ (vertical line passing through $(-2, 0)$). (Proceed as in Projects Exercise 6.)

10. Find the equation of the parabola in the preceding exercise. $y^2 = 8x$

11. Let k be a given positive number. Consider the parabola with focus

at $(k, 0)$ and directrix $x = -k$. Show that the equation of this parabola is

$$y^2 = 4kx.$$

Exercise 12 will probably call for teacher direction. It is another good exercise for general class discussion.

12. A man standing on the edge of a cliff throws a rock horizontally with velocity v.

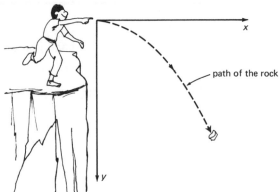

path of the rock

We have placed an "upside down" coordinate system on the picture, with origin at the point where he released the rock. Then x measures the horizontal distance the rock travels, and y measures the vertical distance the rock falls.

If the force of gravity were absent, the rock would travel horizontally without falling. The distance x covered in time t would be given by

$$x = vt,$$

because of the relationship

$$v = \text{velocity} = \frac{\text{distance}}{\text{time}} = \frac{x}{t}.$$

If the man had simply dropped the rock, without imparting a horizontal velocity v, and the rock had fallen straight down under the force of gravity, then the distance y it would fall in time t is given by

$$y = \frac{1}{2} gt^2,$$

where g is a constant called *gravitational acceleration*. From physics it is known that the path of the rock will be such that *both* of these equations hold. In other words at any time t the position of the rock will be such that

$$x = vt \quad \text{and} \quad y = \frac{1}{2} gt^2.$$

Deduce that the path of the rock is a *parabola*. [*Hint:* Find k such that $x^2 = 4ky$ and then use Projects Exercise 8.]

The Hyperbola and LORAN

13. A system of long range navigation called LORAN (*LO*ng *RA*nge *N*avigation) was developed during World War II. LORAN is based on the hyperbola.

Two transmitting stations located at A and B send out simultaneous radio signals (pulses). A ship at point P below receives the signal from A later than the signal from B.

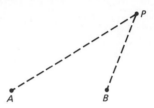

Electronic machinery on the ship records the time lag between the signals and on this basis computes the difference of the distances, $AP - BP$. The ship's navigator then knows that he is somewhere on a branch of a certain hyperbola with foci A and B (why?).

The navigator is on a branch of a hyperbola as the difference $AP - BP$ is some positive constant (derived from the time lag). The navigator must merely find the intersections of the hyperbolas with foci A and B, and C and D. Since the navigator has a rough idea of where he is and since the branches of the hyperbola are generally far apart, the possibility of several intersecting branches can be ignored.

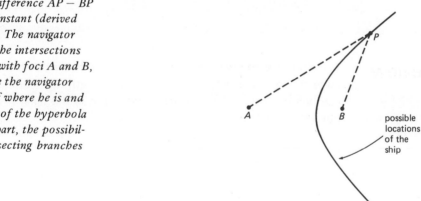

Suppose now that two transmitting stations located at C and D (different from A and B) send out simultaneous radio signals which are received by the ship at P. Discuss how the ship's navigator can use this to fix the location of the ship. (The locations of A, B, C, D are known to the navigator. Assume also that he has a rough idea of where he is.)

The figure on the following page demonstrates the situation.

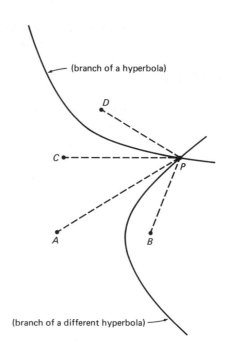

Cumulative Review

This Cumulative Review contains interesting problems on different aspects of the course.

1. Suppose you want to build a closed cardboard box in the shape of a cube to hold a fishing pole 6 feet long (without bending the pole). If cardboard costs 10¢ per square yard, what is the least it would cost to build the box?

2. A ladder leans against a house. The base of the ladder is 5 feet from the house and the ladder makes a 60° angle with the ground. How long is the ladder?   *30°-60° right triangle*

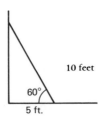

The distance d from the center of the bottom to the top rim is the hypotenuse of a right triangle whose legs are the radius of the bottom and the height of the can. $d = \sqrt{2^2+5^2}$

3. A certain tomato can is 5″ high, and the top has diameter 4″. What is the distance from the center of the bottom to any point on the top rim?

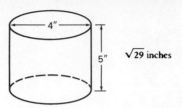

4. Find the area of each polygon.

(a): The height is the long leg of a 30°-60° right triangle whose hypotenuse is 2.
(b): It is a right triangle.
(d): area (5×12 rectangle) + area (4×12 right triangle)

(a)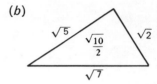

(b)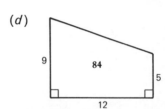

(c)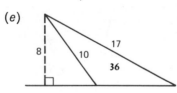

(d)

(e): area (large right triangle) − area (small right triangle)

(e)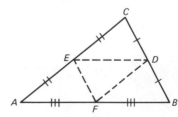

$\dfrac{s^2\sqrt{3}}{4} = 25\sqrt{3}$

5. What is the length of each side of an equilateral triangle of area $25\sqrt{3}$?

6. (a) Prove that $\triangle ABC \sim \triangle DEF$.

$\triangle ABC \sim \triangle DEF$ since the ratio of corresponding sides is constant $\left(\dfrac{1}{2}\right)$.

(b) What is the ratio of similarity of $\triangle DEF$ to $\triangle ABC$?

(c) What is the ratio of the perimeter of $\triangle DEF$ to the perimeter of $\triangle ABC$?

(d) What is the ratio of the area of △DEF to the area of △ABC?

7. If AS = BQ, prove that the quadrilateral PQRS has half the area of rectangle ABCD.

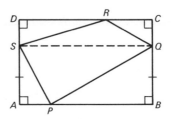

8. Find x.

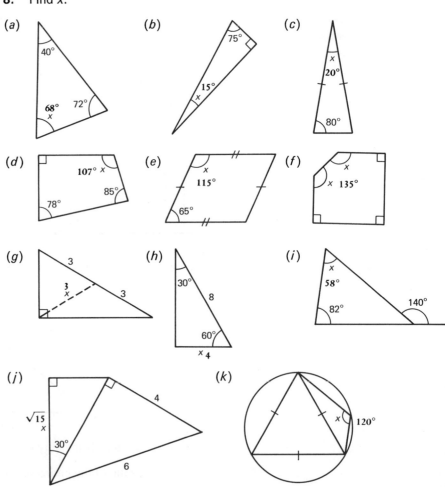

(l) 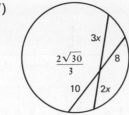 (m)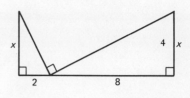

9. If the area of the square is 100 square units, what is the area of the circle?

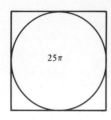

10. Explain why $(AE)(BC) = (BD)(AC)$.

$$\text{area} = \frac{1}{2}(AE)(BC)$$
$$= \frac{1}{2}(BD)(AC)$$

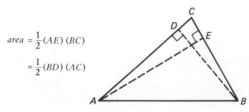

11. The picture shows a rhombus circumscribed about a circle. If the rhombus has area 24 square units, what is the area of the circle?

height of rhombus
= 4
= 2·(radius of circle)

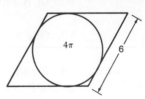

12. Find x.

similar triangles:
$$\frac{x+5}{6} = \frac{6}{x}$$

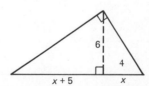

13. Find the length of the shortest path from A to B touching line ℓ in the following picture.

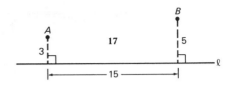

14. *Given:* ∡1 = ∡2 and ∡3 = ∡4
Prove: △ABC ≅ △BAD

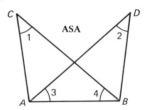

15. Prove that △ABC ≅ △FEG.

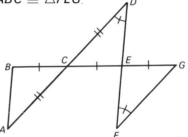

16. A 60-foot flagpole stands in the center of a square field of area 800 sq. ft. Find the distance from the top of the pole to a corner of the field.

17. One angle of a parallelogram is 73°. Find the other angles.

18. A carpenter's square can be used to bisect an angle as follows. Mark points B and C on the sides of ∡A so that AB = AC. Now slide the carpenter's square, keeping the edges on B and C, until BP = CP.

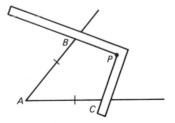

Finally, draw the line through A and P. Prove that this line bisects ∡A. [*Hint:* Given the information in the figure below, prove that ∡1 = ∡2.]

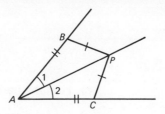

19. Write the converse of each statement. State if the converse is true.

(*a*) If two triangles are congruent, then they have the same area.

(*b*) If two squares are congruent, then they have the same area.

(*c*) If a quadrilateral is a parallelogram, then its diagonals bisect each other.

(*d*) The base angles of an isosceles triangle are equal.

20. A string with a weight is attached to a protractor at point *B*. Let *C* be the point where the string intersects the semicircle. What theorem tells us that the line through *A* and *C* is horizontal?

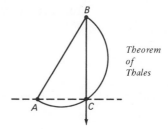

Theorem of Thales

21. If the area of circle I is equal to the area of circle II, then explain why the area of region III is also equal to the area of circle I.

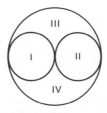

22. A racetrack is built in the shape of a rectangle with semicircles fitted on opposite sides, with the dimensions shown:

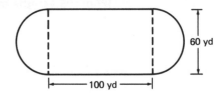

(a) Find the perimeter of the racetrack.

(b) Find the area enclosed by the track.

23. Each side of the triangle is tangent to the circle:

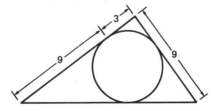

(a) Find the perimeter of the triangle.

(b) If the circle has radius 3, find the area of the triangle. *The triangle is a right triangle.*

24. The picture shows a pyramid with square base $ABCD$ and altitude $\overline{PQ}$.

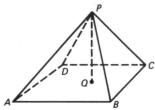

If $AB = \sqrt{8}$ and $PQ = 2$, find

(a) the volume of the pyramid $\quad \frac{1}{3} \cdot 8 \cdot 2$

(b) the surface area of the pyramid (assume Q is the center of the square base). $\quad 8 + 4\left(\frac{1}{2}\right)$ (base)(height)

25. Find the distance between the points in the (x, y)-plane with coordinates

(a) $(0, 0)$ and $(-3, 4)$ \qquad (b) $(2, -1)$ and $(8, -1)$

(c) $(2, 8)$ and $(-3, 1)$ \qquad (d) $(0, 2)$ and $(4, -5)$.

$(AB)^2 + (AC)^2 = (BC)^2$

26. Use the converse of the Pythagorean Theorem to prove that the triangle in the (x, y)-plane with vertices (0, 1), (2, 0), and (3, 7) is a right triangle.
   A   B   C

27. Graph both equations and locate the intersection point of their graphs, then solve the equations algebraically.

(a) $\begin{cases} 2x - 3y = 1 \\ 5x + 2y = 4 \end{cases}$ (b) $\begin{cases} 3x - 6y = 2 \\ 5x + 2y = 7 \end{cases}$

$AB = \sqrt{5} = AC$

28. Prove that the triangle with vertices (3, 4), (2, 2), and (5, 5) is isosceles.
   A   B   C

29. Find the distance between the two points in (x, y, z)-space.

(a) (0, 0, 0) and (1, 2, −1)   (b) (1, −1, 2) and (−3, −2, 2)

(c) (0, −2, 1) and (3, 5, −1)   (d) (5, 4, −2) and (5, 1, 2)

Ratio of similarity of $\triangle CDE$ to $\triangle CAB$ is $\frac{1}{4}$: $\frac{10-x}{10} = \frac{1}{4}$

30. If the triangle has altitude 10 inches, and if $\dfrac{\text{area}(\triangle CDE)}{\text{area}(\triangle CAB)} = \dfrac{1}{16}$, what is x?

$x = 7\frac{1}{2}$ inches

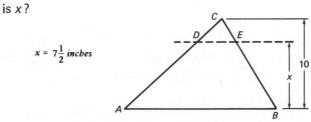

31. (a) Find the ratio of the circumference of the larger circle to that of the smaller circle.

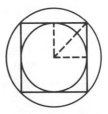

(b) Find the ratio of the area of the larger circle to that of the smaller.

32. Using only straightedge and compass,

(a) construct on a separate piece of paper $\triangle EDF \cong \triangle ABC$

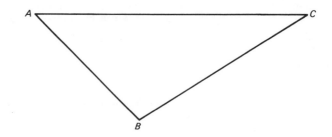

(b) construct point P on $\overline{DE}$ such that

$$\frac{\text{area }(\triangle DPF)}{\text{area }(\triangle EPF)} = \frac{1}{2}.$$

33. Prove that there *cannot* be any circle passing through A, B, C, and D.

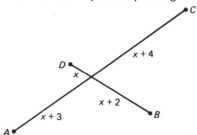

[*Hint:* Imagine a circle passing through the four points: $\overline{AC}$ and $\overline{BD}$ would be intersecting chords.]

34. Suppose P and Q are similar polyhedra, and suppose the ratio of similarity of P to Q is 2.

(a) If P has volume 32 cubic units, what is the volume of Q? $32\left(\frac{1}{2}\right)^3$

(b) If P has surface area 80 square units, what is the surface area of Q?

$80\left(\frac{1}{2}\right)^2$

35. The picture shows an ellipse with foci A and B. From the information given in the figure, find the length of the major axis and the minor axis of the ellipse.

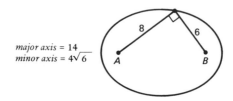

major axis = 14
minor axis = $4\sqrt{6}$

Answers

1. 80¢

2. 10 feet

3. $\sqrt{29}$ inches

4. (a) $3\sqrt{3}$ (b) $\dfrac{\sqrt{10}}{2}$ (c) $4\sqrt{3}$ (d) 84 (e) 36

5. 10

6. (b) $\frac{1}{2}$ (c) $\frac{1}{2}$ (d) $\frac{1}{4}$

7. area $(\triangle SQR) = \frac{1}{2}(CQ)(SQ) = \frac{1}{2}$ area $(SQCD)$, area $(\triangle SQP) = \frac{1}{2}$ area $(ABQS)$

8. (a) 68° (b) 15° (c) 20° (d) 107° (e) 115°
   (f) 135° (g) 3 (h) 4 (i) 58° (j) $\sqrt{15}$ (k) 120°
   (l) $\dfrac{2\sqrt{30}}{3}$ (m) 4

9. 25π sq. units

10. $\frac{1}{2}(AE)(BC) = $ area $(\triangle ABC) = \frac{1}{2}(BD)(AC)$

11. 4π sq. units

12. $x(x + 5) = 36$, therefore $x = 4$

13. 17

14. Since $\angle CAB = \angle DBA$, the triangles are congruent by ASA.

15. $\triangle ABC \cong \triangle DEC$ by SAS and $\triangle DEC \cong \triangle FEG$ by ASA

16. $20\sqrt{10}$ ft.

17. 107°, 73°, 107°

18. $\triangle ABP \cong \triangle ACP$ by SSS

19. (a) If two triangles have the same area, then they are congruent. *false*

    (b) If two squares have the same area, then they are congruent. *true*

    (c) If the diagonals of a quadrilateral bisect each other, then it is a parallelogram. *true*

    (d) If two angles of a triangle are equal, then it is isosceles. *true*

20. Theorem of Thales

21. area of I = area of II = $\frac{1}{4}$ area of big circle

22. (a) $200 + 60\pi$ yd. (b) $6000 + 900\pi$ sq. yd.

23. (a) 36 (b) 54

24. (a) $\frac{16}{3}$ (b) $8(1+\sqrt{3})$

25. (a) 5 (b) 6 (c) $\sqrt{74}$ (d) $\sqrt{65}$

26. Use the distance formula.

27. (a) $x = \frac{14}{19}, y = \frac{3}{19}$ (b) $x = \frac{23}{18}, y = \frac{11}{36}$

28. Use the distance formula.

29. (a) $\sqrt{6}$ (b) $\sqrt{17}$ (c) $\sqrt{62}$ (d) 5

30. $7\frac{1}{2}$ inches

31. (a) $\sqrt{2}$ (b) 2

32. (b) Trisect $\overline{DE}$.

33. There is no $x > 0$ such that $x(x+2) = (x+4)(x+3)$.

34. (a) 4 cubic units (b) 20 square units

35. length of major axis $= 14$, length of minor axis $= 4\sqrt{6}$

TABLE OF SQUARES AND APPROXIMATE SQUARE ROOTS

N	N^2	$\sqrt{N}$	N	N^2	$\sqrt{N}$
1	1	1.000	41	1681	6.403
2	4	1.414	42	1764	6.481
3	9	1.732	43	1849	6.557
4	16	2.000	44	1936	6.633
5	25	2.236	45	2025	6.708
6	36	2.449	46	2116	6.782
7	49	2.646	47	2209	6.856
8	64	2.828	48	2304	6.928
9	81	3.000	49	2401	7.000
10	100	3.162	50	2500	7.071
11	121	3.317	51	2601	7.141
12	144	3.464	52	2704	7.211
13	169	3.606	53	2809	7.280
14	196	3.742	54	2916	7.348
15	225	3.873	55	3025	7.416
16	256	4.000	56	3136	7.483
17	289	4.123	57	3249	7.550
18	324	4.243	58	3364	7.616
19	361	4.359	59	3481	7.681
20	400	4.472	60	3600	7.746
21	441	4.583	61	3721	7.810
22	484	4.690	62	3844	7.874
23	529	4.796	63	3969	7.937
24	576	4.899	64	4096	8.000
25	625	5.000	65	4225	8.062
26	676	5.099	66	4356	8.124
27	729	5.196	67	4489	8.185
28	784	5.292	68	4624	8.246
29	841	5.385	69	4761	8.307
30	900	5.477	70	4900	8.367
31	961	5.568	71	5041	8.426
32	1024	5.657	72	5184	8.485
33	1089	5.745	73	5329	8.544
34	1156	5.831	74	5476	8.602
35	1225	5.916	75	5625	8.660
36	1296	6.000	76	5776	8.718
37	1369	6.083	77	5929	8.775
38	1444	6.164	78	6084	8.832
39	1521	6.245	79	6241	8.888
40	1600	6.325	80	6400	8.944

N	N^2	$\sqrt{N}$	N	N^2	$\sqrt{N}$
81	6561	9.000	91	8281	9.539
82	6724	9.055	92	8464	9.592
83	6889	9.110	93	8649	9.644
84	7056	9.165	94	8836	9.695
85	7225	9.220	95	9025	9.747
86	7396	9.274	96	9216	9.798
87	7569	9.327	97	9409	9.849
88	7744	9.381	98	9604	9.899
89	7921	9.434	99	9801	9.950
90	8100	9.487	100	10000	10.000

Credits

facing page 1	The Shortest Path—*Day Passage* (detail), by Rockne Krebs (American, born 1938). Courtesy, Los Angeles County Museum of Art, *Art and Technology*, May 10–August 29, 1971. Project executed in collaboration with Hewlett-Packard Corporation.
page 32	Donald Dietz
page 80	M.I.T. Lincoln Laboratory
page 136	Donald Dietz
page 178	Frank Siteman
page 230	Alan P. Cohen
page 272	Stock, Boston
page 316	Stock, Boston
page 364	Donald Dietz
page 398	Farrell Grehan/Photo Researchers
page 462	Stock, Boston
page 506	Sandra Manheimer

Index

AAA (angle-angle-angle) condition for similar triangles, 326
Absolute value, 467
Acute angles, 46
Addition-subtraction method, 227, 479
Almagest, 341
Alternate interior angles, 39
Altitude(s), 206
   meet in a point, 205, 206
   of a cone, 452
   of a parallelogram, 233
   of a triangle, 204–207, 222
Angle(s)
   acute, 46
   alternate interior, 39
   bisector of, 190, 206
   central, 400
   complementary, 47
   corresponding, 37
   exterior, 42
   inscribed, 402
   notation for, 10, 399, 430
   obtuse, 46
   of a triangle, 36
   opposite interior, 42
   right, 46
   sides of, 45
   straight, 10
   supplementary, 52
   vertex of, 45
   vertical, 11
Angle bisectors of a triangle, 191, 207, 418, 429
Annulus, 455
Apex
   of a cone, 452
   of a pyramid, 244

Apollo, 389
Apollo 11, 68
Apollo 15, 377
Arbelos, 448
Arc
   bend of, 400–401
   intercepted, 402
   length of, 421
   notation for, 401, 430
   of a circle, 187, 206
Archaeologist, 201
Archimedes of Syracuse, 208, 217, 430, 448, 450, 451
   tomb of, 431
Area, 231
   fundamental properties of, 247
   of a circle, 422–424, 430
   of an equilateral triangle, 278, 284
   of a lattice polygon, 309
   of a parallelogram, 233, 248
   of a polygon, 237
   of a rectangle, 232
   of a sector, 425, 430
   of similar polygons, 367–370, 376, 391
   of a trapezoid, 239, 248
   of a triangle, 235–236, 248
   properties of, 232
   surface, 373
Arithmetic mean, 442
ASA (angle-side-angle) property of congruent triangles, 92, 109
Associative property, 77
Auxiliary lines, 102–103, 109
Average, 442
Axiom(s), 38, 41, 110, 156–157
Axiomatic method, 110, 157
Axis

   coordinate, 464, 485
   of a cone, 452

Babylonians, 69, 274
Ball
   cross section of, 427
   definition of, 427
   volume of, 428, 430, 454–455
Base
   of a parallelogram, 233
   of a pyramid, 244
   of a triangle, 235
Base angles of an isosceles triangle, 104, 109, 127
Basic constructions, 192
Bell, E. T., 68
Bend of an arc, 400–401
Bhāskara, 304
Billiards, 13, 14, 17, 25, 518, 527
Biology, similar figures in, 390
Bisect
   a line segment, 182
   an angle, 190
   an arc of a circle, 188
Bisecting, 106
Bisector of an angle, 190, 206
Bolyai, Janos, 158
Branches of a hyperbola, 520, 522

Calculus, 430
Carpenter's square, 385, 408, 452
Cavalieri's Principle, 393
Center
   of a circle, 180
   of an ellipse, 510, 512
   of gravity, 224
Central angle, 400

Centroid, 223
Chord(s), 206
 extended, 410, 429
 of a circle, 187, 402
 product of segments of, 409, 429
Cicero, 431
Circle(s), 399
 arc of, 187, 206
 area of, 430
 center of, 180
 chord of, 187, 402
 circumference of, 420, 429, 430
 circumscribed, 202–203, 206
 concentric, 401
 definition of, 180, 206
 diameter of, 206, 402
 inscribed, 418
 length of arc of, 421
 radius of, 180
 sector of, 425
Circular reasoning, 360
Circumcenter, 203, 206
Circumference of a circle, 420, 429, 430
Circumscribed
 circle, 202–203, 206
 polygon, 422–423
Coefficient, 130
Commutative property, 77
Complementary angles, 47
Completing the square, 269, 312
Concentric circles, 401
Cone, 452
Congruent, 82, 109
 symbol for, 85, 110
Congruent triangles
 ASA property of, 92
 conditions for, 87
 in space, 125
 SAS property of, 91
 SSS property of, 88
Conic section(s), 507–509, 519
 degenerate, 520
Construction(s)
 bisecting an angle, 190
 bisecting an arc, 188
 bisecting a segment, 182
 copying an angle, 192
 copying a triangle, 193
 dividing a line segment into equal segments, 195
 impossible, 189
 of circumscribed circle, 202

of inscribed circle, 418
of line parallel to a given line, 194
of perpendicular bisector, 181–182
of perpendicular from a point to a line, 203
of perpendicular through a point on a line, 184
of regular hexagon, 186–187
of regular pentagon, 446
of regular polygons, 185, 220
of square root, 441
rules of, 183
trisecting a segment, 195
trisecting the angle, impossibility of, 197–200
Contact, point of, 512, 522
Converse(s), 44, 105, 144, 157
Convex polygon, 69–70, 499
Coordinate(s), 463–504
 axis, 464, 485
 in space, 485
 of a point, 464, 486
 plane, 485
 system, 463–467
Copernicus, 341
Corner reflector, 68
Corresponding, 86
Corresponding angles, 37
Corresponding parts
 of congruent triangles, 85–87, 109
 of similar triangles, 330
Cube, 170, 500
Cuboctahedron, 171, 263
Cylinder, 428, 453

da Vinci, Leonardo, 303
Degenerate conics, 520
Degree of a polynomial, 130
Dehn, Max, 249
Delphi, oracle at, 389
Descartes, René, 463, 491
Diagonal(s)
 of a box, 281–282, 284
 of a convex polygon, 69–70
 of a parallelogram, 140
 of a quadrilateral, 48
Diameter of a circle, 206, 402
 definition of, 188
Dioptra, 16
Directrix of a parabola, 520, 522
Dissection problems, 258
Distance

 between two points, 2
 from earth to moon, 68
 from a point to a line, 3, 4
 on the number line, 468
Distance Formula, 468–471, 488, 490
Distributive
 identity, 71
 property, 77
Domino, 65
Double column form, 100, 109
Dudeney, Henry, 24

Edge
 of a half-plane, 483
 of a polyhedron, 500
Elements of Euclid, 110, 249
Ellipse
 center of, 522
 construction of, 507–508
 cross section of a cone, 531
 cross section of a cylinder, 511, 514
 definition of, 509, 521
 major axis of, 509, 522
 minor axis of, 510, 522
 reflection property of, 515–519
Endpoint(s)
 of a line segment, 34
 of a ray, 45
Equation, 471
 linear, 473
 quadratic, 132, 311–313
Equidistant, 98, 109
Equilateral triangle
 area of, 278, 284
 definition of, 49
Euclid of Alexandria, 110, 157, 249
Euclidean geometry, 157
Euler, Leonhard, 500
Euler's Formula, 500
Exponents, 27
Extended chords, 411
Exterior Angle Theorem, 43
External segment, 411

Fables, similar figures in, 389
Face of a polyhedron, 374, 500
Factoring, 129
Fermat, 492
 prime, 221
Fibonacci numbers, 446–447
Focus (foci)
 of a hyperbola, 521

of an ellipse, 508, 509
of a parabola, 520, 522
Fractions, 171
Fundamental Dissection Theorem, 260

Galilei, Galileo, 377
Garfield, James A., 302
Gauss, Carl Friedrich, 159, 189, 207, 208
   Theorem on constructible polygons, 221
General Theorem for Intercepted Arcs, 455–458
Geoboard, 308
Geometric mean, 441–442
Geometry
   analytic, 463
   Euclidean, 157
   non-Euclidean, 157, 158
Golden
   ratio, 284, 444–448
   rectangle, 444–448
   section, 444–448
Graph(s)
   of a condition, 481
   of a linear equation, 491
   of a linear inequality, 483, 491
   of an equation, 471–472
Graphical method, 478
Gravitational acceleration, 536
Greek cross, 65
   3-dimensional, 67
Gulliver's Travels, 389

Half-line, 45
Half-plane, 483
Heptagon, 35
Heron of Alexandria, 15
   formula for area of triangle, 310
   problem of shortest path, 23–24
Hexagon, 35
Hilbert, David, 249
Hipparchus of Nicaea, 341
Hyperbola, 519–521
   definition of, 521
Hypotenuse, 95, 109

If-then statements, 137
Impossible constructions, 189, 197
Inequalities, 224–226
Inequality
   linear, 483
   of the arithmetic and geometric means, 442

Triangle, 8, 128
Inscribed
   angle, 402
   circle, 418
Inscribed angles, 402
Inscribed Angle Theorem, 403, 405, 429
Intercepted arc, 402, 455–458
Irrational numbers, 285, 305
Isoperimetric Theorem, 443
Isosceles triangle
   definition of, 104, 109
   base angles of, 105
   right, 240, 248, 283

Kepler, Johannes, 447, 523

Lagrange, J. L., 491
Lambert, J. H., 450
Laser, 68
Lattice
   points, 308
   polygons, 308–309
Legs of a right triangle, 95
Leibniz, 430
Length of an arc, 421
Lindemann, Ferdinand, 266
Linear equations, 226
Line dividing sides proportionally, 333–334
Line segment
   definition of, 34
   endpoints of, 34
   midpoint of, 97, 109
   notation for, 34
   perpendicular bisector of, 97, 109
Lobatchevsky, Nicolai Ivanovitch, 158
LORAN, 537

Major
   axis of ellipse, 509, 522
   diagonal of box, 487
Marcellus, 431
Medians of a triangle
   definition of, 205, 206
   meet in a point, 205–207, 222
Midpoint(s)
   of a line segment, 97, 109
   of sides of a quadrilateral, 149–150
   of sides of a triangle, 147
Minor axis of an ellipse, 510, 522
Mirror
   elliptical, 518
   parabolic, 533
   problems, 26, 350

Mosaics, 67

Napoleon, 95
Nappe, 519
Nearest point, 3
Newton, Isaac, 208, 430, 523
n-gon
   definition of, 35
   regular, 49
   sum of the angles of, 49
Non-Euclidean geometry, 157, 158

Obtuse angle, 46
Octagon, 35
Octahedron, regular, 170, 246, 308, 500
Operations, order of, 28
Opposite
   angles of a parallelogram, 139
   sides of a parallelogram, 140
Origin, 464

Pantograph, 354
Parabola
   definition of, 520, 522
   directrix of, 520
   equation of, 534–536
   focus of, 520
   reflection property of, 531–533
Paradox, 241
Paragraph form, 102, 110
Parallel
   lines, 36
   Postulate, 41, 156–158
   rulers, 141
Parallelogram
   area of, 233, 248
   definition of, 51
   properties of, 139, 157–158
   tiles the plane, 53
Parts of a triangle, 86
Penta-, 65
Pentagon
   definition of, 35
   sum of angles of, 48
Pentomino, 65, 352
Perigal, H., 303
Perimeter
   definition of, 366
   of a circle, 418
   of a polygon, 376
Perpendicular
   definition of, 4
   from a point to a line (construction), 203

through a point on a line (construction), 184
Perpendicular bisector
  construction of, 181–182
  definition of, 97, 109
  in space, 498
  of sides of a triangle, 184, 202, 206–207
  some properties of, 200
Pi (π), 419–420, 429, 450
  approximations to, 450
Pick's Theorem, 309
Plane
  coordinate, 485
  line perpendicular to, 498
  reflection across, 498
  tangent to sphere, 512, 522
Planetary motion, Kepler's three laws of, 523
Plato, 151
Platonic Solids, 169
Plumb
  level, 167
  line, 223
Point of contact, 512, 522
Polygon
  area of, 237
  circumscribed, 422–423
  definition of, 34
  diagonals of, 69, 70
  Lake, 56
  regular, 49
  sum of angles of, 48
Polygonia, 244, 258, 525
  Sultan of, 258, 371, 426
Polyhedron (polyhedra), 169, 374
  convey, 499
  faces of, 169
  regular, 169
Polynomials, 72
Polyominoes, 66
Postulate, 41, 156–157
Prime, 129, 220
  polynomial, 130
Prism
  triangular, 169
  volume of, 242–244
Proof(s), 96, 157
  wrong, 154–156
Proportional
  sequences, 321, 340
  sides, 324
Ptolemaic system, 377

Ptolemy, 341
Pyramid
  altitude of, 245
  apex of, 244
  base of, 244
  volume of, 244–246, 394–397
Pythagoras of Samos, 274, 284
Pythagoreans, Order of the, 284
Pythagorean Theorem, 273–277
  converse of, 280, 283
  proofs of, 275, 289–290, 302–304
  proof using similar triangles, 332
Pythagorean Triples, 304

Quadratic
  equations, 132, 311–313
  formula, 313–314
Quadrilateral
  definition of, 35
  skew, 168
  sum of angles of, 48
Q.E.D. (Quod erat demonstrandum), 97, 110

Radicals, simplifying, 266
Radius (radii), 180, 426
Ratio of similarity, 338
  definition of, 339, 340
Ratio of two numbers, 320, 339
Rationalizing the denominator, 267
Ray, 45
Rectangle
  definition of, 51
  diagonals of, 142
Reflection
  across a line, 6
  across a line in the (x, y)-plane, 484
  across a plane, 498
  property of an ellipse, 518, 522
  through the origin, 499
Regular
  dodecahedron, 170, 284, 500
  five-pointed star, 284
  hexagon (construction of), 186–187
  icosahedron, 170, 448, 500
  n-gon (sum of angles of), 50
  octahedron, 170, 246, 308, 500
  pentagon (construction of), 446
  polygon, 49
  polyhedron, 169, 374, 503
  tetrahedron, 126, 127, 170, 263, 306, 500

Rep-tiles, 351
Rhombus, 152–153, 157
Right angle, 46
Right circular cone, 452
Right circular cylinder, 428, 453
Right triangle
  definition of, 46
  midpoint of hypotenuse, 150
  sum of angles, 47
  30°–60°, 151, 158, 283, 277
  45°–45°, 279
Rigid, 89
Rulers, parallel, 141
Rules of construction, 183

Saccheri, Girolamo, 158
SAS (side-angle-side) condition for similar triangles, 337
SAS (side-angle-side) property of congruent triangles, 91, 109
Satisfies, 471
Scott, David R., 377
Secant, 413
Sector, 425
Segments, constructible, 440
Semicircle, 407
Shortest Path Problem, 1
  in space, 499
Sides
  divided proportionally, 333, 334
  of an angle, 45
  of a polygon, 35
Similar, notation for, 324
Similar figures, 317
  in biology, 390
  in fables, 389
Similar polygons
  areas of, 367–370, 376, 391
  definition of, 324, 340
  perimeters of, 367, 376
Similar right triangles, 329
Similar solids
  surface areas of, 375, 377
  volumes of, 371–373, 377
Similar triangles, 326–333, 335–338
  AAA condition for, 326
  proofs of conditions for, 355
  SAS condition for, 337
  SSS condition for, 335
Simultaneous linear equations, 227, 478
Six-Cornered Snowflake, The, 523

Slope
  formula, 476, 491
  of a line, 473
  undefined, 474
Solid Geometry of Wine Barrels, The, 523
Soma cubes, 67
Sphere
  cross section of, 426
  definition of, 426
  radius of, 426
Spider and fly, 25, 291, 292
Square root table, 549
Squaring the circle, 264, 266
SSS (side-side-side) condition for similar triangles, 335
SSS (side-side-side) property of congruent triangles, 88, 109
Straight angle, 10
Straightedge, 179
Substitution method, 228, 480
Super-tongs, 104
Supplementary angles, 52
Surface area, 373–376
  definition of, 374
  of a polyhedron, 376
Swift, Jonathan, 389

Tâbit ibn Qorra, 303
Tangent line
  of a sphere, 512–513, 522
    perpendicular to radius, 414, 458
  to a circle, 413
Tangent plane, 512, 522

Tangent segments
  to a circle, 414–415
  to a sphere, 514, 522
Thales of Miletus, 55
  Theorem of, 407, 429
Theorem, 38, 137
Therefore (∴), 101, 110
Theta (θ), 421
Tiling
  space, 67, 171
  the plane, 33
Trapezoid
  area of, 239, 248
  definition of, 153, 157
Triangle
  altitudes of, 204, 207, 222
  angle bisectors of, 191, 207, 418, 429
  area of, 235–236, 248
  base of, 235
  circumscribed circle of, 202–203, 206
  corresponding parts of, 87
  definition of, 35
  equilateral, 49
  exterior angle of, 42
  Inequality, 8, 128
  inscribed circle of, 418
  isosceles, 104, 109
  medians of, 205–207, 222
  notation for, 85, 110
  parts of, 86
Triangular prism, 500
Trigonometry, 340
Triplication problem, 352
Tripling any angle, 211

Trisecting a segment, 195
Trisecting the angle, impossibility of, 197–200
Tunnel of Eupalinus, 16
Tycho Brahe, 523

Vertex
  of an angle, 45
  of a polygon, 35
  of a polyhedron, 500
Vertical angles, 11
Volume
  fundamental properties of, 241, 247
  of a ball, 428, 430, 454–455
  of a box, 242
  of a cone, 453
  of a cylinder, 428, 454
  of a prism, 242–244
  of a pyramid, 244–246, 394–397
  of similar solids, 371–373, 377

Wantzel, P. L., 189, 198, 207
Whisper chamber, 519

x-axis, 464, 485
x-coordinate, 464
(x, y)-plane, 29, 466, 485
(x, z)-plane, 485

y-axis, 464, 485
y-coordinate, 464
(y, z)-plane, 485

z-axis, 485
Zero product rule, 132